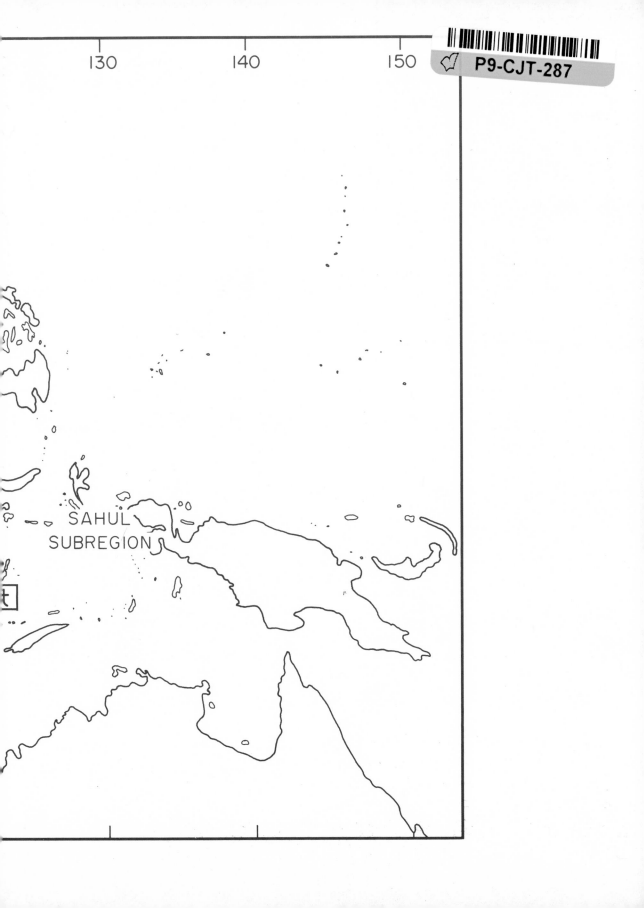

130   140   150

P9-CJT-287

SAHUL
SUBREGION

KEY ENVIRONMENTS

General Editor: J. E. Treherne

MALAYSIA

The International Union for Conservation of Nature and Natural Resources (IUCN), founded in 1948, is the leading independent international organization concerned with conservation. It is a network of governments, non-governmental organizations, scientists and other specialists dedicated to the conservation and sustainable use of living resources.

The unique role of IUCN is based on its 502 member organizations in 114 countries. The membership includes 57 States, 121 government agencies and virtually all major national and international non-governmental conservation organizations.

Some 2000 experts support the work of IUCN's six Commissions: ecology; education; environmental planning; environmental policy, law and administration; national parks and protected areas; and the survival of species.

The IUCN Secretariat conducts or facilitates IUCN's major functions: monitoring the status of ecosystems and species around the world; developing plans (such as the World Conservation Strategy) for dealing with conservation problems, supporting action arising from these plans by governments or other appropriate organizations, and finding ways and means to implement them. The Secretariat co-ordinates the development, selection and management of the World Wildlife Fund's international conservation projects. IUCN provides the Secretariat for the Ramsar Convention (Convention on Wetlands of International Importance especially as Waterfowl Habitat). It services the CITES convention on trade in endangered species and the World Heritage Site programme of UNESCO.

IUCN, through its network of specialists, is collaborating in the Key Environments Series by providing information, advice on the selection of critical environments, and experts to discuss the relevant issues.

KEY ENVIRONMENTS
MALAYSIA

Edited by

EARL OF CRANBROOK

Glemham House, Great Glemham, Saxmundham, Suffolk, IP17 1LP, U.K.

Foreword by

HRH THE DUKE OF EDINBURGH

Published in collaboration with the

INTERNATIONAL UNION FOR CONSERVATION OF
NATURE AND NATURAL RESOURCES

by

PERGAMON PRESS

OXFORD · NEW YORK · BEIJING · FRANKFURT
SÃO PAULO · SYDNEY · TOKYO · TORONTO

U.K.	Pergamon Press plc, Headington Hill Hall, Oxford OX3 0BW, England
U.S.A.	Pergamon Press, Inc., Maxwell House, Fairview Park, Elmsford, New York 10523, U.S.A.
PEOPLE'S REPUBLIC OF CHINA	Pergamon Press, Room 4037, Qianmen Hotel, Beijing, People's Republic of China
FEDERAL REPUBLIC OF GERMANY	Pergamon Press GmbH, Hammerweg 6, D-6242 Kronberg, Federal Republic of Germany
BRAZIL	Pergamon Editora Ltda, Rua Eça de Queiros, 346, CEP 04011, Paraiso, São Paulo, Brazil
AUSTRALIA	Pergamon Press Australia Pty Ltd., P.O. Box 544, Potts Point, N.S.W. 2011, Australia
JAPAN	Pergamon Press, 5th Floor, Matsuoka Central Building, 1-7-1 Nishishinjuku, Shinjuku-ku, Tokyo 160, Japan
CANADA	Pergamon Press Canada Ltd., Suite No. 271, 253 College Street, Toronto, Ontario, Canada M5T 1R5

Copyright © 1988 Pergamon Press plc

All Rights Reserved. No part of this publication may be reproduced, stored in a retrieval system or transmitted in any form or by any means: electronic, electrostatic, magnetic tape, mechanical, photocopying, recording or otherwise, without permission in writing from the publishers.

First edition 1988

Library of Congress Cataloging in Publication Data

Malaysia.
(Key environments)
1. Forest ecology—Malaysia. 2. Ecology—Malaysia.
I. Cranbrook, Gathorne Gathorne-Hardy, Earl of, 1933- . II. International Union for Conservation of Nature and Natural Resources. III. Series.
QH185.M35 1988 574.5′2642′09595 87-35819

British Library Cataloguing in Publication Data

Malaysia.
I. Malaysia. Natural environment
I. Cranbrook, Gathorne Gathorne-Hardy, *Earl of, 1933-* II. International Union for Conservation of Nature and Natural Resources
508.595
ISBN 0-08-028866-9

Printed in Great Britain by A. Wheaton & Co. Ltd., Exeter

The general problems of conservation are understood by most people who take an intelligent interest in the state of the natural environment. But if adequate measures are to be taken, there is an urgent need for the problems to be spelled out in accurate detail.

This series of volumes on "Key Environments" concentrates attention on those areas of the world of nature that are under the most severe threat of disturbance and destruction. The authors expose the stark reality of the situation without rhetoric or prejudice.

The value of this project is that it provides specialists, as well as those who have an interest in the conservation of nature as a whole, with the essential facts without which it is quite impossible to develop any practical and effective conservation action.

Philip

1984

Preface

The increasing rates of exploitation and pollution are producing unprecedented environmental changes in all parts of the world. In many cases it is not possible to predict the ultimate consequences of such changes, while in some, environmental destruction has already resulted in ecological disasters.

A major obstacle, which hinders the formulation of rational strategies of conservation and management, is the difficulty in obtaining reliable information. At the present time the results of scientific research in many threatened environments are scattered in various specialist journals, in the reports of expeditions and scientific commissions and in a variety of conference proceedings. It is, thus, frequently difficult even for professional biologists to locate important information. There is consequently an urgent need for scientifically accurate, concise and well-illustrated accounts of major environments which are now or soon will be, under threat. It is this need which these volumes attempt to meet.

The series is produced in collaboration with the International Union for the Conservation of Nature. It aims to identify environments of international ecological importance, to summarize the present knowledge of the flora and fauna, to relate this to recent environmental changes and to suggest where possible, effective management and conservation strategies for the future. The selected environments will be reexamined in subsequent editions to indicate the extent and characteristics of significant changes.

The volume editors and authors are all acknowledged experts who have contributed significantly to the knowledge of their particular environments.

The volumes are aimed at a wide readership, including: academic biologists, environmentalists, conservationists, professional ecologists, some geographers as well as graduate students and informed lay people.

Although, for simplicity and consistency with the series, we have titled this volume simply "Malaysia", in fact it deals with one family of key environments in one part of the country of that name. A long title for the work would be, "The Forests of Peninsular Malaysia".

Malaysia, the nation, sprung to being in 1963 as a union of the former Federation of Malaya (generally known as "Malaya") with the two former British colonies of Sabah and Sarawak. Malaya, or Peninsular Malaysia as the eleven component States are now known, is a natural biogeographical unit and, as a one-time national entity, has been the focus of much important research into the aspects of the environment covered in this book. To extend the treatment to include the two Bornean States (collectively known as "East Malaysia") would have created unproductive complications.

The terminology is further confused by the fact that, before 1963, the name "Malaysia" had been established for three decades in two distinct usages by zoogeographers and phytogeographers. In zoogeography the term denoted the humid tropical region of the Sunda shelf, delimited by Wallace's Line (dividing Borneo from Celebes and Bali from Lombok) and comprising the southern extremity of peninsular Burma and the peninsular provinces of Thailand south of 10°N, Malaya, Sumatra, Java and Borneo, with intervening islands. For botanists, for whom Wallace's Line represents a less distinctive biogeographical transition, "Malaysia" denoted the entire Indo-Malayan region from Sumatra to New Guinea and beyond, legitimately divided into a western portion (corresponding to the zoological "Malaysia") and an eastern.

Post-1963, the term "Sundaland" (adjectivally, "Sundaic") has found favour as a substitute for "Malaysia" as defined in zoology. Unfortunately, that name also has currency among geologists to denote a more limited unit, the tectonically stable part of Southeast Asia, comprising Peninsular Malaysia, the eastern rim of Sumatra, the tin islands of Bangka and Belitung, the submarine Sunda shelf, western Kalimantan and westernmost Sarawak. Botanists have chosen to standardize an alternative spelling, "Malesia", continuing to recognise the division into East (corresponding broadly with the "Sahul subregion" of zoologists) and West.

We have felt obliged to dwell on these terminological problems because the different usages appear in this book in the writings of authors from the three disciplines involved. The map inside the front and end covers provides a pictorial summary of this explanation.

John Treherne and Earl of Cranbrook

Contents

CONTENTS

CHAPTER 1

The Physical Setting

H.D. Tjia

Department of Geology, Universiti Kebangsaan Malaysia, Locked Bag 13, 43600 Bangi, Selangor, Malaysia

CONTENTS

1.1. LANDFORMS

Peninsular Malaysia projects as the southernmost extremity of the Asian continent between approximately the latitudes 6° 40′ N and 1° 10′ N, and separates the South China Sea from the Strait of Malacca. Its landforms indicate three major trends. Coastlines strike NNW north of the latitude of Kuala Lumpur and northwesterly south of that latitude. The mountain ranges, however, strike northerly and

1

have some less conspicuous NNE trends (Fig. 1.1). The mountainous parts are in the northern two-thirds of the country, where long mountain ranges are capped by summits over 1500 m high, the highest being Gunung (=Mount) Tahan, 2188 m (7174 feet) high. The four principal rivers are the Perak (390 km long) and Muar rivers (190 km) draining into the Strait of Malacca, and the Pahang (500 km) and Kelantan (250 km) draining into the South China Sea. From the disposition of the Pahang and Muar rivers, some geographers suspect that at one time the upper Pahang may have drained via the Muar into the Strait. At present Tasik Bera, a large shallow lake, occupies low ground between the two rivers. Several large tributaries of other rivers are influenced by geological structures. For instance, many left-hand tributaries of the Perak, such as the Korbu and Kinta, seem to follow WNW-striking, curvilinear fracture zones over distances between 20 and 40 km.

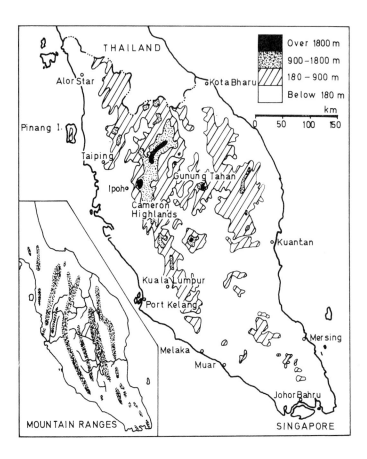

Fig. 1.1. General topography of Peninsular Malaysia. Inset shows the mountain ranges and their trends.

The peninsula is surrounded by coastal plains except where relatively small areas of highland extend to the shoreline. The western coastal plains are wide, in Perak attaining a maximum width of 60 km. Broad estuaries are the rule, and a significant delta exists only at the Kelang river mouth. On the east, wide coastal plains exist only along the Kelantan, Mercang and Pahang courses. The Kelantan delta is of the fan-shaped type; the Pahang delta is cuspate. Elsewhere, the eastern coastal plain is characterized by long, low sandy ridges known as 'permatang', probably former shorelines, that occasionally alternate with lagoons. Three groups of permatang can be distinguished. Nearest to the existing shoreline the ridges rise up 2 m above sea level as a result of present day storm wave action that prevails during the northeast monsoon. Still parallel to the shoreline, but farther inland, is a series of ridges between 2 and 6 m high. A third series of permatang occurs as far as 10 km inland from the present shoreline, with some crests between 9 and 13 m high. At least two events of sea-level lowering are indicated by these inland ridges. Indeed, old maps show that the coastal area south of Kuantan became land only after 1800 A.D. It has been estimated that within the last 1500 years this coastline has accreted at an average rate of 16 m per year.

1.1.1. Morphology of the rock types

Two-fifths of the country consists of granitic rocks and the rest mainly of sedimentary rock with relatively small areas of crystalline metamorphic rocks (Fig. 1.6). Most of the sedimentary rocks, except the sediments on the coastal plains, have been slightly metamorphosed and are actually slate, phyllite, meta-sandstone, and meta-conglomerate. However, intense, prolonged and deep weathering have changed and softened these rocks so that they more closely resemble the rock types before metamorphism occurred.

1.1.1.1. Granitic terrains

Most of the granite forms mountains and mountain ranges. Although deeply dissected, small relatively flat summit areas exist, such as the Cameron Highlands and Penang Hill. Elsewhere steep valley walls, with numerous waterfalls and rapids are the rule. Rectilinear to curvilinear valleys in granitic terrains are caused by major fracture zones. In the southern part of the peninsula and in Singapore, undulating hill topography, rarely exceeding a few hundred metres in height, characterizes granitic terrain. It is possible that this is due to lower initial height of the top of the granite batholith. It is also possible that the subdued topography of the south is caused by the preservation of a thick soil mantle, while elsewhere higher relief prevented such thick accumulations of soil.

Associated with granite are the many quartz ridges in which the material is exposed as sheer cliffs. The largest is the Kelang Gates quartz ridge, 16 km long and jutting 200 m above the surrounding plain northeast of Kuala Lumpur.

1.1.1.2. Metamorphic terrains

Among the larger areas of metamorphic rock are the Taku Schist and the Setong metamorphic complex in Kelantan. The Setong complex has mountainous topography similar to that built by granite. The Taku Schist forms hills that are characterized by steep valley walls and sharp-crested ridges that are 350–400 m high. The rivers generally follow lines of weakness — joints or faults — and so show many straight sections.

1.1.1.3. Areas of sedimentary rocks

Slightly metamorphosed or unmetamorphosed sediments often result in morphologies controlled by bedding. Parallel ridge-and-valley topography reflecting alternating soft and resistant strata is especially

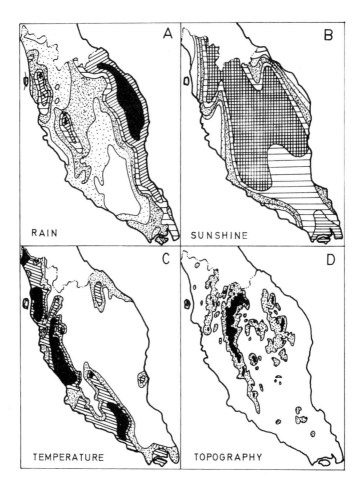

Fig. 1.2. (a) Mean annual rainfall. Contours indicate 2000–2500–2750– 3000–3250 mm of rainfall.After Dale (1959). (b) Mean annual hours of bright sunshine. Countours are 2100–2200–2400– 2500 hours. After Dale (1964). (c) Annual mean temperature. Contours are 26.1°–26.6°–27.2°C.After Dale (1963). (d) General topography of Peninsular Malaysia; blank areas less than 300 m; dotted 300–900 m; black more than 900 m high.

clear in the Triassic and younger rocks. Where fold plunges are present, zig-zag valleys and ridges occur. Drainage is of the trellis pattern.

1.1.1.4. Limestone

Although not extensive in Peninsular Malaysia, limestone exposures are conspicuous landscape formers. The main limestone areas are the Langkawi islands, the northwestern region, the Kinta valley, Perak, and the neighbourhood of Gua Musang, Kelantan. Most limestone hills are tower karst, rising abruptly a few tens to over 500 m above the surrounding terrain. Among the steep hills lie depressions which are called 'wang' if they are rounded in plan-form and do not possess surface drainage. Natural lakes may develop in these depressions. One such isolated habitat on Langgun island of the Langkawi group was studied by van Balgooy *et al.* (1977). Sinoid hills of limestone, the other form of limestone karst in the tropics, have only been observed on the east side of the main island of Langkawi. Caves are common in the limestone; many have been mapped and their origins discussed by Gobbett (1965). Depressions and valleys in limestone terrains are generally linear and often aligned in rows following fractures. In contrast to this rugged topography, the tops of limestone pinnacles and towers often define flat surfaces that are also horizontal to subhorizontal. The lower 20 m or so of the limestone cliffs often display deep, horizontal grooves. These grooves may occur at several levels and are believed to be the result of solution by acid swamp waters, such as in the Kinta valley. In some cases there are now no swamps near the limestone, and the feature is thought to reflect wetter climatic conditions in the past.

1.2. CLIMATE

1.2.1. Temperature

Fluctuations in the warm climate of these low latitudes are further moderated by the surrounding sea and generally low relief of the land, so that Peninsular Malaysia does not experience great extremes in the temperature. Long series of air temperatures measured at stations scattered throughout the country were analysed by Dale (1963), from whom the following details are taken. The maximum shade temperature recorded was 103°F (39.4°C), at Langkawi island, 3.6 m elevation, on 27 March 1931, and again at Segamat, Johor, 11 July 1958. The lowest recorded temperature in the lowlands was 60°F (15.6°C), while the lowest minimum at any elevation was 36°F (1.2°C) at Cameron Highlands. Throughout the lowlands (150 m elevation or lower) the mean annual temperature varied within about 1.6°C of 26.7°C, the highest mean being 27.7°C at Batu Gajah (33 m elevation, lat. 4°N) in the Kinta valley, the lowest at Keluang (64 m, lat. 2°N), 25.4°C. On the west side of the peninsula these annual means were slightly higher (by 0.3–0.8°C) than those on the east side at similar latitudes. Four regions show comparatively high annual means: Perlis, south Kedah and the Krian valley; Kinta valley; Perak valley with southward extensions in south Perak and north Selangor; and a narrow belt on the east side of the Main Range extending northwest and southeast from Segamat. The difference between the maximum and minimum monthly means was everywhere less than 3.3°C. As expected, the diurnal range was usually smaller at the coast than inland. Thus, the annual mean diurnal range is less than 8.9°C for many coastal stations, but along the foot of the Main Range is close to 11°C. On the lowlands the average annual mean relative humidity is 95 per cent. During the day it varies between 55 and 70 per cent and at night it is 95 per cent or higher over almost the entire country.

1.2.2. Sunshine

An analysis of daily hours of sunshine by Dale (1964) showed that most coastal strips, the Kinta valley, Perak, and the Kelantan river valley received more than 50 per cent mean annual sunshine hours. Exceptions were the south coast of Johor and adjacent Singapore island (Fig. 1.2b). This interpretation should only be regarded as preliminary, as it was based on very few recording stations, most of which were in the lowlands.

1.2.3. Rainfall

At any time of the year heavy rainfall may occur anywhere. Equally, there may be dry spells, although these are not regular or long. In most parts of Peninsular Malaysia, four weakly defined seasons may still be recognised (Dale, 1959). Wind direction is the most important governing factor. During the northeast monsoon, from about November until March, northeasterly winds prevail, wind speeds seldom exceeding 40 km h^{-1}. This season is followed by an intermonsoonal or transitional period lasting five to seven weeks, centred on April for the southern region and on May for the northern region. During this period winds are weak and variable. Some time in May light southwesterly winds announce the arrival of the southwest monsoon for the northern part, and light southerly winds begin to blow across the southern part of the country. These winds are weaker than the northwesterlies and are often overridden by the daily land and sea breezes in the coastal regions. A second transitional season occurs in October and early November. Rainfall maximum occurs during the two transitional periods for many parts of the country and a minimum may occur during each of the monsoons. However, other rainfall patterns are also possible elsewhere. The three rain types that occur are orographic, boundary and convectional rain.

Chia (1977) used harmonic analysis to determine seasonal rainfall distribution in the peninsula. He was able to distinguish five regions and five subregions. The Northwest region (A) is characterized by a very low minimum rainfall period during December to February (Fig. 1.3). The West region (B) experiences peak rainfall during the last half of April and the last half of October. The Port Dickson-Muar Coast region (C) has minimum rainfall in February, peaks in March and November, and rather high rainfall during the March-November period. The East region (D) shows considerable variation in the seasonal rainfall patterns; e.g., the minimum rainfall period in the north is from February to April, but in the south it is in June–July. The region is further characterized by pronounced rainfall peaks centred in December. The Interior region (E) is sandwiched between the Main Range in the west and the Boundary Range in the east. High seasonal rainfall occurs in April–May and again and much higher in October–November.

A remarkable feature of rainfall distribution in the peninsula is the variation over short distances. This is due in part to topographic relief (compare Fig. 1.2a with Fig. 1.2d). The moist air streams cross the north-south grain obliquely. Two distinct wet belts occur along the east coast and the northern half of

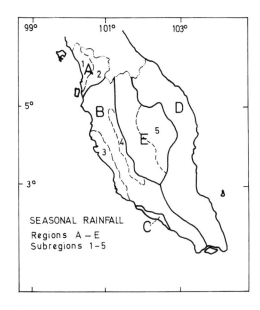

Fig. 1.3. Characteristic regions and subregions of seasonal rainfall. After Chia (1977).

the west coast. In the wettest parts annual rainfall exceeds 2750 mm, with the heaviest fall on the foothills of Terengganu (average 4000 mm per year). The Larut hills (1354 m high), near Taiping, Perak, have the highest recorded annual average total of 5800 mm rain. The axes of the wet belts coincide with the foothills (150–300 m high in the west, 120 m on the east coast) rather than with the highest topography. In the drier areas the average annual total rain amounts to less than 2250 mm. The number of rainy days is also distinctly controlled by relief. The lowlands of the east coast experience considerably more rainy days than those of the west coast. However, the high annual rainfall in the mountains of the eastern part is not reflected by the number of rainy days, a fact probably ascribed to the high-intensity of rainfall during the northeast monsoon and the shelter provided by the Main Range during the southwest monsoon. It is remarkable that eastern Johor has the highest number of rain days but represents the driest part of the eastern wet belt region (Fig. 1.2a).

Dry spells (defined by Dale as less than 0.25 mm of rainfall per day for a specific number of consecutive days) are longest and most frequent in the northwest of the peninsula. For instance, Alor Setar, Kedah, had no rain for 49 days from January to February 1940. Dry spells of 21 days are also rather frequent for Alor Setar (16 occurrences in 23 years).

1.2.4. Past climates

In world history during the Quaternary period, at higher latitudes, glaciers expanded and retreated several times in association with severe climatic changes. There is still no consensus on the reasons for these climatic fluctuations, although it appears that astronomical causes were predominant (see Imbrie and Palmer-Imbrie, 1979). In equatorial regions, the consequences of Quaternary climatic changes were only felt in the higher mountainous areas, such as the summit of Mt Kinabalu (4100 m), Sabah (see Koopmans and Stauffer, 1968; Tjia, 1973).

In Southeast Asia in general three important factors are thought to have influenced past climatic conditions. The first is the position of the Intertropical Convergence zone (ITC), with which most rain in the region is associated, and which itself is affected by the development of anticyclones over Asia and Australia and by changes in atmospheric circulation. During glacial periods the anticyclone over Asia (and perhaps also that over Australia) was more strongly developed than at present, so that the distributional pattern of rainfall must have been markedly different. Secondly, during the glacials global air and water temperatures were lower. Lastly, the Sunda Shelf emerged as dry land during glacial periods. The combined effect of these three factors is thought to have resulted in considerably lower precipitation, a more pronounced dry season and somewhat lower temperatures compared with present conditions. Verstappen (1976) has estimated a 30 per cent reduction in rainfall and a temperature drop of 3–5°C, alternating during interglacial periods and the mid-Holocene climatic optimum with increased humidity and a climate probably more humid and perhaps 1–2°C warmer than present conditions.

The different climatic conditions must have affected the intensity of processes that contribute to landform development. In the interglacial periods, including the Holocene, intense chemical weathering dominated. Weathering resulted in thick soils that contributed mainly clays to the rivers; bed load was relatively unimportant. Mass movements, such as creep and landslides, affected slopes; a dense vegetation kept rivers along linear courses, and vertical erosion was restricted to soft Tertiary rocks. Denudation rates were very high. During the glacial periods lower rainfall, humidity, and temperatures and, particularly, the more pronounced seasonality resulted in a marked reduction of chemical weathering and increased mechanical weathering. A presumed lowering of the tree line and vegetational zones contributed to increased production of coarse debris in the mountains. The regolith tended to become thinner and to consist of coarser material. Mass movements were less important, hill streams became torrential, and sheetwash prevailed on the gentle slopes. The debris was mainly transported by the high discharges of the wet monsoon. During the dry season large tracts of valley-fill marked the river courses where coarser debris were temporarily deposited. Braiding river patterns were the rule. Surface planation through degradation and aggradation must have been important during the glacial periods.

1.3. SOILS

1.3.1. Inland soils

Under hot and humid tropical conditions, weathering is predominantly chemical. This process produces hard, iron-rich masses, or laterite, in the soil, especially in areas of relatively dry atmospheric conditions and on argillaceous rock (Eyles, 1970). Major laterite areas are in Kedah, the coastal regions of Negeri Sembilan and Melaka, and inland also in these two states, in Johor and Pahang, and Kelantan near the border with Thailand. The laterite may occupy flat alluvial areas, or occur at the foot of slopes close to the boundary between dryland soils and alluvial soils, in gently undulating country, and capping hill summits. Under special circumstances, the weathering processes that produce laterite also concentrate aluminium hydroxides or bauxite, as in the Pengerang area of Johor.

In Peninsular Malaysia over 200 soil series have been recorded for land where slopes are less than 20 degrees (Kalpage, 1979). Six out of the ten soil orders known in soil taxonomy are present: Entisols, Inceptisols, Spodosols, Ultisols, Oxisols and Histosols. Entisols do not possess soil horizons and are found on river alluvium throughout the peninsula, on very steep slopes of inland mountains and in the marine clay of the western and in marine sand of the eastern coastal areas. Inceptisols are common in the poorly drained marine clay of the western coastal plan. In the subsoil may be found the sulfic members with acid sulphate layers. Spodosols are found in the depressions of the east coast (among the so-called 'beris' soil landscape) and also at heights above 1370 m at Genting Simpah and Cameron Highlands. Ultisols are widespread on igneous, sedimentary and metamorphic rocks. Oxisols are the extremely weathered soils that are present on flat to undulating surfaces overlying sedimentary, metamorphic, and mafic to intermediate igneous rocks. Histosols consist of peat and organic clay that occur mainly in depressions of the coastal deposits in the central and southern parts of the peninsula. In the river valleys peat may extend inland and is underlain by clay on the west and by sand on the east coast. About six per cent of the total area consists of peat. Inceptisols, Ultisols and Oxisols have excellent to good agricultural potential and are favoured for current development schemes.

Malaysian soils show a wide range of physical and chemical properties. The intense chemical weathering and leaching have caused acidity in most soils, with pH ranging between 4.2 and 4.8. Very low pH (<4 in the topsoil and <3.5 in the subsoil) is encountered in acid sulphate and peat soils.

Soils on alluvium range in texture from pure sand ('beris' soils, e.g., Rudua Series on the east coast) to clay of marine deposits on the west coast. Fluvial alluvium has markedly different textures according to morphology. The levee deposits have more fine sand while back swamp deposits contain more clay. Older alluvial deposits tend to produce soils in which the clay content increases with depth. On the other hand, marine clay soils do not exhibit an increase in clay with depth. On igneous rocks, the soils contain more than 50% clay (e.g., the Renggam Series). Textures of soils developed from sedimentary rock vary according to the parent material; less than 40% clay is found in soils derived from arenaceous rocks and, as expected, more than 50% clay is found in soils upon argillaceous rock.

The content of plant nutrients and capacity for cation exchange are generally low, resulting in poor chemical fertility. In all mineral soils nitrogen content is low and decreases rapidly with depth. In peat, the nitrogen content is high (approximately 1.4%) but due to a high C/N ratio (40), the element is not easily available. Total and available phosphorous are generally low; total P is highest in soils derived from basic and intermediate rocks and from marine clays, but is much lower in other soils. Total K values are more than 5 meq % in most alluvial soils and soils upon clays, but are low in soils on igneous rocks.

1.3.2. Marine and intertidal soils

That part of the South China Sea that adjoins Peninsular Malaysia also covers the Sunda Shelf; with an area of over 1.9 million km^2 it is one of the largest shelves in the world. The sea is about 40 m deep and slopes down towards northeast to about 100 m depth. The large rivers draining the peninsula towards the east were probably headwaters of tributaries to a large drainage system that is now submerged in the Gulf of Thailand. The rivers of southeastern Johor, south of the latitude of Tioman island, were tributaries to the similarly submerged North Sunda River that debouched near the Natuna islands and were fed by tributaries such as the present Kapuas river in western Kalimantan, the Musi and Jambi rivers in Sumatra (Tjia, 1980).

The tidal range is between 3.1 and 3.7 m (between lowest and highest astronomic tides) or generally 1.5 to 2.3. The larger fetch of the South China Sea enables the development of higher and stronger waves on the east coast compared to sea conditions on the west coast. There is strong correlation between the direction of river mouth deflection and the direction of prevailing or stronger currents.

The Strait of Malacca is an 800 km long connection between the Indian Ocean and the South China Sea. Coastal swamps of mangroves are common along both sides of the strait. Coastal accretion of up to 200 m annually in the vicinity of larger rivers on Sumatra has been assisted by the presence of coastal vegetation (Tjia et al., 1968). On the Malaysian side coastal accretion has been estimated at about 10 m per year, but this figure is likely to have increased considerably in the past 20 years as a result of extensive deforestation and land cultivation. For instance, in the forested headwaters the Gombak river contains 1–5 mg suspended particles per litre, but suspended load in the lower reaches may be 100 mg/l (Chapter 17).

1.4. PAST CHANGES IN LANDFORM AND LAND AREA

The last major mountain-building episode occurred in Late Triassic to Early Jurassic time. Since then tectonic movements have been mainly vertical. Younger sedimentary rocks, such as the Jurassic to Cretaceous Gagau Group are continental-type deposits and are not folded. Along major faults strike-slip movements may have persisted throughout the Mesozoic, but by the beginning of the Cenozoic the region had been stabilized tectonically. Radiometric ages of Cretaceous–Early Tertiary time for some granites in the peninsula are interpreted as indications of thermal events (probably metamorphism) and are not considered as indications of forceful intrusions (Burton and Bignell, 1969). During the Tertiary, most of the peninsula apparently remained above sea level, except where small lagoons and lakes have left their characteristic deposits, such as at Batu Arang, Selangor, and a few other isolated localities.

Together with the submarine Sunda Shelf, western Kalimantan and westernmost Sarawak, the eastern rim of Sumatra, the tin islands Bangka and Belitung, Peninsular Malaysia forms an area termed "Sundaland". This geological unit is the tectonically stable part of Southeast Asia. Recent oil exploration in the South China Sea has shown that in the central portions of Sundaland downwarping has resulted in the accumulation of 3000 to 5000 m of Tertiary sediments. There are also some indications of downwarping along the edges of Sundaland, but on the whole the region had been tectonically stable.

This is more so with respect to the 2-million year span of the Quaternary Period. Therefore, the pronounced sea level fluctuations caused by large scale, alternating glaciations and deglaciations of the Quaternary have developed shoreline features that may be regarded as records of eustatic (actual or world-wide) sea level changes.

Various lines of evidence indicate that the last glacial period depressed sea level 100–130 m below its present position (see Flint, 1971). During the interglacials sea levels were higher. Researchers differ in opinion on the levels of Quaternary interglacial seas, and of the postglacial sea. One school of thought maintains that during interglacials sea level was tens of metres (perhaps up to +50 m) above its present position. Others think that there is only evidence for Quaternary sea levels as high as 6 to 10 m above the present. Many of the uncertainties encountered in sea level research are caused by the interference of assorted factors (see Tjia *et al.*, 1977).

That Quaternary sea level had dropped far below its present position is widely accepted. On the Sunda Shelf drowned shorelines are indicated by the presence of river valleys, platforms or terraces on bathymetric maps, and can be deduced from sonograms, from assemblages of pollen and foraminifera, and from peat deposits in the Strait of Malacca. Onshore along the Strait very thick coastal alluvium has been located by drilling in the following localities: Butterworth (137 m thick), lower reaches of the rivers Bernam (135 m), Perak (135 m), and Selangor (66 m). On the east coast, alluvium is at least 30 m thick and 50 m thick at Kuantan, Pahang. In the southern part of the Sunda Shelf, horizontally disposed, surficial sediments and valley fillings have aggregate thicknesses as much as 100 m. The data suggest eustatic Quaternary sea levels at depths of –8, –10, –13, –18, –20 to –22, –30 to –33, –36, –45, –50 to –51, –60, –67, and –82 to –90 metres.

Shorelines between present datum and +10 metres are documented by abrasional platforms, flat and accordant summits, benches, sea level notches, depositional terraces of sand and sometimes of gravel, beachrock, barrier ridges that are beyond the reach of present storm waves, clusters of oysters, algae and coral in growth positions. More than 40 such organic indicators dated by radiometric methods (^{14}C) show that in the past 6000 years eustatic sea level was up to 4 m higher than today in Sundaland (Fig. 1.4). About 1800–1600 years B.P. (or 150–350 A.D.) the sea stood lower by one metre. There are also indications that 300 to 250 years ago, sea level on the Peninsular coasts was 0.5 to 0.7 m higher.

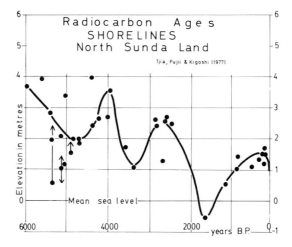

Fig. 1.4. Curve indicating sea level variations in the past 6000 years (B.P. years Before Present) constructed based on *in situ* shoreline indicators. The original data are listed in Tjia *et al.* (1977).

Since 6000 years B.P. sea levels appear to have become progressively lower, every 1000 to 1500 years fluctuating approximately 1 m about the general decreasing trend. Figure 1.5 is an example of an ancient sea level indicator. Other shorelines in the category between the present strandline and +10 m are at +6 m (as abrasional platform, and abrasional as well as depositional terraces), at +7 (as abrasional features), and at +10 m (also as abrasional phenomena).

Fig. 1.5. Raised abrasion platform developed upon hard metamorphic rock. The platform is now covered by grass and is 1 m above high tide. It was abraded when sea level stood 1 to 2 m higher than today. Cenering, Terengganu.

Topographic maps, bathymetric charts, published work, and many field observations in various places in Sundaland suggest former shorelines at +50 m, +34 m to +36 m, +30 m to +32 m, +25 m, +18 m, and +12 to +13 m. Level surfaces and slope breaks above +50 m, such as those at approximately +75 m, may represent pre-Quaternary strandlines that have been raised isostatically. This high shoreline level is represented by breaks in slope and the upper limit of the presumed Early Pleistocene Older Alluvium in the Kinta Valley (Walker, 1956), in Selangor, and also in Johor and Singapore. It is generally accepted that if all present ice melted, sea level could initially rise up to 100 m but, by isostatic adjustment of the crust, the ultimate rise would be approximately 65 metres. Estimates of the maximum glacio-eustatic sea level rise during the Quaternary do not exceed +50 m.

The levels in the +30 to +32 m and +34 to +36 m ranges are indicated by field observations and topographic maps as summit flats, accordant summits, and terraces in the Langkawi island group, on Penang (Pinang) island, along the west and east coasts of the Peninsula, and also in the Riau and Lingga archipelagoes in Indonesia (Table 1.1).

TABLE 1.1. Indicators of Regional Base Level at approximately 30 metres in Sundaland.

Elevation	Locality, type of indicator, reference
+30 to +36 m	West coast, Peninsular Malaysia; coastal bench; Walker (1956)
ca. +30 m	Batu Kulat, Langkawi; accordant summits; field observation (HDT)
+31 to +33 m	Rebak island, Langkawi; flat summits; field observation (HDT)
+29 to +31 m	Selatan, Besar and Angsa islands, off Port Kelang, Selangor; accordant summits, summit flats; field observations (HDT) and British Admiralty Chart 3766
ca. +30 m	Southwest point, Karimun Kecil island, Lingga group, Indonesia; flat surface; field observation (HDT, 1977)
+27.2 m	Cape Pengaruh, southwest Karimun Besar island, Lingga group, Indonesia; accordant summits on granite; field observation (HDT, 1977)
+34 m	Tanjung Balai, southeast Karimun Besar island, Lingga group, Indonesia; field observation (HDT, 1977)
+28 m	Pantai Laut Jauh, Singkep island, Indonesia; abrasion terrace on phyllite; Joesril Hainim (pers. comm., 1977)
ca. +30 m	West of Buding, north-central Belitung island, Indonesia; extensive, very gently undulating surface; field observation (HDT, 1977) and also indicated by topographic map

Haile (1975) has reviewed the evidence used by earlier writers to interpret high Quaternary shorelines in Sundaland and concluded that most of the data was unconvincing and even misquoted and only shorelines up to +6 m can be accepted as proven. On the other hand, certain elevations of summit flats, accordant peaks, and terraces are found throughout Sundaland and strongly suggest base level control of a regional scale. The wide extent of Sundaland further suggests that regional base level is most likely sea level. Recently, Gray *et al.* (1978) found at Seri Medan, Johor, far from the present coast at 15 m elevation young, probably Quaternary marine fossils.

1.5. STRATIGRAPHY AND GEOLOGICAL HISTORY

1.5.1. Palaeozoic

Peninsular Malaysia forms a segment of a Mesozoic mountain belt that extends from Yunnan in China to Belitung and perhaps even into western Kalimantan (Borneo), Indonesia. Lengthwise, Peninsular Malaysia may be subdivided into four zones which constitute a complete geosynclinal couple. From west to east a miogeosyncline is followed by a miogeanticline, now marked by the Main Range (or 'Titiwangsa'), then a eugeosyncline covering most of the east coast states, and finally a geanticline along the eastern border of the east coast (Burton, 1972). According to the distribution of economic minerals, three longitudinal zones can be distinguished: a tin belt in the west and another tin belt in the east flanking a central belt containing gold and base metals.

The oldest rocks are upper Cambrian, crossbedded meta-sandstone with subordinate meta-argillite, estimated to be 1800 m thick. These rocks are exposed only in the Langkawi islands and bear the name Macincang formation. Rocks near Gunung Jerai in Kedah are supposed by some geologists to be time-equivalents, but there is no fossil evidence. The age of the Macincang rocks has been determined by fossils in similar-looking beds on Terutau island in Thailand. The trilobite fossils include *Pagodia thaiensis*, *"Eosaukia" buravasi*, and *Saukiella terutaoensis*. The Macincang rocks were deposited in a deltaic environment. During the Ordovician to Silurian argillaceous carbonates were laid down upon the Cambrian rocks; sedimentation took place in shallow marine surroundings. This rock sequence is also 1800 m thick and its age is indicated by a number of cephalopods *(Robsonoceras)*, gastropods *(Homotoma, Helicotoma)*, conodonts (especially in the limestone in the upper part of the series; *Aconotiodus, Dreponodus, Panderodus, Scolopodus)*, trilobites *(Dalmanitina)*, and graptolites *(Climacograptus, Glyptograptus, Monograptus sedgwicki)*. The lower Palaeozoic rocks in the Langkawi islands and northwestern part of the mainland ended with clastic beds containing lower Devonian fossils.

In central Perak lower Silurian limestone contains diagnostic corals *(Favosites, Heliolites,* and *Halysites)* and is interbedded with graptolite shale *(Glyptograptus* and *Diplograptus)*. Such shales are also well developed in south Kedah and north Perak. In north Perak Ordovician to Silurian limestone and clastic sediments are associated with acid volcanic material. Some serpentinite bands within these rocks probably represent metamorphosed ultrabasic rocks.

In the Kuala Lumpur area, middle and upper Silurian limestone is well developed, containing corals, brachiopods and gastropods *(Ketophyllus, Heliolites, Pentamerus, Atrypella, Euomphalus, Loxonema)*. This limestone rests upon non-fossiliferous phyllite and schist of probable early Silurian age. The lower Palaeozoic sediments are succeeded by geosynclinal development in areas farther to the east. The upper Ordovician to lower Devonian sediments of the Mahang Formation were deposited under euxinic conditions and consist of carbonaceous and often siliceous black shale containing graptolites and/or tentaculites *(Monograptus gregarius; Nowakia* and *Styliolina)*, trilobites and crinoids (Burton, 1967). Still farther eastward, these sediments intercalate with limestone and quartzite of the Baling Group that developed in shallower and better aerated environments. The group contains graptolites of the upper Llandovery *(Climacograptus calaris, Spirograptus grobsdorfensis)*. Towards north, these rocks are succeeded by upper Palaeozoic sediments.

In central Perak, towards the south, the rocks comprise limestone with minor shale that may represent another shelf during Late Ordovician to Early Carboniferous time. Still farther southward, lower Silurian, euxinic-type rocks have been found in Perak and Selangor, while time-equivalent clastics and chert occur in Melaka.

The euxinic-type sediments were succeeded by a sequence of mudstone of flysch facies, the Pokok Sena Formation, and the Singa Formation that locally contain abundant benthos of middle Devonian to early Carboniferous age. The pebbly mudstone of the Singa beds may carry granitic clasts, one of which was dated at 800 to 1100 million years by radiometric methods and thus indicates Precambrian age. Towards south the clastics change into carbonates with subordinate shale, both rock types implying sedimentation in tectonically stable conditions.

In the eugeosyncline east of the Main Range, Silurian to Devonian rocks occur in the foothills and are now known as the Karak Formation. They are accompanied by some serpentinite and amphibolite schist. The formation is at least 4800 m thick and lower Devonian fossils (sponges and brachiopods) are found in the upper part of the sequence (Jaafar Ahmad, 1976). In the central part of the eugeosyncline developed lower Carboniferous limestone, while more to the east clastic sedimentation became more important. Fossil benthos, plants and ripple marks indicate shoaling.

In the northwest and probably also in the central parts of Peninsular Malaysia a major disturbance occurred toward the end of Early Devonian. The rocks in the Langkawi islands and in Perlis were deformed by cleavage and boudinage, and intruded by small granitic bodies during the middle and late Devonian. However, in the Kinta Valley, the middle Devonian limestone forms a conformable part of the entire Devonian limestone sequence. Apparently, a stable shelf persisted throughout the Late Palaeozoic. The area was probably surrounded by deeper water which received detritus from the denudation of folded and uplifted areas. In west Pahang, lower Devonian beds form the top of the lower Palaeozoic geosynclinal rock sequence, but elsewhere in the central and eastern regions the Devonian is not represented. Carboniferous and lower Permian clastics were derived from uplifted and eroded areas in the northwest and were unconformably deposited upon older rocks.

In the central and eastern regions the lower Carboniferous consists of shale and limestone accompanied by tuff. Visean fossils (the brachiopods *Schizophoria, Spirifer, Dielasma,* and the corals *Lithostrotion* and *Amygdalophyllum)* have been determined. There is no coarse clastic material which suggests that if any uplift had occurred in the Devonian, that uplift was probably of minor importance. It appears that the sedimentary conditions persisted until the Late Carboniferous. In the eastern region, sand and grit were produced after the Visean, indicating a change in sedimentation; Carboniferous plant remains in these rocks are *Lepidodendron, Stigmaria, Sphenophyllum.* During the same period, another smaller landmass was also raised in the west-central region. It supplied clastics to the presumably Carboniferous Kenny Hill formation of the Kuala Lumpur area. During the Permian, limestone and shale were again dominant in the eastern region. In the central region a Permian sea with volcanic islands can be interpreted from the sediments. Locally, limestone developed around the islands and upon shoals. Occasional plant fragments drifted into the area, either from the east or from the south of the peninsula where at times the sea withdrew and terrestrial conditions prevailed. In the northwest region, the pure limestone of middle Permian age (fossils: foraminifers *Schwagerina,* corals *Lopophylidium, Caninia,* bryozoans *Fenestella* and *Polypora,* brachiopods: *Reticulatia, Marginifera,* lamellibranchs, scaphopods, gastropods, crinoids and algae) suggests a relatively quiet and shallow sea extending far into Thailand. During the Late Permian widespread uplift took place. Radiometric ages indicate that the uplift was accompanied by granitic intrusions in the northwest area, Kinta Valley, and in the eastern area. In those regions the Triassic rests unconformably upon the Palaeozoic or is absent. However, in some parts of the central region, marine conditions continued without appreciable change in facies from the Late Permian into Early Triassic.

In short, during the Palaeozoic the following geological development took place. The area of deposition was established in the Late Cambrian, and in mid-Palaeozoic time it underwent structural modifications; in the late Palaeozoic it became enlarged and finally the rocks were folded and uplifted as the Thai-Malayan Orogen in early Mesozoic (Jones, 1973).

1.5.2. Mesozoic

The Mesozoic sediments are only preserved in the miogeosynclinal and eugeosynclinal tracts of the peninsula. Similar rocks may have been formed in the geanticlinal zones, but subsequent denudation in those elevated regions may have obliterated any trace. The present form of the peninsula was determined during the Mesozoic. Folding and faulting began in Late Triassic and were accompanied by granitic intrusions that form the Main Range. Many radiometric ages of granitic rocks cluster around Late Triassic and Early Jurassic, suggesting that the main mountain-building event took place during that period. Other dates indicate that some granites experience heating at the end of the Mesozoic; most geologists do not believe that these dates are associated with another period of orogenic movements. After the Early Jurassic, sediments accumulate in thick packets and were deformed into wide folds with wavelengths several km long. This fold style is distinctly different from the tight and sometimes distorted folding that affected older rocks. Gravity and strike-slip faults were also active. Many of the major strike-slip faults show left slip motions, only the Kelau-Karak, approximately north-striking strike-slip fault has dextral sense of movement. The movement pattern of such faults is compatible with regional compression that acted ENE-WSW and that also formed the major structural trends of the Peninsula (Tjia, 1972). See Fig. 1.6.

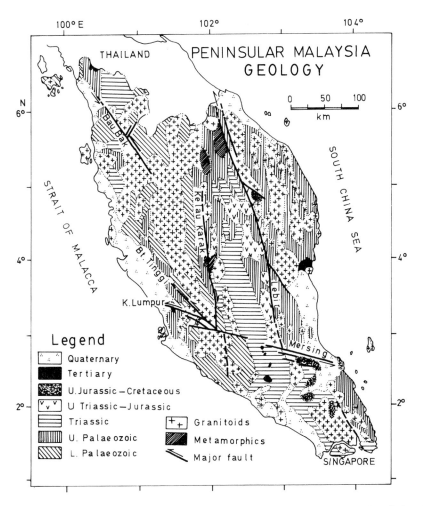

Fig. 1.6. General geological map of Peninsular Malaysia based on various sources. Half-tipped arrows along faults indicate sense of strike-slip movement.

1.5.2.1. Triassic

In general, during Late Permian to Early Triassic the peninsula was an area of non-deposition, except in a few isolated areas where lower Triassic shale, pyroclastics and limestone are now found unconformably overlying Permian rocks (Chung and Yin, 1975).

By the Middle Triassic the sedimentation environment had changed and flysch-type sediments were produced. The presence of Scythian to Norian ammonites makes possible a zonation for the Triassic. During that period sandy shallow water appears to have been restricted to the west of the Main Range, while deep water conditions persisted in the east and alternated with shallow seas in the west. In addition conodonts have been useful in refining the zonation. Within the eugeosyncline the accompanying volcanic tuffs were predominantly silicic.

At the Jengka pass, Pahang, upper Triassic sediments rest unconformably upon upper Permian, tightly folded shale and limestone with diagnostic fossils. Nearby, plant fossils in the uppermost part of the Triassic indicate a change from marine conditions into continental environment.

1.5.2.2. Jurassic–Cretaceous

There are very few fossils indicative of the change from the Triassic into the Jurassic. These fossils indicate continental conditions and among them are *Sagenopteris* flora, *Zamites, Podozamites*, and spores and pollen *Classopollis classoides* and *Circulina* sp. (Chung and Yin, 1975). Upper Jurassic to lower Cretaceous sediments comprise a thick sequence of comparatively undisturbed clastic beds in Kelantan, Pahang and Johor. At some localities these sediments rest unconformably upon older formations, but there is a general belief that these so called Gagau beds are conformable with Jurassic sediments. In the Mt Gagau area the sediments consist of conglomerate, sandstone with minor shale, mudstone, coal and volcaniclastic bands. Towards south the rocks possess a more arenaceous character, the so called Lotong Sandstone that are essentially fluviatile, non-marine sediments. The sediments are about 300 to 360 m thick. The contained flora (*Equisites burchardi, Frenelopsis* among other) indicate a late Jurassic to early Cretaceous age. After their deposition, uplift and tilting took place. The base of the Gagau formation is at 1380 m at Mt Gagau but is only 190 m high in south Johor.

1.5.3. Cenozoic

Land conditions prevailed during the Cenozoic. The known deposits of this period are thin, surficial and appear to be terrestrial. Those in the coastal areas may be partially marine. The Cenozoic geological development of the peninsula has been summarized by Stauffer (1973). One of the rock types was deposited in lakes or lagoons and now outcrops in a number of small basins; the Bukit Arang beds in Perlis, the Enggor coal beds in Perak, the Batu Arang coal beds in Selangor, and in several scattered localities in Johor. The contained plant remains and freshwater gastropod *Viviparus* are not diagnostic for geological age. During this period basaltic volcanism took place: at Segamat, Johor, are dolerite dykes, 60 million years old, and in Kuantan, Pahang basalt flows are 1.6 m.y. old.

In many localities, weathered clastics ranging in size from gravel to clay occur together with peat and partially carbonized plant remains and form low, rounded hills. These deposits are known as "Older Alluvium" and have been laid down at the foot of slopes and along river valleys. Some may be of marine origin. Radiocarbon dates of suitable material all indicate ages beyond the measuring range of the method, but occasional vertebrate fossils suggest a Pleistocene age. There is speculation that the Older Alluvium is the result of higher rates of precipitation and denudation during Quaternary glacial stages.

Other Quaternary deposits consist of terrace sediments along major rivers, shallow marine and lagoonal sediments, sand in barrier and beach ridges, laterite, bauxite and some slope deposits. A group of sediments called the "Younger Alluvium" is distinguished from the Older Alluvium on the basis of weathering. In the Older Alluvium, even the boulders may be thoroughly weathered. Quaternary fluvial sediments also contain tin placers that have made Peninsular Malaysia the world's top producer of the metal. In the Indonesian tin islands a number of good age indicators define the Quaternary age of similar placer deposits. The presence of silicic volcanic ash and tektites are two other interesting features of the Quaternary of the peninsula. The tektites have been collected from opencast tin mines and are believed to belong to the Asia-Australian tektite swarm that rained onto the Southeast Asian and Australian region about 700,000 years ago. There is speculation that the tektites were produced when a large celestial body fell somewhere in the delta region of the Mekong. Acid volcanic ash has been found together with fluviatile and lacustrine sediments in several places in the west as well as in the east coast. The ash deposit in Perak is at least 9 m thick. The nearest probable source for this ash is the Toba cauldron, 250 to 400 km to the west in Sumatra. Fission-track age of zircons from the ash at Serdang, Selangor, has been determined at 30 000 years, similar to the age of one of the ash deposits around Lake Toba.

1.5.4. Volcanic Activity

During Palaeozoic until middle Triassic time, volcanic activity was explosive and most occurred under water contemporaneous with sedimentation. Lava flows are subordinate to pyroclastic products that in composition range from andesitic to rhyolitic. A large portion of the pyroclastics are ignimbrites. Some basic lava may have been produced, now found as serpentinite and metabasic bodies within lower Palaeozoic rocks in Pahang and Negeri Sembilan. Volcanic paroxysm was reached in the Permian and Early Triassic. Then followed decreased volcanic activity during the Late Triassic to Early Jurassic, but it increased again during the Late Jurassic. For the first time in the history of the region were produced basaltic lava flows from fissures. Basalt continued to be produced during the Palaeogene (Segamat, Johor) and also in the Quaternary (Kuantan, Pahang).

Fig. 1.7. A typical metamorphic rock in the peninsula. The phyllite and schist were originally clay and sand, indurated by metamorphism and crumpled by tectonic forces. This rock crops out on the coast; inland, similar rocks have been weathered into softer material.

1.5.5. Igneous Activity

The radiometric ages of granites seem to indicate four periods of graniticactivity. The earliest intrusion took place in late Carboniferous, an other occurred in Early Triassic (approximately 230 m.y.), a third activity was in Late Triassic (about 199 ± 2 m.y.) and a thermal event affected some granite at the beginning of the Tertiary (75 ± 1 m.y.). It is generally accepted that the three earlier dates represent granitic intrusions that accompanied mountainbuilding processes, while the latest age probably indicates re-heating of granite in several parts of the country.

1.5.6. Structural History

The Late Triassic to Early Jurassic orogenic movements produced the general NNW-strike of mountain ranges and valleys of the peninsula. Granitic mountain ranges define a west and an east region, while the central zone possesses lower relief. In detail, structural strikes may deviate from this general NNW grain. Based on the differences in strike, seven tectonic domains can be recognized.

Stratigraphy indicates an interruption in sedimentation sometime in the Middle Palaeozoic as the rocks seem to suggest in the Langkawi islands, the Kuala Lumpur area, and in west Pahang. Detailed fieldwork in the Kuala Lumpur area, near Kuala Terengganu and near Dungun on the east coast, reveal that upper Palaeozoic rocks were subjected to several directions of compressive stress. The earliest stress system that can be identified acted in north–south direction, while the latest compression acted more or less perpendicularly to the regional strike. Post-Triassic to Cretaceous sediments have only been deformed into wide open folds with gently dipping limbs. Faulting appears to have occurred in Jurassic to early Tertiary time. Cenozoic sediments are either flat-lying or have assumed basinal structures, suggesting the influence of gravity without tectonic compressive stresses. The presence of Quaternary basalt near Kuantan, Pahang, supports the interpretation that Peninsular Malaysia has achieved tectonic stability. In granitic rocks, fault and dyke patterns are consistent with a stress system where the maximum principal stress was horizontal and the intermediate principal stress was vertical. This means that after the faults and dykes were formed (at the latest in Early Tertiary) the peninsula has not experienced appreciable tilting. This also supports tectonic stability for the country.

Recently it was found that low-angle thrust structures occur in several localities on the west and the east coast. These structures suggest the probable presence of large overthrust structures. If this is true, a re-evaluation of the geology of Peninsular Malaysia will be necessary.

REFERENCES

Balgooy, M. M. J. van., Kurtak, B. H., Kurtak, D. C., Littke, W. R., Widjojo Parjatmo and Weinheimer, E. A. (1977). A biological reconnaissance of Tasek Pulau Langgun, a sinkhole lake in the Langkawi District, Malaysia. *Sains Malaysiana* 6, 1–30.

Burton, C. K. (1972). Outline of the geological evolution of Malaya. *J. Geol.* 80, 293–309.

Burton, C. K. and Bignell, J. D. (1969). Cretaceous-Tertiary events in Southeast Asia. *Geol. Soc. Am. Bull.* 80, 681–88.

Chia, L. S. (1977). Seasonal rainfall distribution. *Malay. Nat. J.* 31, 11-39.

Chung, S. K. and Yin, E. H. (1975). Brief outline of the geology of Peninsular Malaysia. *Malay. Geol. Surv., Ann. Rept.* 1974, 66–80.

Dale, W. L. (1959). The rainfall of Malaya. Part I. *J. Trop. Geogr.* 13.

Dale, W. L. (1963). Surface temperatures in Malaya. *J. Trop. Geogr.* 17.

Dale, W. L. (1964). Sunshine in Malaya. *J. Trop. Geogr.* 18.

Eyles, R. J. (1970). Physiographic implications of laterite in West Malaysia. *Geol. Soc. Malaysia Bull.* 3, 1–8.

Flint, R. F. (1971). *Glacial and Quaternary Geology*, Wiley, New York.

Gobbett, D. J. (1965). The formation of limestone caves in Malaya. *Malay. Nat. J.* 19, 4–11.

Gray, L. M., Basir Jasin and Tjia, H. D. (1978). Fossils at Sri Medan (Johor). *Warta Geologi* 4(3), 81–4.

Haile, N. S. (1975). Postulated Late Cainozoic high sea levels. *J. Malay. Brch. R. Asiat. Soc.* 48, 78–88.

Imbrie, J. and Palmer-Imbrie, K. (1979). *Ice Ages, Solving the Mystery*. MacMillan, London, 224 pages.

Jaafar Ahmad (1976). Geology and mineral resources of the Karak and Temerloh areas, Pahang. *Geol. Surv. Malaysia, Distr. Mem.* 15, 1–138.

Jones, C. R. (1973). Lower Paleozoic. In *The Geology of the Malay Peninsula* (eds. D.J. Gobbett & C.S. Hutchison), pp. 25–60. Wiley, New York.

Kalpage, F. S. C. P. (1979). *Soils and Fertilisers for Plantations in Malaysia*, Ch.7, pp. 76–81. Incorp. Soc. Planters.

Koopmans, B. N. and Stauffer, P. H. (1968). Glacial phenomena on Mount Kinabalu, Sabah. *Geol. Surv. Malaysia, Borneo Region, Bull.* 8, 25–35.

Stauffer, P.H. (1973). Cenozoic. In *The Geology of the Malay Peninsula* (eds. D.J. Gobbett and C.S. Hutchison), pp. 143–176. Wiley, New York.

Tjia, H. D. (1972). Strike-slip faults in West Malaysia. *24th Session Intl. Geol. Congr.* Montreal, *Sect.* 3, 255–61.

Tjia, H. D. (1973). Geological observations of the Kinabalu summit region, Sabah. *Malay J. Sci.* 2(B), 137–43.

Tjia, H. D. (1980). The Sunda Shelf, Southeast Asia. *Zeits. f. Geomorph.* 24, 405–27.

Tjia, H. D., Asikin, S. and Soeria Atmadja, R. (1968). Coastal accretion in western Indonesia. *Bull. Nat. Inst. Geol. and Mining* 1(1), 15–45.

Tjia, H. D., Fujii, S. and Kigoshi, K. (1977). Changes in sea level in the southern South China Sea area during Quaternary times. *United Nations, ESCAP-CCOP, Tech. Pub.* 5, 11–36.

Verstappen, H. T. (1976). The effect of Quaternary tectonics and climates on erosion and sedimentation in Sumatra. *Proc. Indonesian Petrol. Ass., 4th Ann. Conv., Jakarta 1975*, pp.49–53.

Walker, D. (1956). Studies of the Quaternary of the Malay Peninsula. I. Alluvial deposits of Perak and the relative levels of land and sea. *Fed. Mus. J. (Malaya)* 1–2, 19–34.

Completed June 1985

CHAPTER 2

Forest Types and Forest Zonation

T.C. Whitmore

Department of Plant Sciences, University of Oxford, Oxford OX1 2RB, U.K.

CONTENTS

2.1. INTRODUCTION

Around the equator the climate is cloudy and more or less permanently rainy or perhumid, nowhere more so than in Malesia (see map at end papers). In this belt of wet, cloudy climates the prevailing vegetation is tropical rain forest, the most complex and species-rich vegetation which exists or ever has existed on earth. Close inspection shows there are many different types of tropical rain forest. The major types (called 'forest formations'), defined by particular structure and physiognomy, can be found in all parts of the tropics.

Most rain forest formations are represented in Peninsular Malaysia, as can be seen from the recently published vegetation map of the region (Whitmore, 1984a). These major types are restricted to particular habitats and where, as is often the case, habitat boundaries are fairly sharp, so are the forest formation boundaries. But where habitats change gradually, so too does the forest from one formation to another.

The particular genera and species of tree, i.e., the floristics, of each formation vary from continent to continent, regionally, locally and within forest formations. The last kind of variation is mostly without sharp boundaries. It is manifold and incompletely understood.

Most of the rain forests of western Malesia (see Preface) are dominated by members of the single tree family Dipterocarpaceae. There are no other forests anywhere in the world which have so many genera and species of a single tree family growing together in the same place. Many dipterocarps are very lofty trees; commonly they attain 50 m and occasionally 70 m height. The western Malesian dipterocarp rain forests are taller than any other tropical rain forests, and have more different tree species growing together in a single area than any other forest.

Peninsular Malaysia lies almost entirely within the perhumid equatorial belt but in the extreme northwest slightly seasonally dry climates occur, and these reach southwards in attenuated form as a triangle whose apex lies at about Kuala Lipis, Pahang. A different rain forest formation, comprised of different species whose centres of distribution lie in Burma and Thailand, occupies this area.

Forests once clothed virtually the whole of Peninsular Malaysia. This original vegetation has now become fragmented so that in the lowlands only patches of primary forest remain. Much of the land has been converted to agricultural use, beginning on a large scale on the western plains last century and extending rapidly in the past two or three decades to all flattish lowland areas. Most of the lowland rain forests which have survived have been logged for timber and, except in conservation areas, logging is expected to have affected all by about 1990 (Chapter 10). It follows that much of the still-forested lowlands now carry a production forest, parts of it silviculturally treated, which has little structural resemblance to virgin rain forest and has species present in different proportions. The montane forests have been much less affected, except at their lowest fringe, where upper dipterocarp forest (see below) contains valuable timber, and where roads cross the mountains or penetrate them for hill stations and telecommunications transmitters.

2.2. THE FOREST FORMATIONS

The forest formations and the habitats they occupy are shown in Table 2.1. The most important distinction in Peninsular Malaysia, as globally, is that provided by climatic seasonality. In the perhumid climate of most of the country, a further distinction can be made between drylands and wetlands, the latter bearing various kinds of swamp forest. On dry land there are distinctive forest formations on nutrient-poor, podzolized sands (Chapter 1.3) and on limestone (Chapter 1.1). These occupy only a small area. Most of the country has zonal tropical soils, ultisols and oxisols (Chapter 1), and carries lowland evergreen rain forest in the lowlands and one of two different montane formations, depending on elevation, in the mountains.

2.2.1. Lowland evergreen rain forest

This, the principal forest formation, once occupied all dry land mesic sites where there is no limiting factor and covered most of the country up to about 750 m elevation. It is a lofty forest with the biggest trees reaching 40–50 or a maximum 80 m tall, and a dense undergrowth of small trees of all sizes. The canopy top is a billowing expanse of individual or grouped giant emergent trees. Buttresses are frequent and large. There is much cauliflory. Both bole climbers and big independent woody climbers are often abundant. Pinnate leaves are frequent. Characteristically this forest is very rich, with well over one hundred tree species $\geqslant$ 10 cm diameter at breast height per hectare and sometimes up to 200 or more

TABLE 2.1. The main forest formations of Peninsular Malaysia.

Climate	Location	Soils	Elevation	Forest formation
Ever-wet (perhumid)	dry land, inland	zonal	lowlands to 1200 m	1 lowland evergreen rain forest
			mountains 1200–1500 m	2 lower montane rain forest
			over 1500 m	3 upper montane rain forest
		podzolized sands	lowlands	4 heath forest
		limestone	lowlands	5 forest over limestone
	dry land, coastal			6 beach vegetation
	wetland, coastal			7 mangrove
	wetland, inland	eutrophic, muck & mineral		8 freshwater swamp forest
		oligotrophic peats		9 peatswamp forest
Seasonally dry	dry land, inland in far NW	zonal	lowlands	10 semi-evergreen rain forest

per two hectares (Chapter 9). In most places Dipterocarpaceae are abundant and diverse. Local variations in flora, between localities, between ridges, slopes and valleys and on granite and sedimentary rocks have been detected. Fine patterns of variation are discussed at the end of the chapter.

2.2.2. Heath forest

Heath forest lies mainly along the east coast, including the permatang (Chapter 1.1), but with a small area on the Perak coast at Tanjung Hantu, though probably long ago elsewhere also. This forest formation is very extensive in Borneo where it is called 'kerangas' (Whitmore, 1984a). It is easily degraded by fire after felling and much has been converted to open grassland with scattered shrubs and woodlands. Along the east coast lie a series of permatang, lying parallel to the shore with periodically swampy swales between. Heath forest (still remaining at Mencali and Jambu Bongkok, Terengganu) has rather few dipterocarps, notably *Shorea materialis, S. glauca* and a few *Hopea* species. Tiny areas of heath forest are also associated with quartz and quartzite ridges, such as occur at Kelang Gates and in the Gombak valley, Selangor.

A forest allied to heath forest occurs inland, firstly the summit of the plateau in the Endau-Rompin area and on ridge crests along the east coast, in which a fan palm *Livistona endauensis* (Chapter 4) is conspicuous and also on the summit plateau Gunung Panti, Johor, where there is a very peculiar flora and the ground is periodically waterlogged.

2.2.3. Forest over limestone

The karst towers occurring from Selangor and Pahang northwards (Chapter 1.1) and the low limestone boulder outcrops in Johor provide a great diversity of habitats and consequently a variety of vegetation types, from 30 m tall forest on colluvial footslopes and damp gulley bottoms to low dense scrub on exposed crests. The flora is extremely rich, many species are restricted to limestone and some are endemic to the peninsula. Most species are limited to only one of the very distinct communities. The most characteristic limestone plants are the small and beautiful herbs of rock faces (Chapter 6). On the limestone hills of the northwest a few dipterocarp species occur and are common on limestone but south of Kedah the family shuns these habitats. The forest which develops on acidic peaty soil on the upper slopes of limestone hills has several species in common with heath forest. The drier parts of limestone hills suffer periodic drought and are prone to damage by fire. The limestone forests of Langkawi and Perlis, in the seasonally dry far north west develop 'autumn' yellow and red coloration during January and February and are noticeably deciduous. The limestone accentuates the dryness of the season. In this region is a strong Burma-Thailand floristic element which contributes markedly to the overall floristic richness of Peninsular Malaysia (for example in Euphorbiaceae).

2.2.4. Beach vegetation

Accreting sandy coasts have a zone of herbs and shrubs above high water mark (Chapter 6). This fringes a narrow belt of beach or strand forest of tree species widely spread through the whole Indo-Pacific region, and including *Barringtonia asiatica, Calophyllum inophyllum, Cerbera mangas, Hibiscus tiliaceus* and *Terminalia catappa*. On accreting beaches *Casuarina equisetifolia* forms pure stands. Along the east coast lines of *Casuarina* persist inland on the permatang, mentioned above. After destruction, beach forest is often replaced by scrub with abundant *Rhodomyrtus tomentosa*.

2.2.5. Mangrove forests

Mangrove forests are confined to muddy shores, lagoons and tidal estuaries and are mainly found along the sheltered west coast. Out of a total 140,000 ha, 51,200 ha are found in Perak. Selangor and Johor also have large areas. On the east coast, mangrove occurs as small patches around the mouths of rivers. These are simple forests with the canopy 6–24 m tall, essentially of a single layer of trees, and are floristically poor compared to all other peninsular forests, (though these mangroves are in fact among the most species-rich in the world). Stilt roots *(Rhizophora)* and pneumatophores *(Avicennia, Bruguiera, Sonneratia, Xylocarpus)* occur. Many species are viviparous *(Bruguiera, Ceriops, Kandelia, Rhizophora)*. The composition of mangrove forest varies from place to place, dependent principally on the soil texture, duration and frequency of inundation (hence salinity) and degree of exposure. A regular zonation occurs. *Avicennia* species are pioneers on exposed sites. The mangrove palm *Nypa* forms extensive stands and so does the mangrove date palm *Phoenix paludosa*. Inland, mangrove merges into brackish water forest, known as 'hutan darat', which has its own typical flora including big clumps of nibong palm *Oncosperma tigillarium* (see Chapter 4), and the trees *Dolichandrone spathacea* and *Brownlowia argentata*.

Mangroves have long been utilized for poles, firewood, charcoal and formerly cutch. The regrowth forest is simpler. Some areas have been converted to prawn ponds or saltpans or filled to provide building sites. The importance of mangrove as habitats for fish and shellfish and as breeding grounds for these is increasingly being appreciated.

2.2.6. Freshwater swamp forest

This formation occurs where the soil is inundated with fresh water either permanently (for example at Tasik Bera, see Chapter 16) or periodically. There may be a layer of peat or muck (defined as organic soil with a loss on ignition of 35–65 per cent), but unlike peatswamp forests this is only a few centimetres thick and the water does not come entirely from rain. Inundation may be seasonal or, inland, may occur daily only at spring tides. The soils are highly variable, and include much alluvium of varying textures.

The floristic composition, structure and physiognomy of freshwater swamp forests varies enormously, from low shrublands to 30 m tall forest. Dipterocarpaceae are seldom dominant. This forest formation once occurred in all parts of Peninsular Malaysia but many areas have now been felled and drained for agriculture. Some seral communities developed after felling are dominated by pure stands of *Macaranga pruinosa* or *Campnosperma coriaceum*. Where there has been repeated disturbance forests of gelam *Melaleuca cajuputi* (important for firewood and able to form coppice shoots) or riang-riang *Ploiarium alternifolium* develop.

2.2.7. Peatswamp forest

This formation occurs just inland from the coast on both sides of the peninsula especially in the south east (Fig. 2.2).The peat is semi-liquid under a solid crust of undecayed woody litter and tree roots and is from under one to 10 m thick. The peat is slightly domed. The only incoming water is from rain and is nutrient deficient. In these oligotrophic, permanently waterlogged and anaerobic conditions organic matter does not decay. The drainage water is tea-coloured. The west coast peatswamp forests have formed over clay, those on the east coast over sand. Such forests are even more extensive in Sumatra and Borneo. Their occurrence appears to depend on a particular topographic setting. The forest is about 30 m tall and structurally complex. The flora is fairly rich, and most of its species are restricted to this formation, though kempas *Koompassia malaccensis* is also common on dry land. Dipterocarpaceae are only locally dominant (especially *Shorea teysmanniana*). Other important trees are punah *Tetramerista glabra*, *Shorea platycarpa* and *Durio carinatus* (east coast only). The sealing wax palm *Cyrtostachys renda* is conspicuous (Chapter 4).Much of the shallow peat has been converted to rubber, oil palm or pineapple plantations. After clearing, secondary forest of pure *Macaranga pruinosa* may develop.

2.2.8. Semi-evergreen rain forest

This dry land rain forest formation of mesic sites and zonal soils occurs in a small area of the far northwest in Perlis, Langkawi and north Kedah, where there is annually a marked dry season. In Malesia this formation forms a narrow belt between evergreen rain forest, which covers most of the archipelago, and the true seasonal or monsoon forests of strongly seasonal climates, and its total area is small. In Peninsular Malaysia the boundary of the two formations is approximately at a line connecting Kangar in Perlis to Pattani in south Thailand.

In structure and physiognomy semi-evergreen rain forest differs only slightly from evergreen rain forest. It is tall, 30–40 m, and emergents are rather rare, occurring as individuals rather than as groups. In the top of the canopy some deciduous species occur. Tree species occur in intimate mixture but with a distinct tendency to gregariousness. Big woody climbers tend to be very abundant. Bamboos are present. Epiphytes are frequent to occasional. The experienced eye can at once distinguish this subtly different formation from evergreen rain forest on its structure and physiognomy. It is more conspicuously distinct in its floristic composition. Dipterocarpaceae are abundantly and richly represented. *Shorea* mainly occurs as the White Meranti group (notably *Shorea assamica* f. *globifera*, *S. hypochra*, *S. roxburghii* and *S. sericeiflora*) (see Chapter 3). The Red Meranti and Damar Hitam groups of *Shorea* are totally absent and there are only *S. guiso* and *S. ochrophloia* of the Balau group. *Parashorea stellata* is common and characteristic, as are certain species of *Anisoptera, Dipterocarpus, Hopea* and *Vatica*. There is a complex interdigitation of semi-evergreen and evergreen rain forest, probably ultimately dependent on periodic water stress but interacting with soil nutrient status. Species of evergreen rain forest occur furthest north on granite and semi-evergreen rain forest species penetrate as far south as the upper Tembeling drainage, Pahang, an intermontane valley shielded from both wet monsoons.

Much of the flat lowland plain of the far northwest bears a forest dominated by *Schima wallichii, Shorea roxburghii* and bamboos (principally *Gigantochloa latifolia* and *G. ligulata*, see Chapter 5). This serves to obscure the Kangar-Pattani boundary line. It is a mixture of the more adaptable species of evergreen rain forest with colonists from secondary forest, semi-evergreen rain forest and the monsoon forest formations of Burma and Thailand. *Schima*-bamboo forest is believed to have been derived from evergreen rain forest by intermittent cutting, grazing, cultivation and burning and degraded thereby to this drier forest type.

2.2.9. Montane rain forests

There is a change in structure and physiognomy with increasing elevation shown on Figure 2.1 and summarized in Table 2.2. The approximate extent of the montane forests can be judged from the mountain areas on Fig. 1.1. At about 750 m elevation on the main ranges, but lower on outlying hills, lowland rain forest merges, usually over a broad ecotone, to *lower montane rain forest*. Conspicuous differences are that this has a smoother canopy, only 15–33 m tall and without emergents, buttresses are smaller and less frequent and big woody climbers are rare or absent.

At about 1500 m on the main ranges lower montane rain forest is replaced by the *upper montane rain forest* formation. The boundary is usually sharp. Upper montane rain forest is highly distinctive in its low, even canopy, frequently only 10 m tall, and in its trees which have tiny leaves (microphylls), crowded in dense subcrowns on gnarled limbs. The soil usually has a layer of peat. On exposed peaks and ridges this

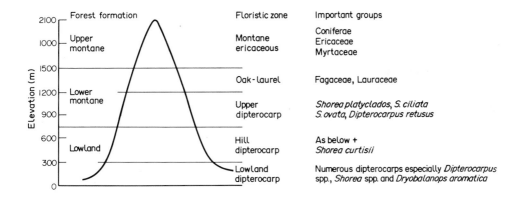

Fig. 2.1. Vegetation zones on the main mountains of Peninsular Malaysia (after Symington, 1943).

TABLE 2.2. Characters used to define the principal montane forest formations.

Formation	Tropical lowland evergreen rain forest†	Tropical lower montane rain forest	Tropical upper montane rain forest
Canopy height	25–45 m	15–33 m	1.5–18 m
Emergent trees	Characteristic, to 60 (80) m tall	Often absent, to 37 m tall	Usually absent, to 26 m tall
Pinnate leaves	Frequent	Rare	Very rare
Principal leaf size class of woody plants‡	Mesophyll	Mesophyll	Microphyll
Buttresses	Usually frequent and large	Uncommon, small	Usually absent
Cauliflory	Frequent	Rare	Absent
Big woody climbers	Abundant	Usually none	None
Bole climbers	Often abundant	Frequent to abundant	Very few
Vascular epiphytes	Frequent	Abundant	Frequent
Non-vascular epiphytes	Occasional	Occasional to abundant	Often abundant

† Included for comparison.
‡ Following Raunkiaer.

Fig. 2.2. Lowland evergreen rain forest in the upper valley of the Batang Kali, Selangor. The protruding giant trees with pale crowns on the ridge crests are seraya *Shorea curtisii*.

forest may be as short as 2 m tall. The stunted facies is sometimes called "elfin forest". The lower limit of upper montane rain forest is at the lower level of prevalent cloud formation. In places the trees are swathed in bryophytes (mostly hepatics) and this facies is often called mossy forest. In especially cloudy places lower montane forest has a mossy facies, too, and a few stunted lower montane forests have been discovered, e.g., on Mt Mandi Angin. On outlying peaks, e.g., Mt Belumut, upper montane rain forest abuts directly on lowland rain forest, and everywhere it descends to lowest elevations along ridge crests, ultimately persisting only on knolls with lower montane forest on the intervening cols.

There is a parallel series of floristic zones with elevation and these are summarized on Figure 2.1. The lower part of lower montane forest is termed *upper dipterocarp forest*. In Peninsular Malaysia, this zone has now been extensively altered by timber extraction. In fact, few dipterocarps reach this elevation, principally *Shorea ovata* and *S. platyclados*, with lesser numbers of *S. ciliata, S. submontana, Dipterocarpus costatus* and *D. retusus*. These species occur up to *c.* 1200 m. Their upper limit appears to be determined by the inability of their seedlings to establish successfully in peat, which first appears at about that elevation. The main floristic zone of the lower montane forest is so-called *oak-laurel forest* because of the prominence of species of Fagaceae (mainly *Castanopsis* and *Lithocarpus*) and Lauraceae. Fraser's Hill lies in this floristic zone.

Upper montane forest is another floristic zone, named *montane ericaceous forest*, because of the conspicuous occurrence of *Rhododendron* and *Vaccinium* species, but there are also other well-represented families, especially Myrtaceae (*Eugenia, Rhodamnia, Tristania*). The Cameron Highlands and Genting Highlands lie in uppermost lower montane rain forest and extend into upper montane. Lower montane seral forests have abundant *Homalanthus populneus*, the tree fern *Cyathea contaminans* and *Symingtonia populnea*. Upper montane forest is commonly replaced by fern brakes with abundant *Dicranopteris* and *Matonia pectinata*.

Altitudinal zonation of rain forest formations on perhumid mountains is found throughout the tropics though on giant cordilleras, such as in Sumatra and New Guinea, the boundaries are considerably higher than in Peninsular Malaysia. There has been much debate as to whether this conspicuous and regular zonation is determined principally by decreasing availability of mineral nutrients with elevation (with much unavailable to plants in the peat) or to rare periods of water stress. It seems likely that the controlling factor is not always the same.

2.3. VARIATION WITHIN LOWLAND EVERGREEN RAIN FOREST

Some of the variants of this, the most extensive forest formation of Peninsular Malaysia, are more distinctive than others. The fragments which remain unaltered will soon be so small that the full diversity that once existed will now never be recorded. Several big trees are commonest on alluvial valley floors, for example *Pentaspadon* spp., *Pometia pinnata* and *Pterygota alata*. *Shorea ovalis* is found in valleys but not on alluvium. Another set of species is most common on ridge crests, this includes damar minyak *Agathis borneensis*, and species of bintangor and kelat *(Calophyllum, Eugenia)*.

Some species have been shown to be commoner on, or exclusive to, granite or sedimentary-derived soils, but the prevalent impression in the peninsula, that most lowland species occur everywhere, arises because several of the commonest and most easily recognized species have been shown to be indifferent to soil type or topography, namely kekatong *Cynometra malaccensis*, jelutong *Dyera costulata*, merbau *Intsia palembanica*, tualang *Koompassia excelsa* (curiously absent from south of a line drawn from Kuala Lumpur to Kuantan), kulim *Scorodocarpus borneensis* and nemesu *Shorea pauciflora*.

Evergreen rain forest occurs on the plains and also up into the hills to about 750 m elevation. A floristic distinction has sometimes been drawn between lowland and hill dipterocarp forest. The only difference however is the presence of the conspicuous, grey-crowned, giant tree, seraya, *Shorea curtisii*, which forms groves along the crests of ridges, commonly in association with the big clump-forming stemless palm bertam, *Eugeissona tristis* (Fig. 2.2). Seraya also occurs almost at sea level on coastal hills (e.g., Cape Rachado). It has been shown that seraya sites are prone to periodic water stress. In topography the hills of Peninsular Malaysia rise sharply from the plain at the so-called steepland or hill-foot boundary. There is no change in forest composition at this boundary, although it is significant in the ecology of mammals (Chapter 12) and birds (Chapter 13).

A broad subdivision was made across the country of the lowland evergreen rain forest into a number of widely defined floristic associations, most of these were said to have particular dipterocarps commoner than others and a few were almost devoid of dipterocarps. These floristic associations and their habitat preferences were not very precisely defined or mapped. They have now largely disappeared because of logging, as the conditions for forest regrowth after massive disruption favour the faster growing species, which were not always abundant before. The approximate former occurrence of these floristic associations is indicated on the map, Figure 2.3.

Red meranti-keruing forests had abundant species of the red meranti group of *Shorea* and of *Dipterocarpus*. *Shorea acuminata, S. curtisii* (locally*), S. leprosula, S. macroptera, S. ovalis, S. parvifolia, Dipterocarpus baudii, D. cornutus, D. grandiflorus, D. kerrii, D. sublamellatus, D. verrucosus, Anisoptera laevis, Canarium* spp., *Dyera costulata, Koompassia malaccensis, Myristica* spp., *Palaquium maingayi, Santiria* spp., *Scaphium affinis* and *Heritiera simplicifolia* were all common.

Balau forests were characterised by the balau group of *Shorea*, of which *S. atrinervosa, S. exelliptica* and *S. maxwelliana* were commonest, sometimes almost replaced by chengal *Neobalanocarpus heimii*. Species of *Dipterocarpus* and red meranti *Shorea* were also frequently present.

Keruing forests occurred on poorly drained land and comprised a mixture of *Dipterocarpus* species often associated with *Dryobalanops oblongifolia, Hopea mengarawan* and *Shorea lepidota*, plus the non-dipterocarps *Koompassia malaccensis* and *Palaquium* species.

Kapur forests, in which the single dipterocarp kapur *Dryobalanops aromatica* dominated (comprising 60-90 per cent of the timber volume), were extensive near the east coast on sedimentary rocks (with tiny outliers in Selangor, mainly at Kanching). In hilly country, where kapur was frequently associated with merpauh *Swintonia penangiana*, the densest stands of *Dryobalanops* were on the upper slopes of ridges, not the crests.

Other forests occurred in which there was an abundance of single dipterocarps, i.e., chengal, nemesu *Shorea pauciflora* or, on coastal hills, damar laut merah *S. kunstleri*, though in none of these was the single-species dominance so marked as in kapur forests.

Two extensive associations occurred in which dipterocarps were rare and locally absent. *Kempas-kedondong forest* was characterized by abundance of *Koompassia malaccensis, Canarium* spp. and *Santiria* spp. *Calophyllum* species were usually common and around Jemaluang, Johor, this genus was dominant. In the main range foothills and along nearby rivers occurred *merbau-kekatong forest (Intsia-Cynometra)*. Both these forest associations had a lower canopy and lesser volume of timber than the dipterocarp-dominated parts of the evergreen rain forest.

The factors which determine predominance of some species in particular locations are not fully understood, though they have been unravelled to a certain extent. The lowland forests poor in dipterocarps may have attained this composition as a result of selective timber extraction over a very long period. In the case of kapur, this species has a limited geographical range in the country and many other examples of restricted range are known. Tualang has been mentioned. The gregarious palm bertam is very patchy in range and mainly western: only three clumps are known in Taman Negara. The conspicuous giant palm sal *Johannesteijsmannia altifrons* is likewise very restricted. For these palms localized extinction has been postulated, perhaps by failure of seedling establishment in a climate which has become slightly more seasonally dry than in the past. There is a group of species restricted roughly to

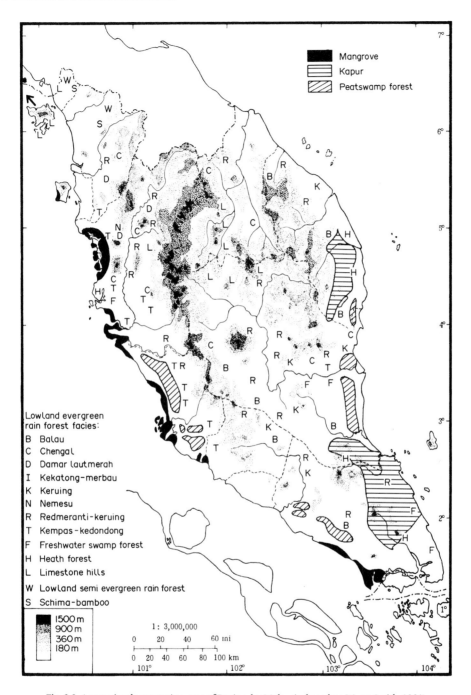

Fig. 2.3. A very simple vegetation map of Peninsular Malaysia (based on Wyatt-Smith, 1964).

the southeast flanks of the Main Range, e.g., *Albizia pedicellata* and *Dialium procerum*. Analysis of the ranges of 1759 tree species (394 genera, 43 families) has shown that, overall, there is no concentration of Bornean species in the east or Sumatran ones in the west of the peninsula, which is a reflection of the close similarity of the Sundaland forests.

Another factor which influences species composition is massive disturbance. A violent storm in 1883 destroyed the forest over a large area of north Kelantan and Terengganu. Eighty years later a forest had redeveloped which bore witness to the catastrophe in its very uneven structure and abundance of meranti tembaga and meranti sarang punai, *Shorea leprosula, S. parvifolia*, many with curved stems. Similar storms are known from elsewhere. One in Melaka in 1917 resulted in a similar *Shorea*-dominated forest, for example in the former Merlimau forest reserve. These are examples where fast-growing, light-demanding trees establish after catastrophic disturbance and will dominate the upper canopy for perhaps a century. It may be conjectured that in the absence of a further catastrophe they will be replaced by more shade-tolerant dipterocarps. Perhaps in this lies at least part of the explanation of the occurrence of areas of, on the one hand, red meranti-keruing forest and, on the other, balau or chengal forest. In the Jengka forest, Pahang, tree positions were mapped on an area of 420 × 640 m. *Dipterocarpus crinitus*, known to be rather light-demanding, was only represented as scattered big trees; *Shorea leprosula* appeared to be invading the plot from the west. At Jengka, and also at Pasoh forest, Negeri Sembilan, several dipterocarps have been observed to occur as family groups, one or a few big trees surrounded by a grove of small ones. Thus we see distribution on a regional and a local scale apparently being determined by ecological response to canopy disturbance and to dispersability of fruits. Detailed study at Pasoh has shown in addition that different groups of tree species are associated with local small differences in relief, drainage and soil.

Variation within the lowland evergreen rain forest occurs at many scales. Some kinds of variation are totally disrupted by logging operations. The composition of the regenerating forest is strongly influenced by the disturbance caused by logging. Ecological studies in regenerating forests will necessarily have to allow for this if they are to continue to elucidate the patterns in the virgin forest.

BIBLIOGRAPHY

There are a few key references which have been drawn on extensively for this chapter and which describe the forests of Peninsular Malaysia in fuller detail. In addition there are papers on particular forest formations.

Burgess, P. F. (1969). Ecological factors in hill and mountain forests in the States of Malaya. *Malay. Nat. J.* 22, 119–28.

Chin, S. C. (1977). The limestone hill flora of Malaya. I. *Gdns. Bull. Singapore* 30, 165–220.

Corner, E. J. H. (1978). The freshwater swamp forest of South Johore and Singapore. *Gdns. Bull. Singapore,* Supplement 1.

Symington, C. F. (1943). Foresters' manual of dipterocarps. *Malay. Forest Record* 16 (reprinted University of Malaya Press, 1974).

Whitmore, T. C. (1984a). A vegetation map of Malesia at scale 1:5 million. *J. Biogeogr.* 11, 561–71.

Whitmore, T. C. (1984b). *Tropical rain forests of the Far East.* Oxford, Clarendon Press. (2nd edition).

Wyatt-Smith, J. (1959). Peat swamp forest in Malaya. *Malay. Forester* 22, 5–31.

Wyatt-Smith, J. (1961). A note on the freshwater swamp, lowland and hill forest types of Malaya. *Malay. Forester* 24, 110–21.

Wyatt-Smith, J. (1963). Manual of Malayan silviculture for inland forests, chapter 7. An introduction to forest types. *Malay. Forest Records* 23.

Wyatt-Smith, J. (1964). A preliminary vegetation map of Malaya with description of the vegetation types. *J. Trop. Geogr.* 18, 200–13.

Completed December 1985

CHAPTER 3

The Dipterocarps

Marius Jacobs†

Late of the Rijksherbarium, Leiden, Netherlands

CONTENTS

3.1. THE FAMILY DIPTEROCARPACEAE

The Dipterocarpaceae comprise about 515 species in 16 genera in three subfamilies. The largest subfamily, Dipterocarpoideae, consists of about 495 species in 13 genera, represented in the Seychelles, Ceylon, India, and Southeast Asia eastwards to New Guinea (not in the Bismarck Islands), most richly in the rain forests of the Sunda subregion and the humid parts of Ceylon. Representation in New Guinea is modest (Jacobs, 1981b). The second subfamily, Monotoideae, occupies tropical Africa between approximately 12°N and 12°S, in the not too dry tropical parts but, remarkably enough, not in rain forest. There is one outlier in Madagascar. The genus *Monotes* has about 15–20 species in the above area and *Marquesia* has three, in the more eastern parts of Africa. The third subfamily, Pakaraimoideae, is monotypic, comprising the one species, *Pakaraimaea dipterocarpacea*, occurring in Guyana and Venezuela. Like the African dipterocarps, it is a savanna tree.

The family is thus pan-tropical in present distribution, with each of the three subfamilies confined to a separate continent. Four observations are consistent with the supposition that the origin of the family lay in the ancient southern continent of Gondwanaland, in a region now part of tropical Africa: primitive features of the Monotoideae; the presence of fossil Dipterocarpoideae in Africa; the richness of the family in Ceylon; and the sudden massive appearance of the family in the pollen record of Sundaland (notably in the northwestern area of Borneo) at a date over 30 million years ago, following the collision of the India-Ceylon tectonic plate with southern Asia (Audley-Charles *et al.*, 1981).

During the subsequent period, the dipterocarps presumably continued to evolve. Their range expanded further into the Philippines and thence to Celebes, the Moluccas and New Guinea. In the

Sunda subregion where, at the time of their arrival, the forests might have been of modest stature, the dipterocarps grew into really big trees. It has been speculated that they have evolved the physiological capacity to recycle mineral nutrients with extreme economy, thus permitting the attainment of large size on poor soils.

3.1.1. Principal references

Foxworthy (1932) treated 122 species in the Peninsular Malaysian flora, setting forth the taxonomy based on anatomy and morphology as can be studied in the herbarium. This book discussed variation, distribution, field characters, pests and diseases, but gave very little information on ecology. Foxworthy included some notes by C. F. Symington. The latter went on to establish himself as the leading worker on the family in Peninsular Malaysia. A true forest botanist, he synthesized herbarium and field knowledge. His *Foresters' Manual of Dipterocarps* (Symington, 1943) is still a classic. Post-war, Peter S. Ashton, a pupil of E. J. H. Corner (Chapter 8), in 1957 came to Brunei, a country which has about the finest dipterocarp forest in the world. Ashton studied both ecology and taxonomy. His Brunei work resulted in two books (Ashton, 1964a, 1964b), followed by a supplement to cover Sarawak forests (1968), and a series of papers describing new species, accounting for changes in nomenclature and revising taxonomic delimitations (Ashton, 1963, 1969). Finally, in a full treatment of the family (Ashton, 1982), he has discussed each species in Malesia (380 in 10 genera) along customary lines. For each genus there is a distribution map, and a full-page plate of the habit with flowers and fruits, supplemented by such other drawings as are needed to illustrate subgeneric diversity, and habitat photographs. The taxonomic part of the work is based on herbarium materiel. In *Shorea* the species have been grouped into 10 botanical 'sections', distinguished on details of the flower, rather than into field groups which to some extent cut across these sections. Accordingly, the keys in *Flora Malesiana* require whole, fertile specimens. To identify slash, fallen leaves or fruits, one needs Ashton's Brunei or Sarawak *Manuals* with their field keys and illustrations for each species of one leaf and one fruit, or Symington's *Manual* for Peninsular Malaysia.

3.2. THE GENERA AND SPECIES OF DIPTEROCARPOIDEAE

For the Dipterocarpoideae over their full range, Ashton has distinguished two groups of genera, namely:

(i) those with scattered resin canals, valvate fruit calyx and lobes, and 2n = 22 chromosomes: *Anisoptera, Cotylelobium, Dipterocarpus, Stemonoporus, Upuna, Vateria, Vateriopsis* and *Vatica*; and

(ii) those with resin canals in tangential bands, imbricate fruit sepals and 2n = 14 chromosomes: *Dryobalanops, Hopea, Neobalanocarpus, Parashorea* and *Shorea*.

Among group (i), *Anisosptera* (11 species) extends from Assam to New Guinea; two sections are distinguished on the number of stamens. *Cotylelobium* has a disjunct range, with one part in Ceylon and another part in Thailand and the Sunda shelf: Peninsular Malaysia, Sumatra, and Borneo. *Dipterocarpus* extends from Ceylon through India and Burma into the Indo-Chinese peninsula as far north as China, and into Malesia as far east as Sumbawa in the Lesser Sunda Islands. Its 69 species form a coherent block; no subgeneric division can be made. *Stemonoporus* is endemic to Ceylon, with 15–20 species. *Upuna* (1 species) is confined to Borneo; taxonomically it is isolated and in many ways primitive. *Vateria* is confined to south India and Ceylon, with one endemic species in each. *Vateriopsis* is endemic to the Seychelles, with one endemic species. *Vatica* (65 species) extends from Ceylon and India through Southeast Asia, including southern China, to the Lesser Sundas; two sections are distinguished, one with equal, one with unequal fruit lobes.

As for group (ii) *Dryobalanops* is another Sundaland genus, homogeneous, with 7 species. In former times, it was more widely distributed: fossils are known from south India, Cambodia, west Java and Ambon. Through the camphor in the trees of this genus the family became first known in the West; Marco Polo in the 13th century made mention of the substance. *Hopea* (102 species) extends from Ceylon through India and south China, through Malesia with the exception of the Lesser Sundas. The one species of *Neobalanocarpus, N. heimii* is confined to Peninsular Malaysia and peninsular Thailand. *Parashorea* (15 species) extends from Burma to Sundaland, and the Philippines. *Shorea* (194 species) is well represented in Ceylon, in India, Burma and Indo-China, with many in Thailand, but centred in Malesia, extending as far as the Moluccas.

Diversity at the generic level is thus summarized as follows: in the Seychelles there is one genus; in Ceylon, 7; in India south of the Godavari river (evidently a major barrier to distribution), 5; India north of the Godavari, 4; the Andamans, 2; the continental Southeast Asian block (i.e., Burma, southern China, Indo-China and Thailand, extending into northern Peninsular Malaysia), 6, excluding *Cotylelobium* and *Neobalanocarpus* which occur only in peninsular Thailand; in the Sunda subregional block (i.e., Sumatra, Peninsular Malaysia, Borneo, the Philippines, Java and Lesser Sunda Islands, Celebes and Moluccas), 9; and New Guinea, 2.

Diversity and endemicity at the species level are illustrated in Table 3.1. Of the total of 495 recognized species of Dipterocarpoideae, 380 (77%) occur in Malesia. All Malesian species which extend to continental Southeast Asia (as defined above) also occur in Peninsular Malaysia. Conversely, of the 135 species occurring outside Malesia, 54 extend to Peninsular Malaysia (9 of them only in the extreme north). The Peninsular Malaysian flora is thus extremely rich, with 156 species, but moderately poor in endemics (26 species 17%).

3.3. ECOLOGICAL NOTES

These present distribution patterns reflect routes of colonization and past climatic conditions. Dipterocarps have limited powers of dispersal, with seeds unable to cross distances of 500–800 m. Dipterocarps therefore must have reached the stations where they now grow overland, in the great majority of cases. The speed of migration is a matter of speculation but if, as a rule of thumb, we assume that a tree is able to get some of its seeds 500 metres, and that at the most 50 years elapse between germination and first fruiting, we arrive at an idea of 1 km in 100 years.

As already noted by Whitmore (Chapter 2), the dipterocarps constitute prominent elements of the lowland rain forest formations in Peninsular Malaysia. All belong to the primary forests, so it is these which contain the entire genetic stock. All species can colonize secondary forest in the course of its succession, but none seems actually to be confined to secondary formations. Dipterocarps are also altitudinally limited (Symington, 1943). From Borneo there are clear indications that natural barriers have obstructed the distribution of species. For instance, the Batang Lupar, Sarawak, a very ancient river, is the edge of the range of 57 species (Ashton, 1969). Across the watershed which forms the international boundary, Indonesian Borneo (Kalimantan) is not so rich in dipterocarp species as are the states of Sarawak, Brunei and Sabah; moreover, Indonesian Borneo is richest towards the west and towards the north. Of those regions which through time have been accessible to dipterocarps, the ever-wet areas are very much richer in species than the seasonal. We see this in Ceylon, with a concentration of species in the southwest quarter, in the Thai-Malaysian transition belt, and from Java to the Lesser

TABLE 3.1. Summary of the genera of dipterocarps of Sumatra, Peninsular Malaysia and Borneo, showing the number of species represented (with the number of endemics in parentheses). Compiled from Ashton (1982).

Genus	Sumatra	Peninsular Malaysia	Borneo*
Anisoptera	4 (0)	6 (0)	5 (2)
Cotylelobium	1 (0)	2 (0)	3 (1)
Dipterocarpus	25 (1)	31 (2)	41 (15)
Dryobalanops	2 (0)	2 (0)	7 (5)
Hopea	14 (3)	32 (9)	42 (22)
Neobalanocarpus	0	1 (0)	0
Parashorea	3 (1)	3 (1)	6 (4)
Shorea	50 (3)	57 (7)	127 (82)
Upuna	0	0	1 (1)
Vatica	11 (4)	21 (9)	35 (23)
TOTALS	106 (11)	155 (28)	267 (155)

* Since nearly all Bornean species are found in East Malaysian territory, this table reveals the very great number of dipterocarps within the political bounds of the nation of Malaysia. This country is in fact richer in dipterocarps than any other in the world.

Sundas. The richness of the Sunda subregion, especially in Peninsular Malaysia and Borneo, surely reflects the long persistence of favourable climatic conditions. Flenley (1979) suggested that the tropical rain forests shrunk very much during the Pleistocene Ice Age and afterwards extended, and Whitmore (1981) assumed that, at the greatest recession of the sea, rain forest refugia on hills were surrounded by seasonal forests. I am convinced that the Sundaland dipterocarp forests are the outcome of at least 30 million years of evolution under equable hot and wet conditions.In Peninsular Malaysia dipterocarps occupy a wide variety of habitats: coastal to inland, riverine to swampy to dry land, undulating to level terrain, ridges, slopes, valley bottoms, soils deeply weathered to shallow, well-drained to poorly drained, rich to poor in nutrients. Symington (1943, p.xiv) tabulated the main habitat types of the dipterocarp species, with indication of their altitudinal zonation: 0-300 m, to 750 m, and to 1200 m (Fig. 2.1).

So much for the individual species. Throughout, we should realise that most rain forest dipterocarps occur in mixed-species forests. There are forests where as a family they dominate, but even in such a 'dipterocarp forest' there may well also be over a hundred tree species belonging to other families. Based on overall species composition, a number of forest types have been distinguished in lowland Peninsular Malaysia: nine dry-land forest types in the 0–300 m zone, six in the 300–750 m zone (the hill dipterocarp forests), and two in the 750–1200 m zone (the upper dipterocarp forests) (see Wyatt-Smith, 1963, and Chapter 2, above).

Of other forest types, the richest are the freshwater swamps. Corner (1978) recorded 35 dipterocarp species from those in southern Peninsular Malaysia, mostly in the drier parts. True peat-swamp is relatively poor in dipterocarps. Equally poor is the limestone flora, in which only *Hopea ferrea* has been listed as a regular component. The forest fringes along rivers are frequented by a few species, of which neram *Dipterocarpus oblongifolius* is the best known. Finally, there is the transition area on the isthmus of Kra, drier than Peninsular Malaysia yet more humid than the rest of Thailand, with outliers from the north and the south, 30 species of dipterocarps finding their limit (see Jacobs, 1981b, pp.176–178).

Fig. 3.1. Hill dipterocarp forest, at 600 m in the Ulu Gombak Virgin Jungle Reserve, looking north towards Mt Bunga Buah, Selangor. (Photograph by Lord Medway).

3.4. RESEARCH ON DIPTEROCARPACEAE

An International Working Group on Dipterocarpaceae was set up in 1977 as an informal club of interested scientists. 'Round Table' meetings were held at Paris in 1977 and at Kepong in 1980 (Jacobs, 1981a). Proceedings of the latter were published in the *Malaysian Forester* 44 (2 and 3) of 1981.

The Forest Research Institute at Kepong is the finest place in the world for further scientific work on the family. The Aberdeen–Kuala Lumpur project ran for six years in the 1970s, elucidating pollination and breeding systems of a handful of species. Work on chromosomes conducted at Aberdeen by K. Jong has improved understanding of genetics in the family. Work on the architecture of trees has investigated the processes by which a sapling and pole develop into a mature tree and also provided guidance to the student of vegetative propagation to the part of a tree from which to take his cuttings (see Chapter 9).

Among major steps forward — apart from the taxonomic work which laid the basis for all proper identification, in field, herbarium, and laboratory — we may regard the following:

1. In phenology: the discovery that dipterocarps do flower regularly, if only with a few flowers, and the interpretation of this and related phenomena (Chapter 9).

2. In vegetative propagation: including the role of mycorrhiza and ways to propagate from leaves.

3. In pollination: the discovery of thrips as pollinators of some red meranti *Shorea*.

4. In genetics; chromosome studies and work on apomixis.

5. On growth phenomena: especially work on galls in certain species of *Shorea*, and on growth in general in relation to this.

Much is to be expected from further work on mycorrhiza. Lines in phytochemistry, too, are still waiting to be pursued, in connection with biological defences and with genetics. I hope for a continued, disinterested cooperation in this and several other fields, in supra-national efforts.

REFERENCES

Ashton, P. S. (1963). Taxonomic notes on Bornean Dipterocarpaceae. *Gardens' Bull. Singapore* 20, 229–83.

Ashton, P. S. (1964a). *A Manual of the Dipterocarp Trees of Brunei State*. Oxford University Press.

Ashton, P. S. (1964b). *Ecological Studies in the Mixed Dipterocarp Forests of Brunei State*. Clarendon Press, Oxford.

Ashton, P. S. (1968). *A Manual of the Dipterocarp Trees of Brunei State and of Sarawak. Supplement*. Sarawak Forest Department, Kuching.

Ashton, P. S. (1969). Speciation among tropical forest trees: some deductions in the light of recent evidence. *Biol. J. Linn. Soc.* 1, 155–96.

Ashton, P. S. (1982). Dipterocarpaceae. *Flora Malesiana* (1) 9, 237–552.

Audley-Charles, M. G., Hurley, A. M. and Smith, A. G. (1981) Continental movements in the Mesozoic and Cenozoic. In Whitmore, T. C. (ed.), *Wallace's Line and Plate Tectonics*, pp.10–23. Clarendon Press.

Corner, E. J. H. (1978). The freshwater swampe-forest of South Johore and Singapore. *Gardens' Bull. Singapore,* Suppl. 1, 1—266, 18 figs. + 40.

Foxworthy, F. W. (1932). *Dipterocarpaceae of the Malay Peninsula. Malay. Forest Records*, 10.

Jacobs, M. (1981a). Keep the forests, Keep the forests, Keep the forests. The Kepong Round Table Conference on Dipterocarpaceae. *Flora Males. Bull.* 34, 2588–99

Jacobs, M. (1981b). Dipterocarpaceae: the taxonomic and distributional framework. *Malay. Forester* 44, 168–89.

Flenley, J. R. (1979). *The equatorial Rain Forest: a geological history*. London and Boston: Butterworths.

Symington, C. F. (1943). *Foresters' Manual of Dipterocarps*. Reprint, Penerbit Universiti Malaya, Kuala Lumpur (1974). At the end of the reprint edition, B. C. Stone & P. S. Ashton gave corrections. See also the note by T. C. Whitmore (1976) *Taxon* 25, 629–30.

Whitmore, T. C. (1981). Palaeoclimate and vegetation history. In Whitmore, T. C. (ed.), *Wallace's Line and plate tectonics*, ch.5. Oxford: Clarendon Press.

Wyatt-Smith, J. (1963). *Manual of Malayan Silviculture for Inland Forest. Malay. Forest Records, 23:*

Completed August 1982

CHAPTER 4

Forest Palms

John Dransfield

Royal Botanic Gardens, Kew, Richmond TW9 3AE, U.K.

CONTENTS

4.1. THE GLOBAL CONTEXT

The palms form a large, diverse and very ancient family (Palmae or Arecaceae) of mostly tropical and subtropical distribution. Although palms are of great economic importance, they are still rather poorly known taxonomically. This may be a consequence of their neglect by collectors, which follows from characteristically large size, which ill-suits their floral and vegetative parts for reduction to herbarium specimens, and their largely tropical distribution, far from the traditional centres of taxonomic research. At present, about 200 genera and 2700 species are recognized, organized into six subfamilies, (Dransfield and Uhl, 1986). Greatest diversity in terms of number of species, genera and growth forms occurs in the perhumid tropics of South and Central America and Southeast Asia to the West Pacific. Africa on the other hand (even in the most humid areas) has a poor palm flora, perhaps reflecting a long history of pronounced climatic fluctuations and consequent extinction of species adapted to the humid forest environment. The indigenous palm flora of Peninsular Malaysia comprises 31 genera and about 199 species; a few species like the coconut, *Cocos nucifera* L., the betel palm *Areca catechu* L., the oil palm *Elaeis guineensis* Jacq. and the sago palm *Metroxylon sagu* Rottb. are widely cultivated and have become locally naturalized. In a global context, the indigenous palm flora is very rich: 14.6% of all palm genera and about 7.7% of all palm species are found in the small area of Peninsular Malaysia. Guyana, of similar size, has a palm flora of 19 genera and 47 species; Colombia, a much larger country, has 45 genera and 252 species, and Brazil has 33 genera and 387 species (numbers adapted from Glassman, 1972). Malaysian rattan palms are now taxonomically rather well known (Dransfield, 1979). Other palms are less well known, though there is a good general introduction to the flora (Whitmore, 1977).

Of the six subfamilies, four are present in Peninsular Malaysia, and one of these, the Calamoideae which includes bertam *Eugeissona tristis* Griff., rattans (several genera), salak *Salacca* and kelubi *Eleiodoxa*, is very richly developed. There is only one endemic genus, *Calospatha*, a short erect rattan with a good edible fruit. At the species level, 82 species are endemic (i.e., *c.* 41%). The low generic endemism is not very surprising when the close juxtaposition of Borneo and Sumatra is considered. Sumatra has no endemic palm genera, and Borneo only one, *Retispatha*. If, however, the closely related floras of Peninsular Malaysia, southern Thailand, Sumatra, Borneo and Java are combined together, a very different picture emerges — a rather well-defined, extremely rich palm flora, different in many respects from that of monsoonal Thailand to the north and ever-wet Sahul Shelf lands (Moluccas and New Guinea) to the east, with Celebes the transition area of these palm floras (Dransfield, 1981a). Sundaland contains 14 endemic palm genera and is the main centre of diversity of several other genera. Thus, the Peninsular Malaysian palm flora is part of a rich Sunda Shelf palm flora, with some infiltration from genera that are largely Asiatic, such as *Plectocomia* and *Corypha*, and some, such as *Rhopaloblaste* and *Cyrtostachys* that are largely centred on the Sahul Shelf. Sundaland is one of the major palm regions of the World. Peninsular Malaysia lies in the middle and thus the conservation of local palms, viewed in this context, takes on considerable importance.

All the indigenous Peninsular Malaysian palms are plants of forest, though some are adapted to specialized forest types such as mangrove or the low growth on limestone hills. Very few seem capable of tolerating severe disturbance. Tall tree palms spared during land clearance may produce fruit, but no regeneration takes place in the cleared land. The palm flora of second growth ('belukar') is very poor, and native species seem to be slow colonizers of such man-made habitats. By contrast, the introduced oil palm, a native of Africa, does seem to be a quick colonizer of belukar. This fast growing crop plant is becoming naturalized in secondary and primary lowland forest near oil palm estates in Johor, where it may be occupying niches which would otherwise be open to indigenous species. Some native palms are favoured by disturbance short of clear-felling. A well known example is bertam, which seems to dominate the undergrowth where the canopy has been selectively felled.

4.2. TAXONOMIC DIVERSITY AND AFFINITIES

Subfamily Coryphoideae comprises three tribes, all represented in Peninsular Malaya. Tribe Corypheae, perhaps the least specialized group, is present in both Old and New Worlds (although there are only two genera, *Chamaerops* and *Livistona*, present in Africa). The simplest Peninsular Malaysian genus, *Maxburretia*, with two local species, is confined to limestone hills: *M. rupicola* (Ridl.) Furt. to Batu Caves and the Takun hills, Selangor, and *M. gracilis* (Burr.) Dransf. (*Liberbaileya gracilis* (Burr.) Burr. and Potz.) to Langkawi. A third species, *M. furtadoana* Dransf. is known from southern Thailand. The genus seems more closely related to Indochinese *Rhapis* than to any other, but is still rather isolated and may be regarded as a relict of Coryphoid evolution. *Johannesteijsmannia* and *Pholidocarpus*, both relatively close relatives of widespread *Licuala* and *Livistona*, are distinctly Sundaland genera, though one species of *Pholidocarpus* is known from Celebes and Moluccas. *Licuala* is very well represented, but there is also a considerable diversity in the eastern part of its range in New Guinea. *Livistona* on the other hand, with a range from Africa, Arabia and the Himalayas through to Australia, has the greatest diversity in Australia.

Tribe Phoeniceae is exclusively Old World and most of the 17 species of the single genus *Phoenix* are adapted to semi-arid or monsoon climates. The single species in Peninsular Malaysia, *P. paludosa* Roxb., reaches its southern limit here and, perhaps a parallel to the rest of the genus, is confined to mangroves, an area of water stress.

Tribe Borasseae comprises exclusively Old World palms, being confined to Africa, Arabia, Madagascar, Indian Ocean Islands, India and Southeast Asia, with *Borassodendron* in Peninsular Malaysia and Borneo, and one strange little-known *Borassus, B. heineana* Becc. in New Guinea; *B. flabellifer* L. has an

TABLE 4.1. The taxonomic diversity of the palms of Peninsular Malaysia.

	Number of species in Peninsular Malaysia†	Number of species in genus Worldwide
Subfamily Coryphoideae		
Tribe Corypheae		
Maxburretia	2 (2)	3
Johannesteijsmannia	4 (3)	4
Livistona	4 (1)	36
Pholidocarpus	2 (2)	6
Licuala	25 (14)	108
Corypha	1 (0)	8
Tribe Phoeniceae		
Phoenix	1 (0)	17
Tribe Borasseae		
Borassodendron	1 (0)	2
Subfamily Calamoideae		
Korthalsia	9 (2)	27
Eugeissona	2 (2)	6
Salacca	7 (4)	c15
Eleiodoxa	1 (0)	1
Plectocomia	3 (1)	16
Plectocomiopsis	4 (2)	5
Myrialepis	1 (0)	1
Calospatha*	1 (1)	1
Daemonorops	22 (9)	115
Calamus	62 (23)	370
Pogonotium	1 (0)	3
Ceratolobus	2 (1)	6
Subfamily Nypoideae		
Nypa	1 (0)	1
Subfamily Arecoideae		
Tribe Caryoteae		
Arenga	5 (1)	17
Caryota	2 (0)	12
Tribe Areceae		
Orania	1 (0)	16
Cyrtostachys	1 (0)	8
Nenga	3 (2)	5
Pinanga	20 (c10)	120
Areca	4 (2)	48
Iguanura	4 (0)	18
Rhopaloblaste	1 (1)	7
Oncosperma	2 (0)	5

* Endemic genus

† Numbers in parentheses are the numbers of endemic species

artificially wide distribution, related to its usefulness. *Borassodendron* seems to be a relict of a perhaps wider range of the tribe in the past.

Subfamily Calamoideae is found in both New and Old Worlds. Although diversity in the subfamily is overwhelmingly Southeast Asian, two of three African rattan genera *Eremospatha* and *Laccosperma* possess a combination of characters simpler than in any other genus in the subfamily. This suggests, perhaps, that the group has migrated to the Malaysian region and then undergone diversity.

The sole member of subfamily Nypoideae, *Nypa fruticans* Wurmb., has had a much wider distribution in the past, with reliable fossil records from the New World, Europe, Africa and India. Its present distribution extends from the Bay of Bengal to Australia and the Solomon Islands.

Subfamily Arecoideae is represented in Peninsular Malaysia by two tribes. Tribe Caryoteae ranges from India and China to Australia, but there is no doubt that its greatest diversity is in Southeast Asia. Although the Asian genus *Wallichia* is absent from Peninsular Malaysia (it reaches southern Thailand), the local representation of this group is considerable.

Tribe Areceae is pantropic, but the African representation is very poor. Of the Peninsular Malaysian genera, *Cyrtostachys* is largely Papuasian, with only one species in Sundaland; *Oncosperma* belongs to a group best represented in the Mascarenes and Seychelles; *Rhopaloblaste* and *Orania* have strange distributions disjunct across Wallace's Line (see Dransfield, 1981), and *Iguanura*, though a Sundaland endemic, is perhaps most closely related to Papuasian genera. *Areca, Pinanga* and *Nenga*, closely related to each other, are best represented in Sundaland, though *Areca* has a minor centre of diversity in New Guinea.

4.3. GROWTH FORM DIVERSITY

A few Peninsular Malaysian palms are tall trees, contributing to the formation of the forest canopy. Of these, *Pholidocarpus macrocarpus* Becc., a massive, single trunked fan palm of lowland swamp forest is about the tallest in the lowlands; individuals with trunks about 40 m tall have been observed in Johor. The large corky-warted fruits of this species are possibly adapted for elephant dispersal. Very similar, and scarcely distinguishable when sterile, is massive *Pholidocarpus kingianus* (Becc.) Ridley which also grows in the same sort of habitat, but differs in its smooth fruit. Local species of *Livistona* are tree palms but do not reach such great heights. *L. saribus* (Lour.) Merr. is an elegant glaucous-leaved fan palm of the landward fringe of mangrove forest in Selangor, where it is approaching extinction. *L. speciosa* Kurz is a palm of the mid-mountain forest of the Main Range. *L. tahanensis* Ridl. (Fig. 4.1) is, with certainty, confined to Mt Tahan, but quite closely related is a gregarious species abundant on the East Coast Hills of Terengganu, with a closely related population in the Ulu Endau, Johor. The Terengganu taxon has been referred to, incorrectly, as *L. saribus* (Lour.) Merr., and the Johor population has now been described as *L. endauensis* (Dransfield and Wang). All these tree fan palms, with the possible exception of *Pholidocarpus kingianus* and *P. macrocarpus*, are characteristically gregarious. A forest composed of great numbers of these elegant palms has an altogether unusual physiognomy.

Corypha utan Lam. (*C. elata* Roxb.) is rarely gregarious and its distribution seems to have been much affected by man. It is a massive, but not especially tall, monocarpic palm which seems to be adapted to submaritime storm forest. In Peninsular Malaysia it is associated with man-made habitats and may, in fact, be introduced. The remaining fan-leaved palm is the rare *Borassodendron machadonis* (Ridl.) Becc. I know few extant populations in Peninsular Malaysia.

The tallest palm of montane forest is without doubt the majestic doubly-pinnate *Caryota maxima* Bl. (*C. aequatorialis* (Becc.) Ridl.), a single-stemmed monocarpic species with a trunk sometimes reaching 35 m tall, abundant in some mid-montane forest types on better soils (Fig. 4.2). Its habit contrasts with *C. mitis*

Fig. 4.1. The endemic fan palm *Livistona tahanensis* Ridl., in upper montane forest on poor soils on the slopes of Mt Tahan. This beautiful palm occurs in large numbers on this mountain only, at elevations about 900–1500 m. (Photograph by J. Dransfield.)

Lour., a clustering moderate palm of lowland forest, one of the few species present in suburban scrub. In the same tribe is the pinnate-leaved genus *Arenga*, of which three Peninsular Malaysian species are tree palms. One, *A. pinnata* (Wurmb) Merr., a very useful palm, may not be native as it is usually associated with human settlements. The others, kerjim *A. westerhoutii* Griff. and langkap *A. obtusifolia* Mart. — confused by the botanist but easily distinguished by Orang Asli (Chapter 18) and the initiated — are massive sympatric species, abundant especially in the hill forests of the Main Range. The former is monocarpic* whereas *A. obtusifolia* is biologically very different in its clustering pleonanthic* habit and possibly dioecious state. On some hillslopes such as Bukit Lagong, Kepong, and in the Ulu Gombak, Selangor, these two species may occur in great abundance, forming a distinct layer in the canopy and, with their massive litter, partially supressing the development of other species.

Among members of tribe Areceae, only ibul *Orania sylvicola* (Griff.) Moore, an abundant palm especially in the Main Range, and bayas *Oncosperma horridum* (Griff.) Scheff. of inland hills, and nibong *O. tigillarium* (Jack) Ridl. of submaritime habitat, contribute to the forest canopy. The two *Oncosperma* spp.

---

*Monocarpic · a plant which flowers and fruits once only, after which it dies.

*Pleonanthic · a plant (palm) whose stems produce inflorescences over an extended period, inflorescence production not ending in death of the stem.

Fig. 4.2. The giant mountain fishtail palm *Caryota maxima* Bl., in lower montane forest at Genting Highlands. (Photograph by J. Dransfield.)

form large clumps and, like the two *Arenga* spp. effectively suppress regeneration with their massive litter. The Orang Asli at Kepong recognize a third species of *Oncosperma*, but I am still not convinced.

One characteristic feature of most Peninsular Malaysian forest types is the abundance of palms in the undergrowth; in particular, there is usually a great abundance of climbing palms, the rattans. Undergrowth palms, excluding the rattans, are of two main types: those with erect stems, and the so-called 'stemless' palms, in which the stem is procumbent or subterranean. Frequently the two types are found within the same genus. Thus, in *Johannesteijsmannia*, a magnificent genus of palms with large undivided diamond-shaped leaves, three of the four species are stemless, while the fourth, *J. perakensis* Dransf., eventually has an erect trunk to 3 m tall. Another parameter of variation in habit is between single-stemmed palms and clustering palms, in which the most basal nodes produce axillary buds, thus creating a clump. This difference is of some taxonomic value, but occasionally individuals of the same species may be clustered or solitary. Variations in the length of the underground portion of branches gives close or loose clumping. The acaulescent nipah *Nypa fruiticans*, the remarkable mangrove palm, represents a case perhaps unique in the local palm flora, as its subterranean stems appear to branch dichotomously (Tomlinson, 1971). *Salacca* and *Eleiodoxa* are, like *Eugeissona tristis* and *E. brachstachys* Ridl., spiny acaulescent palms tending to form great thickets. *Eleiodoxa conferta* (Griff.) Burr. is hapaxanthic (that is, individual stems flower once only and then die) but, unlike monocarpic palms such as *Corypha utan*, perpetuate themselves by the production of axillary branches. *Salacca*, superficially very similar to *Eleiodoxa*, has strange lateral inflorescences enclosed in the bud within the body of the leaf sheath, emerging through a slit at anthesis (Fisher and Mogea, 1980). Most species of *Salacca* produce sucker shoots in a position opposite the leaf rather than in the axil; in *S. flabellata* Furt. and *S. wallichiana* Mart. new shoots are produced from the tips of the long flagelliform inflorescences, and in *S. graciliflora* Mogea and *S. minuta* Mogea leaf-opposed branches are produced but they are flagelliform and root at some distance from the parent plant before developing into a new shoot. There is thus a rather diverse range of clump building, even in the one growth form of acaulescent palms.Most of the Arecoid undergrowth palms are rather small, with erect stems each bearing at its tip a crownshaft, i.e., a column formed by the tightly sheathing, tubular leaf bases. In *Cyrtostachys renda* Bl. (syn. *C. lakka* Becc.), a rather large undergrowth palm of some facies of peat swamp forest, the crownshaft is brilliant sealing-wax red. It is a clumping species; in the wild it seems to produce rather open colonies, but in cultivation dense suckering occurs close to the main stem. The genus *Pinanga* (Fig. 4.3) is well represented, ranging from undergrowth palms with stems to 4.5 m tall (e.g., *P. malaiana* (Griff.) Scheff.) to very small palmlets scarcely 30 cm tall (e.g., *P. paradoxa* Scheff.). An acaulescent species has been described (*P. acaulis* Ridl.) but is known only from its type. Two species, *P. simplicifrons* (Miq.) Becc. and *P. cleistantha* Dransf., are remarkable and quite aberrant in their flowering behaviour: the inflorescence remains enclosed within its prophyll, leaving no obvious access for pollinators. The floral biology of these remarkable species has not been studied.

Nenga with three Malayan species (Fernando, 1983) and *Areca* with four are superficially similar to *Pinanga* and have more or less the same growth form. In one undescribed species of *Areca* from Terengganu, however, the leaves do not abscise neatly, a crownshaft is scarcely developed, and the inflorescences emerge by bursting through the rotting sheaths.

Iguanura and *Rhopaloblaste* in Peninsular Malaysia have ill-defined crownshafts. *Iguanura* has been studied in detail by Kiew (1976), who recognized four local species. *Rhopaloblaste* is represented by one aberrant clustering undergrowth species. In both genera the leaves do not abscise neatly, a well defined crownshaft is absent and the inflorescences are interfoliar.Apart from those mentioned above, undergrowth palms are found in *Licuala*, *Maxburretia* and *Phoenix*. The only Peninsular Malaysian *Phoenix*, *P. paludosa*, forms dense colonies on the landward edge of mangrove in the north.

The final major growth form represented is the climbing habit. The rattans amount to 106 species, though some can scarcely be called climbers, being more or less stemless forest undergrowth palms. The principal habits of rattans are solitary hapaxanthic (i.e., individual stems monocarpic), clustering hapaxanthic, solitary pleonanthic or clustering pleonanthic (Dransfield, 1979). A further habit is seen in most species of *Korthalsia*, where branching is not confined to the stem base, but also occurs in the canopy to produce vast aerial entanglements.

4.4. ECOLOGY

The major forest types of Peninsular Malaysia (Chapter 2) tend to have distinct palm floras, yet some palm species appear to be remarkably catholic in their ecological requirements and are correspondingly

Fig. 4.3. *Pinanga polymorpha* Becc. forming dense thickets in upper montane forest, Genting Highlands. (Photograph by J. Dransfield.)

widespread. The only strictly mangrove palm is *Nypa*, which grows gregariously forming a very distinct facies. *Calamus erinaceus* (Becc.) Dransf., *Phoenix paludosa* and *Oncosperma tigillarium* grow at the landward margin of the mangrove, and beyond this *Licuala spinosa* Thunb. becomes a regular feature. Peat swamp forest has a characteristic though rather impoverished palm flora consisting of *Licuala paludosa* Griff, *Cyrtostachys renda, Eleiodoxa conferta, Korthalsia flagellaris* Miq., *Calamus scabridulus* Becc. and a few other less constant species. The limestone hill palm flora is very restricted, perhaps the two *Maxburretia* spp. being the only strict calcicole elements. Where soil accumulates in gulleys, other species more characteristic of lowland dipterocarp forest occur on limestone. *Arenga westerhoutii* and *Borassodendron machadonis* occur on pockets of deep soil, but can scarcely be called calcicoles. The majority of Peninsular Malaysian palms occur in lowland and hill dipterocarp forests (Fig. 2.1). At about 1000 m altitude there is usually a rather marked change in the palm flora, and a great reduction in the number of species; some genera (e.g.,

Korthalsia) are wholly absent from montane forest. Certain habitats within primary forest, such as valley bottoms, have particularly rich palm representation, but it is remarkable how many species seem to tolerate a wide range of conditions within primary forest. More remarkable still are instances of disjunct distribution. *Johannesteijsmannia* is a good example: in Selangor, three species may be found growing sympatrically in Sungei Lalang Forest Reserve (Fig. 4.4), yet over the hill in nearby Ulu Langat, in an apparently very similar forest type, the genus is absent. *Rhopaloblaste singaporensis* (Becc.) Moore, very much a feature of the forests of Johor, has an outlier on the Main Range in Perak. Such distributions are at present inexplicable, although a few generalizations can be made. The palm flora of Johor and southeastern Pahang has some species otherwise only known in Borneo. Terengganu and east Kelantan have several endemics, and the palm flora of the area west of the Main Range has species which have also

Fig. 4.4. *Johannesteijsmannia magnifica*, one of the finest and rarest undergrowth palms of the lowland forest, in the virgin jungle reserve, Sungei Lalang, Selangor. Note man. (Photograph by J. Dransfield.)

been recorded in north Sumatra. These trends probably reflect the fact that Sumatra and Borneo were connected to Peninsular Malaysia at periods during the Pleistocene (see Chapter 1.4).

Mention has already been made of the response of *Eugeissona tristis* to forest clearance. Natural clearings caused by landslides or the falling of large trees are the habitat of a specialized group of palms (Fig. 4.5). In the lowlands and uplands, *Plectocomiopsis* spp and *Myrialepis paradoxa* (Kurz.) Dransf. (syn. *M. scortechinii* Becc.) seem to be exclusive to this type of habitat, and it has been suggested elsewhere (Dransfield, 1979) that the hapaxanthic behaviour of these rattans may be associated with the colonization of rare temporary habitats, by the production of large quantities of seed at one time. On poor soils in the lowlands and in the uplands *Plectocomia elongata* Bl. (syn. *P. griffithii* Becc.) behaves similarly.

Fig. 4.5. *Eugeissona brachystachys* Ridl., an undergrowth palm common on the slopes of Mt Tahan and a few hills of Terengganu and eastern Pahang, a close relative of bertam *E. tristis*. (Photograph by J. Dransfield.)

4.5. PALM/ANIMAL INTERACTIONS

Few interactions between palms and animals have been worked out in detail, but casual observations indicate a wonderfully diverse range of relationships for detailed investigation. Some such observations are recorded below to indicate the field for future study.

4.5.1. Pollination

Pollination of very few Peninsular Malaysian palms has been studied. More is known about pollination of the introduced oil palm than of any wild species. Detailed studies will obviously have to be carried out by locally based naturalists. Wind pollination, frequently invoked for the palms in elementary texts, is almost certainly an exception in this region.

Most palm flowers seem to be scented and at anthesis are visited by insects. *Salacca zalacca* (Gaertn.) Voss in Java has been shown to be pollinated by curculionid beetles (Mogea, 1978); similar beetles are abundant in *Salacca* and *Eleiodoxa* inflorescences in Peninsular Malaysia. Many *Daemonorops* spp. and *Pinanga* spp. when in flower attract nitidulid and staphylinid beetles; *Korthalsia* flowers seem to be visited by bees and many *Calamus* spp. by wasps. We need detailed studies.

4.5.2. Dispersal

Our state of knowledge of palm fruit dispersal is similarly incomplete and based on casual sightings of feeding or of faeces with seed.

4.5.3. Ant/rattan relationships

These have been summarized by Dransfield (1979). The gross details of the relationships are probably well established; detailed studies made over prolonged periods are now required. A casual observation in Sabah suggests that ant rattans may be avoided by herbivores (Dransfield, 1981b).

4.5.4. Palms as food for herbivores

Olivier (pers. comm.) regards rattan apices as one of the important food sources for the elephant. My own observations suggest that rattans are frequently grazed by squirrels and possibly pigs. Young rattans, and the basal leaves of tufted rattans are used by pigs to build their nests (Medway, 1963).

4.6. ECONOMIC POTENTIAL AND CONSERVATION

Wild palms are of great significance to the Orang Asli (Chapter 18). Most species are put to some use, though this may not be very specific. Construction of dwellings, basketry, twine, food, ornament, ritual and medicine all involve the use of forest palms. At the other extreme, a few elite wild rattans, such as *Calamus manan* Miq. and *C. caesius* Bl. are the basis of a lucrative export trade of cane for cane furniture, the value of which in 1977 has been estimated to exceed US$2 billion (including rattan from sources other than Peninsular Malaysia). High prices, forest destruction and rather uncontrolled collecting has resulted in many rattans becoming dangerously scarce. Yet as a gene pool for the establishment of rattan cultivation the wild populations are essential.

Mention must also be made of the great value of palms as garden ornamentals, Malaysian *Cyrtostachys renda* being outstanding in this respect. As yet, relatively few indigenous species have been introduced into cultivation. There is an increasing international demand for seed of wild palms, a demand which need not conflict with conservation.

The most serious threat to the palm flora is large scale clearance of lowland forest for agricultural projects. I have made rough estimates on the number of palm species in the major forest types. I estimate that about 140 species (that is, about 70% of the total flora) are found in lowland dipterocarp forest; about 100 species (*c.* 50% of the total) are found in hill dipterocarp forest, and only 20 species are found in montane forest (*c.* 10%). The implications for conservation are obvious. If the montane forest alone survives clearance, then almost 90% of this rich palm flora will become extinct outside nature reserves. Even with some protected forest in reserves, the patchy distribution of palms means that the representation in the very few lowland reserves will be far from complete. Such possible extinction should be totally unacceptable when, quite apart from any aesthetic reasons, due consideration is paid to the great economic potential of the palms of Peninsular Malaysia.

REFERENCES

Dransfield, J. (1979). *A Manual of the Rattans of the Malay Peninsula.* Malaysian Forest Records No. 29. Kuala Lumpur.
Dransfield, J. (1981a). Palms and Wallace's Line. In Whitmore, T. C. (ed.), *Wallace's Line and Plate Tectonics,* pp.43–56. Oxford, Clarendon Press.
Dransfield, J. (1981b). A synopsis of *Korthalsia* (Palmae-Lepidocaryoideae). *Kew Bull.* 36, 163–194.
Dransfield, J. and Uhl, N. W. (1986). An outline of a classification of palms. *Principes* 30, 3–11.
Fernando, E.S. (1983). A revision of the genus *Nenga. Principes* 27, 55–70.
Fisher, J. B. and Mogea, J. P. (1980). Intrapetiolar inflorescence buds in *Salacca*: development and significance. *Bot. J. Linn. Soc.* 81(1), 47–59.
Glassman, S. F. (1972). *A Revision of B. E. Dahlgren's Index of American Palms. Phanerogamarum Monographiae, vol.6.* J. Cramer.
Kiew, R. (1976). The genus *Iguanura* Bl. (Palmae). *Gns. Bull., Singapore* 28(2), 191–226.
Medway, Lord (1963). Pigs' nests. *Malay. Nat. J.* 17, 41–5.
Mogea, J. P. (1978). Pollination in *Salacca edulis. Principes* 22, 56–63.
Moore, H. E., Jr. (1973). The major groups of Palms and their distribution. *Gentes Herbarum* 11(2), 27–140.
Tomlinson, P. B. (1971). The shoot apex and its dichotomous branching in the *Nypa* palm. *Ann. Bot.* 35, 865–879.
Whitmore, T. C. (1977). *Palms of Malaya.* Revised Edition. Oxford University Press, Kuala Lumpur.

Completed August 1985

CHAPTER 5

Forest Bamboos

Soejatmi Dransfield

Royal Botanic Gardens, Kew, Richmond TW9 3AB, U.K.

CONTENTS

5.1. INTRODUCTION

Bamboos are an important group of useful plants, found either cultivated in villages, or wild or spontaneously in forest or forest margins, river banks, etc. Those occurring in Peninsular Malaysia can be divided into village or cultivated bamboos and native or forest bamboos (Holttum, 1958). The former are usually not native; the origin of some species is not known. For instance, *Bambusa vulgaris* Schrad. grows spontaneously along river banks, where culm cuttings easily sprout and produce new shoots and roots in suitable conditions, but is considered to be cultivated, is never found in truly wild conditions, and is of unknown origin. Another village bamboo is buluh galah *Bambusa heterostachya* (Munro) Holtt., cultivated especially in Negeri Sembilan, Melaka and Johor for use as poles for collecting coconuts, and recently begun to be used widely in oil palm plantations for poles. *Dendrocalamus asper* (Schult.) Backer ex Heyne, planted for its excellent edible shoots ('rebung') here as in Indonesia, is without known origin though it may be native in the northern part of the Peninsula or Burma (Holttum, 1958).

Native bamboos grow wild at forest margins along road sides, on hillsides or around clearings, or in the forest (Fig. 5.1). Native forest bamboos in Peninsular Malaysia can be divided into two groups, those which grow in lowland and hill dipterocarp forests and those which grow in montane forest. A few native species have been planted in villages. Some are used by local people. Two or more species are useless and often become a nuisance in cleared or logged forests.

Bamboos belong to the subfamily Bambusoideae of the grass family, the Gramineae, and share some common features with the grasses such as flower structure. The flower or spikelet typically has glumes and one or more florets; each floret consists of a lemma, a palea, lodicules, stamens and an ovary.

Fig. 5.1. Bamboo-rich forest-edge habitat along a stream course on the eastern slopes of the Main Range; the upper Nenggiri river, Kelantan. (Photograph by Lord Medway.)

Traditionally bamboos are separated from grasses by having woody culms, three lodicules, six stamens and three stigmas. From anatomical data now they can be separated from the grasses also by having arm cells and fusoid cells in the leaf-blades. Based on anatomical features, and also some flower structures, many herbaceous grasses are included in the subfamily Bambusoideae, such as the genus *Buergesiochloa* from New Guinea and many genera from tropical and south America and Africa (Calderon and Soderstrom, 1973).

Holttum (1958) found two classes of information on bamboos, one from herbarium botanists (Munro, 1868; Gamble, 1896) who classified bamboos on flower structure, and the other from field botanists (such as Kurz, 1876) who recognized and differentiated bamboo species on vegetative characters. By combining information from both sources Holttum (1958) presented a most useful account, which remains the basic work on bamboos for Malaysia and neighbouring countries.

In modern bamboo taxonomy almost any structure, either vegetative or reproductive, has been employed for classification and identification (McClure, 1966 and 1973). Characters such as growth habit, rhizome, culm, branches, culm-sheath, leaf-blade, inflorescence, flower including ovary and fruit, have been used. However a formal and overall classification of the woody bamboos has not yet been established. Herbarium specimens of bamboos are usually few and incomplete. Because of irritant hairs on culm-sheath and for other reasons (no flowers, planted in villages, etc.) plant collectors usually ignore bamboos. Sometimes, when a bamboo plant produces flowers, the specimen collected consists of leafy branch and a flowering branch which often causes difficulty in naming because the culm-sheath is not available.

There are about 56 genera of woody bamboos in the world (Soderstrom and Calderon, 1979) but only seven genera occur in Peninsular Malaysia, one, *Thyrsostachys*, being introduced from Thailand. In the six remaining genera only 28 species in all may be regarded as native (Table 5.1). This appears to be a poor bamboo flora compared with other areas to the north. Many species of bamboos have very limited distribution, and endemism is not uncommon. There are about 19 species endemic to Peninsular Malaysia. Many native species have restricted ranges in the northern part of the country.

TABLE 5.1. Bamboo species occurring in Peninsular Malaysia (data from Holttum, 1958 and Wong, 1982).

Genus	Total species	Native/forest species	Endemic	Introduced/cultivated
Bambusa	13 + 1(?)	7	3–4	6 + 1(?)
Dendrocalamus	8	5	4	3
Gigantochloa	9 + 2	5 + 2	2 + 2	4
Schizostachyum	9	9	3	
Dinochloa	1 + 1(?)	1		
Racemobambos	1	1	1	
Thyrsostachys	1			1

5.2. GROWTH FORM DIVERSITY

A bamboo plant has no central trunk. The basic frame consists of a ramifying system of segmented vegetative axes which can be differentiated as rhizome, culm and branches (McClure, 1966; Wong, 1986). Each axis has nodes and internodes; each internode is enveloped by a sheath originating from the node. These enveloping sheaths are alternate at the successive nodes along the axis.

The rhizome is generally subterranean and forms the foundation of the bamboo plant. Two types of rhizomes are recognized in bamboos, pachymorph and leptomorph (McClure, 1966). Pachymorph rhizomes are short, thick and curved, and have short segments; the apex of each rhizome grows upwards and becomes a new culm. Thus the rhizome here is determinate and sympodial. Leptomorph rhizomes have long segments and elongate indefinitely. Peninsular Malaysian bamboos have pachymorph rhizomes; the culms grow close to each other and form a clump (Fig. 5.2). In an established clump, the subterranean part is often a collection of massive rhizomes. Once the bamboo clump is established, it becomes difficult to eradicate. Weedy bamboos, such as *Schizostachyum grande* Ridl. or *Dinochloa* spp., can be a serious problem, as the rhizome systems can be large. On the other hand, bamboos can be planted to protect the soil from erosion.

The culm is usually hollow or with a small lumen, except in some *Dinochloa* species where it is solid. Some species have erect and straight culms with pendulous tips, as in some *Bambusa* spp., *Gigantochloa* spp., *Dendrocalamus* spp., and some *Schizostachyum* species. Some have scrambling culms; in these, the internodes are usually long with thin walls and branches are many at each node, so that the culms cannot support themselves and lean on trees or scramble between trees, as in *Schizostachyum grande, Bambusa wrayi* Stapf, *B. montana* Ridl. and *Racemobambos setifera* Holtt.. Some have climbing culms, as in the genus *Dinochloa*. In these, the internodes are straight, but each is produced at an obtuse angle following

Fig. 5.2. A clump of *Gigantochloa scortechinii* in the Ulu Gombak Forest Reserve, western slopes of the Main Range. The man has his foot on a young culm. (Photograph by Lord Medway.)

previous internode, and thus the culm becomes zig-zag. This structure and the rugose base of culm-sheath, which easily clings to trees, give support to the culm in climbing (Dransfield, 1981).

A bamboo culm possesses branch buds emerging just above the node on alternate side of the culm. In many *Bambusa* species all branch buds, one at each node, will develop horizontally from the lower nodes up to the upper ones. The primary branch will elongate and remain dominant. It will produce secondary lateral branches, one at each node. The pattern will be repeated to several degrees and the result is that the branches of all mature culms of a clump will entangle and form inpenetrable thickets. In *Gigantochloa* and *Dendrocalamus* branches will develop and elongate from the mid-culm nodes upwards. Each is usually accompanied by few to several short branches at the lowermost nodes. In *Schizostachyum*, branches will develop and elongate also from the mid-culm nodes upwards, but the process of rebranching happens at the earlier stage of successive orders, and the result is a dense tuft of short subequal branches. The primary branch is indistinguishable from other secondary branches. The genus *Dinochloa* has a peculiar branching habit, structurally similar to that of *Gigantochloa* and *Dendrocalamus*, but with the primary branch bud dormant. This bud will develop if the tip of the culm is damaged and will then elongate, behaving as the main culm (Dransfield, 1981). In this way the culm will continue to grow.

5.3. BAMBOOS OF MONTANE FORESTS

In Peninsular Malaysia there are at least three species of bamboo found only in the mountain forest, all of the genus *Bambusa*: *B. pauciflora* Ridl., *B. magica* and *B. wrayi*. Other species are found at lower altitudes, growing at the forest margins at the foot of the Main Range: *Schizostachyum grande*, *Dendrocalamus hirtellus* Ridl. and *D. pendulus* Ridl.. In addition, *B. montana* (Ridl.) Holtt. is endemic in Penang, growing on Penang Hill; *Racemobambos setifera* is found on two hills of lower altitude.

B. pauciflora, B. magica, B. wrayi, and also *B. montana*, have thin-walled culms of about 2–3 cm in diam., and many short branches at each node from mid-culm upwards, as in the genus *Schizostachyum*. *B. pauciflora* is native, very rare, and grows in forest at Fraser's Hill. *B. magica* is endemic, and grows in open and exposed ridges on the Main Range. The culms are about 5 m tall, erect first then drooping with the internodes of about 60 cm long. *B. wrayi* has very long internodes, up to 200 cm long, and the culm can be as long as 18 m (Holttum, 1958). Young culms are erect, light green, with green culm-sheaths covered with black hairs; when mature they are pale green to dark green, straight, becoming very small with shorter internodes towards the apex. The long internodes are used for blowpipes by Orang Asli (see Chapter 19). Along roads in the mountains, the hanging tips of culms can often be seen among the trees even though the clump may remain invisible.

These four species of *Bambusa*, together with *B. klossii* Ridl., endemic to Kedah Peak, *B. cornuta*, very rare from Java, and *B. griffithiana* Munro from Burma, form a group which is characterized by a reduced number of florets in the spikelet. They are not necessarily closely related to each other but may have arisen separately from different species of *Bambusa* with many flowers (Holttum, 1958).

Schizostachyum grande often occupies large areas where the forest has been cut, establishing itself as a weed which is difficult to eradicate. Young shoots are erect, up to 5 m tall, like spears, whitish to pale green because of a wax covering. Mature culms are first erect; when branches at each node have developed the culm cannot support itself, but leans on nearby trees or (if there are no trees) curves or droops to the ground.

Dendrocalamus hirtellus has relatively thick culms forming a dense clump, and is common in open places on hill slopes, often forming a small bamboo forest. It is not difficult to find a clump of this species producing flowers, usually on leafless small culms. Another species of *Dendrocalamus* growing on hills is *D. pendulus*, usually found at forest margins, but also within the forest. It has erect greenish culms with long pendulous tips and branches.

The genus *Racemobambos* contains species which are found in mountain forests. There are 15 species, all in Malesia (Dransfield, 1983). *R. setifera* is endemic to Peninsular Malaysia, growing on low altitude hills (G. Pulai, type locality; G. Angsi; Endau-Rompin). This species differs from typical *Racemobambos* in having two florets in the spikelet. It seems that *Racemobambos* spp. have a definite flowering periodicity, but it is not known whether the plant dies after flowering.

5.4. BAMBOO OF LOWLAND AND HILL FORESTS

Although bamboos are usually found at forest margins or in secondary growth, it has been shown that *Dinochloa* species are true forest bamboos and found in primary forest (Dransfield, 1981). Nonetheless, *Dinochloa* species may be seen at forest margins along roads and, like many other members of the Gramineae, they will occupy cleared areas and will survive outside forest. *D. scandens* (Bl.) O. Kuntz is the most common species of the genus, especially in western Malesia. There is probably a second species in Peninsular Malaysia (Holttum, 1958). To recognize *Dinochloa* species, at least three features are needed: culm-sheath and its blade, leaf-blade, and fruit. When a plant flowers, it produces masses of fruits. Because of its peculiar characters, a climber, producing plenty of fruits which germinate quickly and branching habit, *Dinochloa* species can survive in forests and logged forests, and in Sabah (where there are seven species) become weeds at forest margins.

In Peninsular Malaysia *Schizostachyum* species grow wild. All member species except one have thin-walled culms. Because of the ready availability and of the thin-walled culm which is easily split, *Schizostachyum* species (except *S. grande* and related species) are among the most useful bamboos for local people. One, *S. brachycladum* Kurz, has been brought into cultivation in villages. The plant with yellow culms is usually found only in cultivation; its origin is not known, but it was possibly introduced from Sabah. This species and its relative, *S. zollingeri* Steud., are used locally for cooking rice and glutinous rice. *S. zollingeri* often can be seen growing in open areas in the forest on hill slopes. *S. gracile* (Munro) Holtt., also related to these two species, grows on the forest margins or river banks in southern Peninsular Malaysia. *S. jaculans* Holtt. has a long lanceolate blade of the culm-sheath and long internodes; it grows in the forest in the north, sometimes planted in villages. The internodes are used for making blowpipes, hence its local name, buluh sumpitan (also buluh temiang).*S. latifolium* Gamble (*S. longispiculatum sensu* Holttum, 1958) grows wild in forest or forest margins all over Peninsular Malaysia and is also found in Sumatra and Borneo (Dransfield, 1983).

The genus *Gigantochloa* contains some wild species and some cultivated species. Wild species appear to be confined to the region of tropical mainland Asia (Burma, Thailand, Indo-China and Peninsular Malaysia). *G. scortechenii* Gamble is the commonest local species, usually growing along forest margins at the foot of the Main Range. The culms are up to 20 m tall and about 10 cm in diameter; culm-sheaths of young shoots are light orange but often covered with black hairs. This species flowers frequently, but not gregariously, and the flowers are borne on a culm without leafy branches. The other species, *G. ligulata* Gamble, *G. latifolia* Ridl. and *G. wrayi* Gamble, are found in the north extending into southern Thailand. *G. scortechenii* is used for making baskets.

Dendrocalamus is a genus related to *Gigantochloa*, often very difficult to differentiate without both culm-sheaths and flowers. The genus has a centre of diversity in the regions between northeast India and Thailand. Wild species in Peninsular Malaysia are found growing in hill or lowland forest in the northern part of the country, but are rather rare. *D. pendulus* which is found commonly in hill forests of the Main Range, is also found growing wild in lowland forest, and here and there along the Tembeling River in the Taman Negara. *D. giganteus* Munro, native in Burma and Thailand is one of the world's biggest bamboos.

TABLE 5.2. Alphabetical list of named bamboo species occurring in Peninsular Malaysia.

Bambusa	Dendrocalamus	Dinochloa	Gigantochloa	Racemobambos	Schizostachyum	Thyrsostachys
arundinacea (Retz.) Willd.	asper (Schult.) Backer ex Heyne	scandens (Bl.) O. Kuntz	apus (Schult.) Kurz hasskarliana (Kurz) Backer ex Heyne	setifera Holtt.	aciculare Gamble brachycladum Kurz, var. auriculatum Holtt.	siamnesis Gamble
blumeana Schult.	dumosus (Ridl.) Holtt.		holttumiana Wong		gracile (Munro) Holtt., var erectum Holtt.	
burmanica Gamble	elegans (Willd.) Holtt.		latifolia Ridl., var. efimbriata Holtt., var. alba Holtt.		grande Ridl.	
glaucescens (Willd.) Sieb. ex Munro	giganteus Munro		levis (Blanco) Merr.		insulare Ridl.	
heterostachya (Munro) Holtt.	hirtellus Ridl.		ligulata Gamble		jaculans Holtt.	
klossii Ridl.	pendulus Ridl.		pseudoarundinacea (Steud.) Widjaja, var.		latifolium Gamble	
magica Ridl.	sinuatus (Gamble) Holtt.		viridis Holtt.,		terminale Holtt.	
montana (Ridl.) Holtt.	strictus (Roxb.) Nees		ridleyi Holtt.		zollingeri Steud.	
pauciflora Ridl.			rostrata Wong			
ridleyi Gamble			scortechenii Gamble, var. albovestita Holtt.			
ventricosa McClure			wrayi Gamble			
vulgaris Schrad.						
wrayi Stapf						

5.5. CONCLUSIONS

As many species are very useful they seem at the moment to be in little danger from a conservation point of view. Some species, however, are very restricted in their distribution and it is these rare, often more or less useless species which are of greatest botanical interest. *Racemobambos setifera* with its three extant localities (Wong, 1987) is of interest and could easily disappear if both are cleared. Similarly *Bambusa montana, B. pauciflora* and *B. klossii*, have few extant localities. Although there seems to be little danger of extinction at the moment, it is important to draw attention to these species of very restricted distribution.

ACKNOWLEDGEMENTS

Professor R. E. Holttum kindly read and commented on a draft version of this chapter.

REFERENCES

Calderon, C. E. and Soderstrom, T. R. (1973). Morphological and anatomical considerations of the grass subfamily Bambusoideae based on the new genus *Maclurolyra*. *Smiths. Contr. Bot.* 11, 1–55.
Dransfield, Soejatmi (1981). The genus *Dinochloa* (Gramineae-Bambusoideae) in Sabah. *Kew Bull.* 36(3), 613–33.
Dransfield, Soejatmi (1983a). The genus *Racemobambos* (Gramineae- Bambusoideae). *Kew Bull.* 37(4), 661–79.
Dransfield, Soejatmi (1983b). Notes on *Schizostachyum* (Gramineae- Bambusoideae). *Kew Bull.* 38(2), 321–32.
Gamble, J. S. (1896). The Bambuseae of British India. *Ann. R. Bot. Gard., Calcutta* 7, 1–133.
Holttum, R. E. (1958). The bamboos of the Malay Peninsula. *Gdns. Bull., Singapore* 16, 1–135.
Kurz, S. (1876). Bamboo and its use. *Indian Forester* 1, 219–69; 335–61.
McClure, F.A. (1966). *The bamboos - A fresh perspective.* Harvard University Press, Cambridge, Massachusetts.
McClure, F. A. (1973). Genera of bamboos native to the New World. In T. R. Soderstrom (ed.) *Smiths. Contr. Bot.,* Vol.9, 1–148.
Munro, W. (1868). A monograph of the Bambusaceae, including all of the species. *Trans. Linn. Soc. Lond.* 26, 1–157.
Soderstrom, T. R. and Calderon, C. E. (1979). A commentary on the bamboos (Poaceae: Bambusoideae). *Biotropica* 11(3), 161–72.
Widjaja, E. A. (1987). A Revision of Malesian *Gigantochloa* (Poaceae–Bambusoideae). *Reinwardtia* 10, 291–380.
Wong, K. M. (1982). Two new species of *Gigantochloa* (Bambusoideae) from the Malay Peninsula. *Malays. Forester* 45, 345–353.
Wong, K. M. (1986). The growth habits of Malayan bamboos. *Kew Bull.* 41, 703–720.
Wong, K. M. (1987). The bamboos of the Ulu Endau Area, Johore, Malaysia. *Malay. Nat. J.* 41, 249–256.

Completed September 1985

CHAPTER 6

Herbaceous Flowering Plants

Ruth Kiew

Universiti Pertanian Malaysia, 43400 Serdang, Selangor, Malaysia

CONTENTS

6.1. INTRODUCTION

Herbs are sometimes simply defined as small non-woody plants but many of the longer lived herbaceous plants in the tropics become woody with age. They may still be recognized as herbs because they have soft leaves, in contrast with shrubs and trees which in general have leathery leaves. But the distinction between herbaceous and woody plants is not clear cut. Some large woody genera, such as the peppers *Piper* or *Ardisia*, include species that can be considered herbaceous (*P. stylosum* and *A. villosa*, for example), while some species of herbaceous genera are as branched and woody as some shrubs (e.g., *Cyrtandromoea acuminata, Didymocarpus corchorifolia*). Over emphasis of the distinction between herbaceous and woody plants has resulted in distancing the temperate herbaceous families from their largely woody close tropical relatives, for instance, Umbelliferae and Araliaceae, Scrophulariaceae and Bignoniaceae or Labiatae and Verbenaceae.

There is great variety among herbaceous plants in Peninsular Malaysia. Large examples include the bananas, gingers and aroids. Other life forms include an abundance of epiphytes (Section 6.3), parasites or saprophytes (Section 6.5), ant plants and insectivorous plants (Section 6.5), water plants (Section 6.3) and climbers.

With some 2580 native species recorded, in 551 genera and 94 families, the species of herbaceous flowering plants are quite as numerous as the trees in Peninsular Malaysia. This count is undoubtedly an underestimate. Many large herbaceous families (such as Acanthaceae or Araceae) have yet to be studied in detail and it is likely that many species remain undescribed. Keng (1970) estimated that for such families, when properly known, the number of species will increase by about 28 per cent.

More than half the species belong to just eight families (Table 6.1). Four of the five largest families are monocotyledons, the orchids (Orchidaceae) grossly outnumbering all other families and also including three of the four largest genera of herbaceous plants (Table 6.2). Most native species of herbaceous monocotyledons live as epiphytes (notably the Orchidaceae) or in the undergrowth of rain forest, as do members of the largest dicotyledonous genera, *Didymocarpus*, *Begonia*, *Sonerila* and *Argostemma*. Most grasses and sedges grow in open conditions, except for the grasses *Leptaspis* and *Lophatherum* and the sedge *Mapania*, which are forest plants.

TABLE 6.1. Herbaceous families with more than a hundred recorded indigenous species in Peninsular Malaysia.

Family	No. Genera	No. Species
Orchidaceae	100	846
Gramineae	83	205
Gesneriaceae	21	162
Cyperaceae	24	154
Zingiberaceae	23	150
Rubiaceae	14	128
Araceae	23	120
Acanthaceae	26	116

Many of the species of weeds of disturbed areas are South American or African in origin. Although these exotic herbs have run wild in Malaysia and are now abundant and widespread, they are light-demanding and unable to invade the deeply shaded ground layer of the forest.

In a comprehensive introduction to herbaceous plants, Henderson (1954,1959) described and illustrated the commoner species. Holttum (1977) provided interesting information on the morphological diversity of Malaysian plants. For the largest families fairly recent taxonomic revisions are available: Orchidaceae (Holttum, 1964), Gramineae (Guilliland, 1971), Cyperaceae (Kern, 1974; Kern & Nooteboom, 1979) and Zingiberaceae (Holttum, 1950). For the other large families the standard reference remains Ridley's flora (1922-1925) although it is now outdated and in some instances unreliable.

TABLE 6.2. Herbaceous genera with more than forty recorded indigenous species in Peninsular Malaysia.

Genus	Family	No. Species
Bulbophyllum	Orchidaceae	127
Dendrobium	Orchidaceae	107
Didymocarpus	Gesneriaceae	68
Eria	Orchidaceae	61
Begonia	Begoniaceae	50
Sonerila	Melastomataceae	47
Argostemma	Rubiaceae	42

Among herbs of economic interest, the wild bananas hold great importance as breeding stock, especially for disease resistance, as S. E. Asia is the centre of banana evolution. In Peninsular Malaysia alone 30 to 40 distinct varieties have been selected over the years and are available in local markets (Simmonds, 1955).

Other herbs have won fame for their beauty: rajah's begonia *Begonia rajah* and mountain aeschynanthus *Aeschynanthus longicalyx* var. *superba* won medals at the Royal Horticultural Society of Britain. The slipper and moth orchids (*Paphiopedilum* and *Phalaenopsis*), in particular, are sought in their native habitats by collectors and growers, presenting a threat to their survival (Section 6.6).

6.2. HABITATS OF FOREST HERBS

In lowland rain forest, the herbaceous layer experiences deep shade (receiving only 1–3 per cent of incident sunlight) and high relative humidity (rarely below 80%). Most of the forest floor is bare but the appearance of emptiness is illusory. Beneath the leaf litter, the surface layer of soil is completely occupied by a tangled mass of roots and fungal mycelia competing for water and nutrients. In some habitats, herbs predominate. In permanently wet areas there may be lush thickets of bemban *Donax grandis* or *Hanguana malayana*. In shallow rocky streams species of *Elatostema*, begonias and aroids are common. On inhospitable, damp vertical rock faces cling species with a rosette habit, such as the wavy-leaved begonia *Begonia sinuata* (Fig. 6.1) or the primrose-yellow didymocarpus *Didymocarpus primulina*, or with hanging stems such as aeschynanthus species. In this rocky habitat, herbs can escape root competition. Some, such as *Neckia serrata* or the begonia-leaved sonerila *Sonerila begoniaefolia*, are found only on steep slopes where leaf litter does not collect and so does not cover them.

In montane forests herbaceous plants are conspicuous, especially in mossy or elfin forests (Chapter 2) where the tree canopy is open and lower and the abundant epiphytes are accessible to the naturalist. Here the white star-like flowers of argostemmas and the pink-flowered sonerilas live on the moss carpet. Above, orchids vie with ferns for a foothold on the branches of the trees which are also festooned with mosses and leafy liverworts. In the mountains species of several temperate genera are found, such as violets (*Viola*), balsams (*Impatiens*) and anemones (*Anemone*).

On Mt Tahan, from 1700 m to the summit, the vegetation is of particular botanical interest as it harbours a large number of endemic and unusual plants. For example it is only here that the Malayan

Fig. 6.1. The wavy-leaved begonia *Begonia sinuata* commonly grows on damp rock faces where it avoids being covered by leaf litter or competing with other plants. One of the fibrous-rooted begonias, it produces many seedlings wherever it grows.

gentian *Gentiana malayana* is found. The flat, open plateau below the summit, known as the Padang, has poor acid soil and exposed granite boulders. Here the vegetation is dwarfed to 1 m in height, except in the sheltered watercourses with peaty soil where trees may attain 15 m. The ground layer consists largely of sedges, ferns and orchids, as well as endemic species such as the giant xyris *Xyris grandiflora*.

Several edaphic habitats, such as sandy beaches, the margins of still or slow-flowing freshwater lakes or limestone hills, do not support forest. Here herbs predominate, but of different species from those found in the forest. Sandy beaches support relatively few species and these are widespread throughout the Indo-Pacific region, their seeds being dispersed by sea currents (Section 6.4). The fleshy purple-flowered sea morning glory, *Ipomea pes-caprae*, spreads itself along the beach by runners. The coarse porcupine grass *Spinifex littoreus* forms clumps on some beaches. The crinum lily *Crinum asiaticum*, reputed to be an antidote for the dreaded arrow poison from the upas tree, grows above the high tide mark.

The margins of Tasik Bera show gradation from floating plants (the bladderworts, *Utricularia* spp.) in the open, acid, tea-coloured water, to rooted plants with floating leaves (the water gentian *Nymphoides indica*) in shallower water, to the beds of the large sedge *Thoracostachyum bancanum* and rushes *Lepironia articulata* on the muddy margins (Merton, 1962 and Chapter 15).

Limestone hills support a special flora, of which almost a quarter of the species are endemic to Peninsular Malaysia (Chin, 1977). From the shaded base and narrow valleys where monophyllaeas abound, to the scorching vertical cliffs and jagged summits, herbs are an important component of the flora. Several genera are characteristic of limestone, such as begonias, balsams, chiritas and paraboeas (the latter formerly included in *Boea*; Burtt, 1984). The paraboeas are often the only plants that can survive on the sheer cliff faces exposed to heat and desiccation. Their adaptation to these severe conditions is not understood. Some species that grow on the summit have xeromorphic features: fleshy leaves (such as the peperomias) or have fleshy stem bases and are deciduous in the dry season (Section 6.3). A few paraboea species accumulate calcium in their epidermal cells (*P. caerulescens* and *P. verticillata*) or have calcium-excreting glands (*P. acutifolia*) (Burtt & Tan, 1984) but most show no anatomical adaptations, apart from a dense indumentum that may serve to reduce transpiration. Most paraboea species grow well in garden soil, showing they are not calcium demanding. A few species, such as the common epithema *Epithema saxatile* or the one-leaf plant *Monophyllaea horsfieldii*, are also found on other types of rocks, such as granite, although the majority are confined to limestone. The reasons for this are not yet understood, but the absence of herbaceous limestone species from forest in some cases may be explicable by factors other than edaphic, such as intolerance of shading or competition with other plants.

Man has created several types of habitat where herbaceous plants predominate; for example, forest fringes, 'lalang' (*Imperata cylindrica*) grasslands or tin tailings. Herbaceous plants of the forest fringes include wild bananas and giant gingers with stems up to 5 m tall. Under natural conditions they colonize landslips or riverbanks.

Unfortunately we understand very little about the ecological requirements of herbaceous species. Although it is well-known that epiphytic orchids taken from the mountains languish and eventually die in the lowlands, it is not known whether this is caused by higher temperatures and/or lower relative humidities. Some species appear to have strict ecological requirements and are found only in very narrow microhabitats; for example some species of the miniature orchid, *Corybas*, grow only on banks of sphagnum moss.

6.3. MORPHOLOGICAL DIVERSITY

6.3.1. Growth forms

Most of the herbaceous plants are perennial and many are long-lived. The broad-leaved didymocarpus *Didymocarpus platypus* (Fig. 6.2) develops a woody stem and is estimated to live for up to 20 years (Kiew, 1986). In addition, many monocotyledons are potentially immortal due to their evergrowing rhizomes or other means of vegetative reproduction. The perennial habit is in tune with the slow tempo of the rain forest where trees can live for hundreds of years and the conditions in the undergrowth are extremely stable. In the deep shade on the nutrient-poor soil of the forest, many of these plants cannot but be slow growing.

Growth patterns of herbs are as diverse as those of trees. Growth in monocotyledons is limited by the root system which has no cambium and so cannot produce secondary tissues to keep pace with growth of the shoot (Holttum, 1955). In many monocotyledons the root system is adventitious, individual roots are relatively small and the number produced at a single node is limited, restricting the potential size of the

Fig. 6.2. The broad-leaved didymocarpus *Didymocarpus platypus*, with pure white flowers (foreground), growing alongside another didymocarpus on the forest floor.

erect stem. This restriction can be overcome to some extent by crowded nodes at soil level (as is seen in taro *Colocasia esculenta*), or by the prostrate growth of the stem (as in the rhizomes of the gingers). The success of these adventitious roots can be judged by, for example, the size of ginger stems (some grow to 5 m in height) and leaves of the aroid *Alocasia macrorrhiza*. The prostrate stem is also well-suited to the epiphytic way of life where the adventitious roots can function to attach the plant to its support. This in part accounts for the success of orchids as epiphytes. The climbing aroids, such as *Philodendron* species, also cling to trees or rocks by their adventitious roots.

Unbranched herbs are rare among monocotyledons; many have sympodial growth and produce suckers at the base, e.g., bananas. In theory, the production of basal buds or the continuous growing tip of rhizomes allows the plant to grow indefinitely. The rhizomes of gingers and the arrow-roots are thick, fleshy and often contain starch. In everwet conditions they do not function as perennating organs but rather as stores of sufficient starch for the production of the next shoot. Among limestone plants that experience a dry season in the north, such as *Arisaema* and *Amorphophallus* species, the rhizome or fleshy stem base does have a perennating function when the leaves die back.

The dicotyledonous herbs are less striking in size and form. Some are unbranched plants with a tuft of leaves at the top (e.g., *Neckia serrata* or the gesneriads in *Didymocarpus* section *Heteroboea*). Plants that grow on rock faces such as the gesneriads *Paraboea* and *Loxocarpus* species often have a rosette habit. Others, with longer internodes, such as scorpion's tail, *Pentaphragma* species, produce taller stems which often become decumbent and root at the nodes (Burtt, 1977). Some species with branching prostrate stems, e.g., elephant's footprint *Phyllagathis rotundifolia*, form clumps in the forest.

Among flowering epiphytes, the orchids predominate and show several morphological adaptations to the exposed conditions under which they live. The succulent leaves and stems store water (those with a pronounced swollen base are known as pseudobulbs). The velamen (the lignified layer that surrounds the roots) prevents water loss and does not, as some textbooks suggest, absorb and store water. Species with succulent leaves are CAM plants (i.e., they have Crassulacean Acid Metabolism). Their stomata are closed during the day, which reduces water loss from the leaf (Winter *et al.*, 1983). Very light, wind-dispersed seeds are also common among epiphytes (Section 6.4).

Orchids are also unusual in having green roots that can photosynthesize. *Taeniophyllum* is a bizarre epiphytic orchid which depends entirely on its green roots as its leaves are brown and scale-like and its stem is very short. At the other extreme, the largest epiphytic orchid, the tiger orchid *Grammatophyllum speciosum*, has stems 3 m long and produces a magnificent inflorescence 2 m or more long with large brown-spotted flowers 10 cm across.

The leaves of the epiphytic *Dischidia* species (Asclepiadaceae) range in shape from lens-shaped to concave to those with ascidia, i.e., flask-shaped leaves, into which adventitious roots grow. Whatever their form, all species have succulent leaves and *D. nummularia* has been shown to be a CAM plant (Winters *et al.*). It is probable that their striking leaves are in part adaptations to conserving water rather than in providing a living place for ants (Section 6.5). Ants do help, however, in the establishment of *Dischidia* seedlings on the branches.

Water plants are morphologically very varied. Those that are rooted include species that produce leaves above the water surface, such as the reeds, wild rice, sedges, the lotus and water mimosa *Neptunia oleracea* (the latter two have edible swollen rhizomes). The waterlilies and water gentian are rooted plants with floating leaves; while the aroid *Cryptocoryne affinis* and the forest waterlily *Hydrostemma* (formerly known as *Barclaya*), which are both important in the aquarium trade, have submerged leaves. *Indotristicha malayana* (Podostemaceae) grows submerged on rocks in torrential streams and only came to light recently when an exceptionally dry period exposed it (Dransfield & Whitmore, 1970). Without its flowers, it is difficult to recognize *Indotristica* as a flowering plant as its thallus resembles a green seaweed. Water plants that are not rooted include species of duck-weed (*Lemna* and *Wolffia*) and the introduced water hyacinth, which float on the water surface, and some species of bladderwort which are submerged. All these grow in fresh water, but a few species, such as the aroid *Cryptocoryne ciliata*, are capable of extending into the brackish water of estuaries.

The rheophytes form an interesting group. They are found in torrential streams and are able to withstand the rush of flood water. They are strongly rooted and have narrow, willow-like leaves. Most rheophytes are shrubs but a few are herbaceous such as the aroid *Piptospatha perakensis* or the gesneriads *Didymocarpus densifolia* and *D. salicifolia* (see Chapter 16).

6.3.2. Leaf size

Herbaceous monocotyledons have some of the largest leaves in the forest, outclassed only by palms. Bananas have oblong leaves, up to 7 m by 1 m in *Musa truncata*, that tatter in the wind. The aroid *Colocasia gigantea* has a sagittate leaf with a petiole more than 1 m long and a lamina 2 m long. The aroid *Amorphophallus* produces a massive mottled petiole 2 m tall and as thick as a man's arm, bearing a wide compound lamina at the top. These large leaves contain very little lignified tissue and are supported by turgor.

Leaves of dicotyledonous herbs in Peninsular Malaysia are never so big as those of monocotyledons, although those of the one-leaf plant can grow to more than a metre long. *Monophyllaea* is also remarkable as its single leaf develops from one of the cotyledons. Of course, some herbs also have small leaves, such as the pick-a-back plants, *Phyllanthus* species, with leaves 5 mm long. Compared with the more uniform leaves of trees and shrubs (glossy, lanceolate and leathery), leaves of herbs are broader and softer and may be variegated (Section 6.3.4). The only herbs with thorny leaves are the aroids *Lasia, Podolasia* and *Cyrtosperma*, which have thorny petioles.

The leaves of forest herbs have several attributes of shade leaves. For example, among grasses of comparable size, the forest grass *Leptaspis urceolata* has broader leaves than those that live in the open. The anatomy of forest herbs also shows several features associated with shade leaves (Hughes, 1965), such as epidermal cells with a sinuous outline, a single layer of very short palisade cells which are tapered to the base and a few layers of spongy mesophyll which combine to give a very thin leaf about 50 μm thick. Another anatomical feature seen in many species is the large size of the upper epidermal cells or the presence of a hypodermis, which by turgor support the lamina which otherwise has very little supporting tissue.

6.3.3. Life span of leaves

A few herbs of the limestone hills in the north are deciduous in the dry season: gouty balsam *Impatiens mirabilis,* the ginger *Kaempferia pulchra*, some orchids and begonias. In the vast majority, herbs are evergreen. Leaves may be produced continously at a rapid rate (bananas produce a new leaf every 7 to 10 days; Purseglove 1972), more slowly (the broad-leaved didymocarpus produces leaves at 2 to 3 monthly intervals) or seasonally (elephant's footprint and scorpion's tail *Pentaphragma horsfieldii* produce new leaves between April and June and between August and December; Kiew, 1986).

The life span of leaves is still largely an unexplored field. Banana leaves live for about 3 to 5 months. In the rain forest leaves of broad-leaved didymocarpus live for about 22 months and those of scorpion's tail for 33 months (Kiew, 1986), while the single leaf of a *Monophyllaea* lives for the entire life of the plant, continuously growing from its base while the tip gradually moulders away.

6.3.4. Leaf colour

The colour of leaves of forest herbs is a source of fascination. There are blue leaves (the peacock begonia *Begonia pavonina*), black leaves (in some argostemmas), golden leaves (the orchid *Malaxis*), leaves with white or reddish spots or stripes (sonerilas), or with veins outlined in white or gold (the batik orchids, *Anectochilus*), marbled leaves (the gesneriad *Cyrtandra pendula*), velvety leaves that change colour from blue-green to golden depending on the angle of the light (the begonia *Begonia thaipingensis*) and many leaves red or purple underneath. The beauty of the coloured foliage was much appreciated by the Victorians and is coming into vogue again. For instance, wild aglaonemas with variegated leaves are popular house plants, able to tolerate deep shade.

The causes of the colours are easier to understand than their function. The blue leaves of the fern-ally *Selaginella willdenowii* are not caused by pigmentation but by the reflective properties of the cuticle and outer epidermal wall which differentially reflect blue light (Hébant and Lee, 1984). When these plants are grown in full sunlight, the leaves do not develop the reflective layer and so do not look blue. The black leaves of argostemmas, too, are not caused by pigmentation but by the total absorption of light by the compact palisade layer which has cells that are completely filled with large chloroplasts.

Red or purple coloration is due to the presence of pigments but there is often considerable variation within a species, or even within a single population. For example, in a single population of the variegated begonia *Begonia phoeniogramma* there are plants with leaves that are completely green, while others are purple underneath, and yet others are purple underneath with white spots on the upper surface. This variability casts doubt on any putative function of variegation.

6.3.5. Epiphylls

Small plants which colonize leaf surfaces (epiphylls or epiphyllae) are abundant in rain forest and herbaceous plants are not immune. Macroscopic epiphylls include crustose lichens (which are most common), green algae (usually *Trentepohlia* species, which are thread-like and grow from the edge of leaves, and *Phycopeltis* species, which form velvety cushions) and leafy liverworts, such as species of *Cololejeunea* and *Drepanolejeunea* (Kiew, 1982). Leaves provide a horizontal surface for epiphylls to grow where they will not be covered by leaf litter or have to compete with soil organisms (Fig. 6.3). They may also obtain nutrients which leach out from the leaf. At the same time, in the early stages the leaf may gain nutrients when colonized by microscopic nitrogen-fixing bacteria or blue green algae. Some crustose lichens, on the other hand, penetrate the leaf tissue and may be partly parasitic. In all cases, eventually epiphylls shade the leaf surface and prevent photosynthesis.

Fig. 6.3. A five-year-old leaf of the undergrowth palm *Iguanura wallichiana* covered with epiphylls including white lichens, thread-like algae and leafy liverworts.

Initial establishment is slow. A few colonies of lichens may appear within the first six months and increase steadily thereafter. Colonies of *Phycopeltis* and the leafy liverworts appear after about two years (Kiew, 1982). Epiphylls thus only become established on long-lived leaves. Leaves of elephant's footprint are spotted with lichens after about 18 months but algae and leafy liverworts do not become established. Leaves of scorpion's tail aged between 18 and 24 months are conspicuously colonized by leafy liverworts, and by 36 months half the leaf surface is covered (Kiew, 1986).

6.4. THE BIOLOGY OF HERBACEOUS PLANTS

6.4.1. Vegetative growth

Although herbaceous plants are more convenient to work with than trees on account of their size and lifespan, few experimental studies have been made. Bananas are an exception, being a commercial crop. We know they grow quickly, producing 35 to 50 large leaves (roughly one every ten days) for a period of

7–9 months, after which the inflorescence emerges and puts a stop to vegetative growth (Purseglove, 1972). The productivity of the water hyacinth (150 tonnes per hectare per year) is perhaps the highest of any plant in the world; its closest rival is sugar cane (90 tonnes ha^{-1} per year; Westlake, 1963). Some species of weeds can grow rapidly e.g., *Hedyotis dichotoma* (Rubiaceae) produces mature fruit within 11 weeks of germinating (Kiew, 1975).

In the rain forest, where the deep shade and competition with other plants limit growth, herbs are perennial and grow at a slower rate, for example the broad-leaved didymocarpus produces four leaves a year with a total leaf area of 792 cm^2, and scorpion's tail just two leaves with a leaf area of 175 cm^2 (Kiew, 1986).

Richards (1952) suggested that reproduction of rain forest herbs depends on vegetative means rather than seed production. Some are rarely found in flower or fruit, for instance, species of arrow-root, some aroids and gingers or the common purple-leaved climbing pepper *Piper porphyrophyllum*. Most, however, especially dicotyledons, do flower regularly. Fruit production may still be low: of 41 flowers produced in five years by one plant of broad-leaved didymocarpus only five set fruit (Kiew, 1986). The common spanglewort *Lepidagathis longifolia* has never been found in fruit (Henderson, 1959) although it is a common forest plant, easily identified as its leaves fall off when the stem is shaken. A possible explanation is that there is a failure in pollination (van Steenis, 1969).

Whether it follows that vegetative reproduction is more important is doubtful. In the vast majority, rain-forest herbs are perennial and some, such as the broad-leaved didymocarpus (which has no means of vegetative reproduction), may live as long as 20 years. The quantity and frequency of seed production therefore needs only to replace an individual once in 20 years to maintain the species. Vegetative reproduction may serve an important function in extending the life of the plant, besides providing a more certain method of spreading itself in a suitable habitat.

The sympodial growth habit of monocotyledons lends itself to spread by means of suckers, rhizomes or runners — the walking ginger *Hornstedtia scyphifera*, which has a runner that 'walks' on metre-long adventitious roots, is a bizarre example. Many of these monocotyledons can therefore form clumps in the forest. Clumps of dicotyledonous plants are formed by species with erect stems that become decumbent and root (Burtt, 1977). Still others produce bulbils, such as the wavy-leaved begonia or the polyploid species of the ginger *Globba*. The tiger's sireh *Phyllagathis griffithii* and elephant's footprint regularly produce plantlets from severed leaves under natural conditions (Kiew and Latipah, 1981).

6.4.2. Flowering

Visitors from temperate regions are disappointed by the lack of lurid flowers and the overwhelming greenness of the rain forest. It is extremely rare to find a show of flowers, though a few species, such as the gregariously flowering Malayan cowslip *Didymocarpus malayanus*, periodically covers slopes with its golden-yellow flowers. The common ever-flowering argostemmas and sonerilas have small flowers.

Other species of *Didymocarpus* flower more regularly than the Malayan cowslip but their flower buds, which develop in sequence, then wait before opening gregariously. The phenomenon that controls this is not understood. Another gregariously flowering species, the pigeon orchid *Dendrobium cruminatum* in Kuala Lumpur flowers nine days after a sudden drop of 4 °C induced by rain storms (Wycherley, 1973). Some flower seasonally: species of *Phyllagathis* (Melastomataceae) flower between March and May and in September and October (Kiew, 1986); the tiger orchid flowers either in January or July (Holttum, 1964). In these cases, however, flowers are not always produced every season or even every year.

Richards (1952) noted that many herbs produce flowers near ground level, though whether such species predominate numerically is doubtful. Several gingers, such as *Achasma* and *Amomum*, produce inflorescences among the leaf litter while the gesneriad *Cyrtandra pendula* produces a pendant inflorescence that hangs to the ground. These, however, are exceptions. Other gingers bear inflorescences at the tips of their long leafy shoots, and most *Cyrtandra* species flower from their leaf axils.

Flower colours include pinks (*Begonia, Phyllagathis, Sonerila*), white (*Argostemma, Pentaphragma*) and dull purple (the black lily *Tacca* and bananas). Yellows are rather rare (the yellow forest-star *Acrotrema costatum* and a few *Didymocarpus* species) and blues even more so (the common spiderwort and the crested day-flower, Commelinaceae). Orchid flowers cover a range from greens and browns (often accompanied by a strong unpleasant smell) to purples, magenta, orange, yellow to white, and may be scentless or smell sweet or spicey. In contrast, most other herbs have scentless flowers, apart from those which smell of carrion, such as species of *Rafflesia* and some aroids (*Amorphophallus*); *Balanophora reflexa* smells strongly of small rodents.

6.4.3. Pollination

Pollination studies are few and many merely record visitors rather than supplying evidence that the agent actually carried pollen. Bat and bird pollination is well-known for bananas. Bats visit the species with pendant inflorescences with purple bracts and birds visit those with erect inflorescences with pink or pale purple bracts (Holttum, 1977). McClure (1966) observed three species of spiderhunter that probed the long tubular scarlet flowers of the climbing gesneriad *Aeschynanthus* for nectar and green leaf birds, nectar 'thieves', that punctured the base of the tube to reach the nectar.

Pollination studies on forest herbs are discouraging as insect visitors are scarcely ever seen. I have only once seen flies at flowers of scorpion's tail and have rarely seen an insect visiting *Argostemma, Begonia, Didymocarpus* or *Sonerila* flowers. Most of these flowers offer nectar. Begonia is a pollen flower, the stigmas of the female flower acting as a pollen dummy to deceive the potential pollinator.

Only in open conditions has the Malayan honey bee *Apis cerana* been seen at work, collecting pollen from the flowers of the sensitive plant *Mimosa pudica* and nectar from *Asystasia intrusa* (Acanthaceae), a common weed (Mardan and Kiew, 1985). The white flowers of the pigeon orchid are only rarely visited by *A. cerana* (Burkill, 1919). Sweat bees *Trigona* visited a white-flowered balsam *Impatiens opinata* (Fig. 6.4) and the large solitary bee *Xylocopa rufescens* the pale purple flowers of the gesneriad *Paraboea treubii*, both plants growing on limestone (Kiew and Yong, 1985).

Flies are attracted to the putrid smelling flowers of *Rafflesia hasseltii* (Meijer, 1983). Its flowers take 18 months to develop and then open for only five days before rotting away. Many aroid inflorescences are designed to trap flies to effect pollination but the gigantic *Amorphophallus titanum* from Sumatra with inflorescences 2.5–3 m tall is visited by carrion beetles (Meeuse and Morris, 1984), apparently attracted by its overwhelming stink.

Most water plants are pollinated by insects, and their colourful flowers open above the water surface (the lotus, waterlilies and the bladderworts). Very few are pollinated by water. *Cryptocoryne griffithii* exhibits the typical aroid mechanism of trapping beetles in the 30 cm long spathe, which projects above the water surface, and allowing them to gain entry to 'overnight in an underwater cabaret', as Corner (1964) put it. Wind pollination is only effective in open conditions and is largely confined to grasses.

Strange phenomena still have to be explained. Do the hinged parts of *Bulbophyllum* flowers really throw the insect into the flower and so effect pollination? Does the purple cloud of pollen released from the explosive stamens of *Poikilosperma* (Urticaceae) powder the pollinating agent or is it a mechanism for wind pollination? And what part do the 30 cm-long whiskers of the black lily *Tacca* flowers play in pollination? What animals are attracted to the crimson flowers of *Balanophora reflexa* by its overpowering smell? And why should the inflorescences of some aroids heat up by as much as 15°C? Is it really to make

Fig. 6.4. A sweat bee *Trigona* (visible on right) visits a white-flowered balsam *Impatiens opinata*, growing on limestone.

the inflorescence stink the better? More than a life time's study is needed to answer questions like these.

Self-pollination occurs in some cleistogamous orchids, for example *Spathoglottis microchilina* (Holttum, 1964), but in others the relative importance of self- or cross-pollination is not known. Natural hybridization, as judged from plants showing a combination of characters intermediate between species, is known in a few instances: hybrids are recorded between the gingers *Globba cernua* and *G. patens* (Lim, 1972), and between *Begonia decora* and *B. longicaulis*. Airy Shaw (1954) suggested that *Pentaphragma xellipticum* (Pentaphragmataceae) is a hybrid between scorpion's tail *P. horsfieldii* and *P. acuminatum*, although the latter is a Bornean species not (now) occuring in Peninsular Malaysia. Simmons (1955) found that the bananas *Musa acuminata* and *M. trunctata* form hybrid swarms where their populations overlap. One putative parent of the commercial sterile banana is *Musa acuminata*, one of the commoner wild species in Malaysia.

The classic instance of hybridization occurred in Miss Agnes Joaquim's garden in Singapore in 1893 between the Malayan orchid *Vanda hookeriana* and the Burmese *Vanda teres*. This hybrid, 'Vanda Miss Joaquim' became immensely popular as a cut flower and laid the foundation for the successful orchid industry in the region which is based on hybrids. The ease with which hybrids are produced in orchids by artifical pollination indicates that they are self-compatible to the extent that trigeneric hybrids can be made, such as x *Holttumara*, a hybrid between species of *Arachnis* and *Renanthera* (x *Aranthera*) hybridized with a species of *Vanda*. Under natural conditions, orchid species are probably isolated by specificity of pollinator.

The genus *Globba* (a small ginger which produces bulbils) has been studied in detail cytologically. Malaysian species are polyploid (tetraploid, hexaploid and decaploid) and the hexaploid and decaploid species produce fruits rarely if at all (Lim, 1972).

6.4.4. Fruiting and dispersal

Seeds of herbaceous plants range in size from minute (orchids) to those of the aroid *Amorphophallus*, which are about 1 cm long. Orchid seeds are produced in great quantities and are dispersed by wind.

Many other epiphytes are wind dispersed, a method that allows the seeds to be swept from one treetop to another: *Dischidia* belongs to a family with plumed seeds (Asclepiadaceae); the two climbing genera of the Gesneriaceae, *Aeschynanthus* and *Agalmyla*, are the only ones in their family with plumed seeds. The other group of herbs dispersed by wind are ground plants of open places, such as lalang grass *Imperata* or the reed *Phragmites* (which have plumed fruits) or porcupine grass, where the infructescences form balls 20 cm across that roll along the beaches in the wind dropping their fruits as they go.

Fleshy fruits are characteristic of some families (aroids and gingers) and some genera of others (bananas, *Labisia* in Myrsinaceae, *Pentaphragma* and *Cyrtandra*); fleshy arils occur in other genera (*Forrestia*, Commelinaceae). Bananas and perhaps the gesneriad *Cyrtandra* are dispersed by bats. Bananas are also dispersed by birds or mammals including monkeys (Simmons, 1955) and by civets (Ridley, 1930). Brockleman (1984) suggested that blue fruits are attractive to pheasants and, indeed, several plants of the forest undergrowth have striking blue fruits, such as *Pollia* species (Commelinaceae), *Labisia longistyla* (Myrsinaceae) and *Peliosanthes violacea* (Liliaceae). Birds probably disperse the red or orange berries of aroids and the orange fruits of *Nertera granadense* (Rubiaceae), which is popular as a diminutive pot plant in Europe. In his monumental work, Ridley (1930) also suggested that brownish fruits which ripen at ground level, such as species of *Curculigo* (Hypoxidaceae), the sedge *Mapania* and some gingers, are dispersed by forest rats. Lumbah *Curculigo latifolia* is interesting as its fruit, although tasteless when eaten, gives a sweet taste to any drink taken for about an hour afterwards. This property is caused by an unstable compound and so has no commercial value.

In the still conditions of the undergrowth of the rain forest, dispersal by wind for the several genera that have dry capsules which contain minute seeds is extremely unlikely. More likely is dispersal by rain splash. The gesneriad *Didymocarpus* has long narrow capsules that split along the upper side to form a gutter along which a rain drop can run carrying with it the minute seeds. Other genera (*Argostemma*, *Phyllagathis* and *Sonerila*) have cup-shaped capsules and their seeds may be dispersed by rain drops bouncing out of the cup. The forest dwelling begonias have minute seeds,too, but their capsules hang from a thin stalk so that they are easily shaken out, perhaps again by rain.

Balanophora seeds are minute, weighing about 7 μg each and are produced in vast quantities (female inflorescences are estimated to produce up to ten million flowers). In the still air of the forest floor, they are unlikely to be dispersed by wind. Perhaps they are carried on the beetles that lay eggs in the fleshy part of the inflorescence.

A few forest herbs have explosive fruits (e.g., the balsams) or adhesive, burr-like fruits carried by animals (the grasses *Centotheca* and *Leptaspis* and the small forest legumes, *Desmodium*).

Water plants frequently have floating seeds or fruits buoyed by corky structures such as the seed coat of the sea morning glory or the receptacle of the lotus. The fruits of seashore plants, such as the crinum lily *Crinum asiaticum*, will float for one to two weeks; those of *Tacca pinnatifida* for many months (Ridley, 1930). A few species such as *Hygrophila saxatile* (Acanthaceae), which lives on rocks in the river, have tenaciously sticky seeds which may be dispersed on the feet of water birds. Ridley concluded that wading birds introduced *Enhydris* and the bladderworts that appeared in the artificial pond in the Botanic Gardens, Singapore. Some water plants, such as duck-weeds, are dispersed by fragmentation. Ridley (1930) considered *Potamogetum* species likely to be dispersed by fish and the hooked seeds of the forest waterlily (*Hydrostemma* species) by wild pigs, though what evidence he had for this is not clear.

Seedlings of herbaceous plants are uncommon in forest (with the exception of a few fibrous rooted begonias, e.g., *Begonia sinuata, B. phoeniogramma*) and yet under artificial conditions percentage germination is usually high. This suggests that seedling establishment may be the critical phase in the life of herbaceous plants in the forest. Some of the smallest seeds belong to genera of forest plants with the largest number of species, e.g., *Argostemma, Begonia, Didymocarpus* and *Sonerila*. Small seed size is not confined to genera that are dispersed by rain splash; they are also a feature of species such as scorpion's tail that have fleshy berries.

The minute seeds of the orchids, which weigh only a few micrograms, require a mycorrhizal association for the establishment of their seedlings. This is apparently not the case in other groups, among which the seeds germinate rapidly to produce minute green cotyledons and leaves. Although small seeds facilitate dispersal, they carry with them the disadvantage of a tiny seedling which risks being covered by leaf litter on the forest floor. It is significant that those species whose seedlings are locally abundant (*Begonia phoeniogramma* and *B. sinuta*) grow on rock faces where leaf litter does not accumulate. Perhaps the advantage of minute seeds is that a vast number are produced from a single fruit, sufficient to compensate for the number of seedlings that perish during the establishment phase.

Germination is usually rapid. In the epiphyte *Dischidia nummularia* (Asclepiadaceae) the root is produced within 12 hours of the seed being wetted. Under artifical conditions percentage germination is high. In fact, if the seeds are not dispersed, they will sometimes germinate in the capsule as I found in the case of the gesneriad *Loxocarpus incana* and the melastome *Sonerila rudis*. This, however, is not the same as vivipary as there is no way for the germinated seedling to be dispersed and it would eventually perish, trapped in the capsule. True vivipary is seen in the aroid *Cryptocoryne ciliaris* which grows on tidal mud and therefore parallels *Rhizophora* (Chapter 2). Its root develops before the seed is freed. Once dispersed, the root can thus quickly anchor the seedling when stranded on mud by the tides.

While seeds of the bananas *Musa gracilis* and *M. violascens* germinate after a month, other banana species will only germinate after more than six months (Chin H.F., pers. comm.) and therefore provide one of the few examples of tropical forest herbs with seeds that remain viable in forest soil for a period of years. Once the soil is exposed to full sunlight, their seeds germinate rapidly, probably in response to the higher soil temperatures which break their dormancy. Seeds of *Dianella ensifolia* (Liliaceae) germinate after a three-month period, not because they are dormant, but because the embryo of fresh seed is still immature (Garrard, 1955).

It is not known how the saprophytic species establish themselves from seed. Seeds of the parasite *Balanophora* produce tubular cells which rapidly attach the embryo to rootlets of the host tree after which a haustorium develops. *Rafflesia* apparently can only enter its host through wounds in the host's stem.

There are still many unanswered questions about the biology of herbs, apart from the simple lack of observations on pollination or dispersal. What, for example, contributes to the success of the broad-leaved didymocarpus *Didymocarpus platypus*, which is common and widespread throughout Peninsular Malaysia, while other *Didymocarpus* species have more restricted distributions, some (*D. primulina*) even being confined to an area a few hundred metres square?

6.5. NUTRITIONAL DIVERSITY

6.5.1. Parasites and saprophytes

Two groups of herbs have opted out of the struggle for light by dispensing with photosynthesis. These are the saprophytes and the parasites. The saprophytes are mostly monocotyledons and the parasites are all dicotyledons (Table 6.3). Their number, compared with the 2500 odd species of green herbs, is very small: 37 species of saprophytes and 15 species of parasites.

The root system of the saprophytes is not extensive nor is it associated with other roots (i.e., they are not root parasites). Nor do saprophytes decompose dead vegetable remains themselves. Anatomical study shows that these species contain endomycorrhiza in their cortical cells. The fungus therefore does the decomposing and these so-called saprophytes are really dependent on the fungus. Perhaps it is not surprising that about half the saprophytic species are orchids which have a mycorrhizal association from

TABLE 6.3. Herbaceous parasites and saprophytes in Peninsular Malaysia.

Family	Genus	No. Species
PARASITES		
Balanophoraceae	*Balanophora*	6
	Exorhopala	1
Lauraceae	*Cassytha*	1
Monotropaceae	*Cheilotheca*	1
Orobanchaceae	*Christisonia*	1
	Aeginetia	1
Rafflesiaceae	*Rafflesia*	2–3
	Rhizanthes	1
Scrophulariaceae	*Striga*	1
SAPROPHYTES		
Burmanniaceae*	*Burmannia*	5
	Gymnosiphon	1
	Thismia	8
Orchidaceae*	*Aphyllorchis*	3
	Cystorchis	1
	Didymoplexis	2
	Epipogum	1
	Eulophia	1
	Galeola	3
	Gastrodia	2
	Lecanorchis	2
	Stereosandra	1
Petrosaviaceae*	*Petrosavia*	1
Polygonaceae	*Epirixanthes*	1
Triuridaceae*	*Sciaphila*	5

* Monocotyledons

germination onwards. From this first beginning, some are able to harness the fungus throughout their life.

These saprophytes are found in the deep shade of the primary forest. Most are small, leafless, often with a reduced shoot system with an inflorescence to 30 cm tall. *Galeola kuhlii*, an orchid, is exceptional in producing a climbing stem 15 m or more in length. The saprophytic way of life has obviously evolved from the autotrophic, as several genera possess both green and saprophytic species (*Burmannia, Cystorchis* and *Eulopia*), while *Epirixanthes* is closely related to, and sometimes regarded as synonymous with *Salomonia*, a genus of green plants.

Saprophytic orchids are commonly encountered but the other saprophytes are more rarely seen. Perhaps they are overlooked because of their small size and ghostly colouring — pallid purples, yellows or deep leaf-litter brown — or, as in the case of *Thismia*, because their flowers are ephemeral. Sometimes, several genera occur together, for example, in Taman Negara *Sciaphila secundiflora, Epirixanthes cylindrica* and *Thismia aseroe* have been found within a few metres of each other. Perhaps this association reflects the sharing of the same species of endomycorrhizal fungus. Study of endomycorrhiza in the tropics is a new field and none of the species associated with these saprophytes has yet been identified.

The parasites, in contrast, can be larger and more colourful. They tap the host's food and water through haustoria. The hemiparasites are green and tap only the host's water supply. *Viscum articulatum* belongs to the mistletoes, Loranthaceae, a family of shrubby hemiparasites. It is almost herbaceous and, being parasitic on other species of mistletoes, gets its water third hand.

The parasites of the primary forest have an extremely reduced shoot system and are conspicuous only when they flower. The two species of broom-rape produce delicate inflorescences. Both are rare: *Aeginetia pedunculata* has been collected only twice in Peninsular Malaysia; *Christisonia scortechinii*, which is endemic, has only been collected five times although the bamboo it parasitizes is plentiful and widespread.

Species of *Balanophora* are also root parasites and are commonly encountered at the hill stations (Kiew, 1978). Their stout fleshy inflorescences 3–15 cm long are scarlet, crimson or pale brown and grow from a tuberous vegetative body composed of the cells of both the parasite and host. Species of *Balanophora* parasitize a wide range of tree species and even occasionally the stems of the vine *Tetrastigma*.

The pride of the parasites is *Rafflesia* which holds the world record for flower size — the Sumatran species *R. arnoldii* is a metre across and weighs 7 kg. The vegetative body of all *Rafflesia* species is a thread-like structure that lives inside the trailing stems of its host, the vine *Tetrastigma*. Its presence only becomes known when its flower buds, which develop in the tissues of the host, burst through the bark. They attain the size of a small cabbage before opening. The common species in Peninsular Malaysia is *R. hasseltii* which has a flower about 50 cm across, but a larger species has been found in the north (E.J.H. Corner, pers.comm.).

Rhizanthes lowii is much rarer than *Rafflesia* and has only been collected from three places in Peninsular Malaysia. It is more modest: its flower measures 10–15 cm across and has the appearance of an earthstar with 14–16 pointed lobes. It opens a delicate pink and is covered by sugary-looking hairs, but darkens to mahogany red and becomes woody. Like *Rafflesia*, it becomes visible only when the buds burst through the bark of the vine stem. It parasitizes both *Kadsura* and *Tetrastigma* vines. Why *Tetrastigma* should be so appetising to these parasites is not known.

6.5.2. Insectivorous plants and ant plants

Several groups of plants which live in nutrient-poor habitats are supposed to benefit from gaining additional nutrients, particularly nitrogen and phosphorus, from insects. Insectivorous plants derive nutrients directly from the bodies of insects; ant plants from waste material from ant nests. These plants possess special structures related to their unusual mode of nutrition, which have attracted the attention of naturalists. Unfortunately, experimental evidence is lacking to show whether the additional nutrient source in fact significantly increases plant growth.

Of the insectivorous plants two genera are cosmopolitan herbs, the sundews *Drosera* and the bladderworts. In Peninsular Malaysia, the sundews are represented by two species and the bladderworts by 15 — four are aquatic, the rest terrestrial. Members of the third genus, *Nepenthes*, the pitcher plants, although fascinating are not considered here as they are woody vines. There is a growing literature about life in the pitchers and Shivas (1984) has provided an illustrated account of the Peninsular Malaysian species.

Ant plants are a tropical phenomenon and Peninsular Malaysia has its share of them. Some are trees (*Macaranga*, Euphorbiaceae); others are epiphytes, such as the shrubby baboon's head *Hydnophytum formicarium* and *Myrmecodia tuberosum* (Rubiaceae), and the herbaceous dischidias with about 23 species in Peninsular Malaysia. Some other species are uncritically considered ant plants because ants nest in them. However, many tropical ants nest above ground and plants offer one of several convenient sites.

Reassessment of the ant-plant relationship in the genus *Dischidia* (Weir and Kiew, 1986) shows a series of increasingly specialized association with ants. The common *Dischidia nummularia*, which has small lens-shaped leaves, has creeping stems which root superficially on the branch of its supporting tree and are not associated with ant nests. *Dischidia parvifolia* has the same type of small lens-shaped leaves, but it is associated with one genus of ants, *Crematogaster*. In the mountains at the base of gelam bukit trees *Leptospermum flavescens* these ants may nest and make connecting tunnels in the wood of the trunk and branches with exit holes in the branches. The plumed seeds of *Dischidia parvifolia* which settle on the branches are taken by the ant into the tunnels where they germinate and where their roots have access to nutrients from insect frass. There is a constant association between the presence of *D. parvifolia* on gelam bukit trees with *Crematogaster*. However, it is not a true ant plant since it does not provide a nesting place for the ant nor a source of food — although, initially, oil in the seed coat may attract the ant, as it may in several other species of dischidia, such as *D. hirsuta*, also found growing from ant nests.

Dischidia astephana grows together with *D. parvifolia* on gelam bukit trees and its roots also penetrate into the ant tunnels. *Dischidia astephana* produces two types of leaves — lens-shaped ones and larger concave leaves appressed to the branch, which are erroneously thought to offer shelter to ants (Holttum, 1977). Ants neither live underneath these leaves nor keep their broods there; the *Crematogaster* nest is at the base of the tree. If the branch is shaken ants pour out from underneath these leaves, but they have come from the nest at the base of the tree via the tunnels and small holes in the tree trunk and branches. This species, therefore, is not an ant plant.

At the most specialized, *Dischidia major* (formerly *D. rafflesia*) and *D. complex* have two types of leaves, the lens-shaped ones and flask-shaped ascidia with adventitious roots growing into them. Eggs and larvae as well as adult ants of the genus *Iridomyrmex* are found inside the ascidia which therefore offer some protection to the ant, although the main nest is located elsewhere. These two species can validly be called ant plants, because the ants do obtain shelter for their brood and the plant may gain nutrients from earth and frass that collect in the ascidia.

6.6. CONSERVATION

Ng and Low (1982) have shown the endemism in tree families is about 30 per cent, the ivy family (Araliaceae) having the highest value with 60 per cent. In contrast, much higher levels are found in herbaceous plants: the African violet family (Gesneriaceae) 80 per cent and Begoniaceae 90 per cent (Kiew, 1983). There are at least six endemic herbaceous genera: *Acrymia* (Labiatae), *Codonoboea* and *Orchadocarpa* (Gesneriaceae), *Exorhopala* (Balanophoraceae), *Klossii* (Rubiaceae) and *Stenothyrsus* (Acanthaceae).

Many endemic species are rare and have extremely narrow distributions. For example, the primrose-yellow didymocarpus is known from a single population of less than 200 plants, most of which grow on a rock face 10 m by 4 m (Fig. 6.5). *Acrymia ajugiflora* (Fig. 6.6), the only representative of its genus, is considered one of the ten most endangered plant species in Malaysia by virtue of its narrow distribution and small population (Kiew *et al.*, 1985). It is found only in forest at the base of one quartzite ridge. Many other herbaceous species are confined to a single mountain top, *Loxocarpus semitorta* is endemic to Mt Ophir, *L. holttumii* to Mt Panti and *L. minima* to Bukit Tangga.

The extremely restricted distribution of these endemic species makes them very vulnerable to threats such as exploitation of the forest, quarrying of limestone hills or commercial collecting. Forest exploitation takes several forms, such as clear felling and logging. Logging, which may destroy up to 40 per cent of the tree stand by the extraction of timber and the building of logging tracks, is disproportionately injurious to herbs which cannot withstand full sunlight or may be covered by silt washed into the streams by erosion. *Rafflesia* is particularly vulnerable, depending as it does on the

Fig. 6.5. The primrose-yellow didymocarpus *Didymocarpus primulina*, one of Peninsular Malaysia's rare herbs.

Fig. 6.6. *Acrymia ajugifolia*, one of Malaysia's ten most endangered plant species, grows in a restricted locality which is not in a national park.

survival of its host. As a silvicultural practice, vines are cut to encourage growth of sapling timber trees and in so doing *Rafflesia* is destroyed (Meijer, 1983).

While montane forest is generally unsuitable for logging or agriculture, it is nevertheless vulnerable to destruction and disturbance from the building of microwave stations, trigonometrical points and the development of hill stations. As the mountain tops carry a disproportionate number of endemic herbaceous species, many of these are in danger of extinction.

Specimen collection is particularly serious for orchids, both from commercial collectors and amateurs. The latter take any attractive orchid in flower they find, only to watch the montane species die in their lowland gardens. Collecting without permission is prohibited in Taman Negara (Fig. 6.7) but at present no Peninsular Malaysian plant is banned from export or import under CITES (Convention for International Trade in Endangered Species of Flora and Fauna) regulations.

In Peninsular Malaysia there is no convenient legislative process to protect small areas, some of which may support the entire population of a herbaceous species. Besides, the preservation of herbaceous plants falls outside the traditional interests of either the Department of Wildlife and National Parks or the Forest Department. It is for such reasons that there is no protected area for *Rafflesia hasseltii* in Peninsular Malaysia, whereas in Sumatra areas with *Rafflesia* are protected and are a tourist attraction.

Although it is easy to grow and propagate herbaceous plants in artificial conditions, they are

Fig. 6.7. The large leaved aroid *Alocasia lowii* grows on limestone in Taman Negara.

nevertheless vulnerable to neglect and can easily be lost. *Begonia eiromischa*, a Penang endemic illustrated in Ridley's flora, is no longer in cultivation and is probably extinct in the wild. *Begonia rajah*, which won the Royal Horticultural Society's medal in 1894, is no longer in cultivation in either Singapore or Kew Botanic Gardens. This species was collected only once from an unknown locality in Terengganu and has not been refound since (Kiew, 1986). Without doubt, the surest way to maintain a species is to protect it in its natural habitat.

REFERENCES

Airy Shaw, H. K. (1954). Pentaphragmataceae. *Flora Malesiana* 4, 517–528.

Brockleman, W. Y. (1984). Blue Berries. *Malay. Naturalist* 38 (2), 13.

Burkill, I. H. (1919). Some notes on the pollination of flowers in the botanic gardens, Singapore, and in other parts of the Malay Peninsula. *Gdns'. Bull. Straits Settlements* 2, 165–176.

Burtt, B. L. (1977). Notes on rain-forest herbs. *Gdns'. Bull. Singapore* 29, 73–80.

Burtt, B. L. (1984). Studies in the Gesneriaceae of the Old World: XLVII. Revised generic concepts of *Boea* and its allies. *Notes R. Bot. Gdn Edinburgh* 41, 401–452.

Burtt, B. L. and Tan, K.(1984). Studies in the Gesneriaceae of the Old World. XLVIII. Calcium accumulation and excretion in *Paraboea*. *Notes R. Bot. Gdn Edinburgh* 41, 453–456.

Chin, S. C. (1977). The limestone hill flora of Malaya I. *Gdns'. Bull. Singapore* 30, 165–219.

Corner, E. J. H. (1964). *The Life of Plants*. Weidenfeld & Nicolson, London.

Dransfield, J. and Whitmore, T. C. (1970). A Podostemacea new to Malaya: *Indotristichia malayana*. *Blumea* 18, 153–155.

Garrard, A. (1955). The germination and longevity of seeds in an equatorial climate. *Gdns'. Bull. Singapore* 14, 534–545.

Guilliland, H. B. (1971). *Flora of Malaya III. Grasses*. Government Printers, Singapore.

Hébant, C. and Lee, D. W. (1984). Ultrastructural basis and developmental control of blue iridescence in *Selaginella* leaves. *Am. J. Bot.* 71, 216–219.

Henderson, M. R. (1954). *Malayan Wild Flowers. Monocotyledons*. Malayan Nature Society, Kuala Lumpur.

Henderson, M. R. (1959). *Malayan Wild Flowers. Dicotyledons*. Malayan Nature Society, Kuala Lumpur.

Holttum, R. E. (1950). The Zingiberaceae of the Malay Peninsula. *Gdns'. Bull. Singapore* 13, 1–249.

Holttum, R. E. (1955). Growth-habits of monocotyledons — variations on a theme. *Phytomorphology* 5, 399–413.

Holttum, R. E. (1964). *The Flora of Malaya. I. Orchids*. Government Printing Office, Singapore.

Holttum, R. E. (1977). *Plant life in Malaya*. Longman Malaysia, Kuala Lumpur.

Hughes, A. P. (1959). Effects of the environment on leaf development in *Impatiens parviflora*. *J. Linn. Soc. (Botany)* 56, 161–165.

Keng, H. (1970). Size and affinities of the flora of the Malay Peninsula. *J. Trop. Geogr* 31, 43–56.

Kern, J. H. (1974). Cyperaceae I. *Flora Malesiana* 7, 436–753.

Kern, J. H. and Nooteboom, H. P. (1979). Cyperaceae II. *Flora Malesiana* 9, 107–187.

Kiew, B. H., Kiew, R., Chin, S.-C., Davison, G. and Ng, F. S. P. (1985). Malaysia's 10 most endangered animals, plants and areas. *Malay. Naturalist* 37 (1), 2–5.

Kiew, R. (1975). The reproductive biology of some tintailing weeds. *Malay. Nat. J.* 29, 52–57.

Kiew, R. (1978). The genus *Balanophora* in Peninsular Malaysia. *Malay. Nat. J.* 30, 539–549.

Kiew, R. (1978). Floristic components of the ground flora of a tropical lowland rain forest at Gunung Mulu National Park, Sarawak. *Pertanika* 1, 112–119.

Kiew, R. (1982). Observations on leaf color, epiphyll cover and damage on Malayan *Iguanura wallichiana*. *Principes* 26, 200–204.

Kiew, R. (1983). Conservation of Malaysian plant species. *Malay. Naturalist* 37 (1), 2–5.

Kiew, R. (1986). Phenological studies of some rain-forest herbs in Peninsular Malaysia. *Kew Bulletin* 41, 733–746.

Kiew, R. (in press). The role of botanic gardens in the conservation of Malaysian plant species. In *Botanic Gardens of the Tropics — Their Roles and Future in a Changing World* edited by Leong, Y.-C.

Kiew, R. and Latipah, M. (1981). First record of leaf cuttings under natural conditions in *Phyllagathis* (Melastomataceae). *Malay. Nat. J.* 34, 131–134.

Kiew, R and Yong, G.-C. (1985). The limestone flora of the Batu Luas area, Taman Negara. *Malay. Naturalist* 38 (3), 30–36.

Lim, S. N. (1972). Cytogenetics and taxonomy of the genus *Globba* (Zingiberaceae) in Malaya I. Taxonomy. *Notes R. Bot. Gdn. Edinburgh* 31, 243–269.

Mardan, M. and Kiew, R. (1985). Flowering periods of plants visited by honeybees in two areas of Malaysia. In *Proceedings of Third International Conference on Apiculture in Tropical Climates* ed. P. Walker. International Bee Research Association, Gerrards Cross, England.

McClure, H. E. (1966). Flowering, fruiting and animals in the canopy of a tropical forest. *Malay. For.* 29, 182–203.

Meeuse, B. and Morris, S. (1984). *The Sex Life of Flowers*. Faber and Faber, London.

Meijer, W. (1983). *Rafflesia* rediscovered. *Malay. Naturalist* 6 (4), 21–27.

Merton, F. (1962). A visit to Tasek Bera. *Malay. Nat. J.* 16, 103–110.

Ng, F. S. P. and Low, C. M. (1982). Check list of endemic trees of the Malay Peninsula. *Research Pamphlet* 88, Forest Research Institute, Kepong, Malaysia.

Purseglove, J. W. (1972). *Tropical Crops. Monocotyledons*. Longman, London.

Richards, P. W. (1952). *The Tropical Rain Forest*. Cambridge University Press, England.

Ridley, H. N. (1922–1925). *Flora of the Malay Peninsula*. Vols. 1–5. Reeve & Co., England.

Ridley, H. N. (1930). *The Dispersal of Plants Throughout the World*. Reeve & Co., England.

Shivas, R. G. (1984). *Pitcher Plants of Peninsular Malaysia and Singapore*. Maruzean Asia, Singapore.

Simmonds, N. W. (1955). Wild bananas in Malaya. *Malay. Nat. J.* 10, 1–8.

Steenis, van C. G. G. J. (1969). Plant speciation in Malesia, with special reference to the theory of non-adaptive saltory evolution. *Biol. J. Linn. Soc., Lond.* 1, 97–134.

Westlake, D.F. (1963). Comparisons of plant productivity. *Biol. Rev.* 38, 385–425.

Wier, J. S. and R. Kiew (1986). A reassessment of the relations in Malaysia between ants (*Crematogaster*) on trees (*Leptospermum* and *Dacrydium*) and epiphytes of the genus *Dischidia* (Asclepiadaceae) including 'ant-plants'. *Biol. J. Linn. Soc., Lond.* 27, 113–132.

Winter, K., Wallace, B. J., Stocker, G. C. and Roksandic, Z. (1983). Crassulacean acid metabolism in Australian vascular epiphytes and some related species. *Oecologia* 57, 129–141.

Wycherley, P. R. (1973). The phenology of plants in the humid tropics. *Micronesia* 9, 75–96.

Completed December 1985

CHAPTER 7

Ferns

R.E. Holttum

Royal Botanic Gardens, Kew, Richmond TW9 3AB, U.K.

CONTENTS

7.1. INTRODUCTION

Ferns are most abundant, and most diversified, in the wet tropics. The cause of this is the necessity of a moist environment for the gametophyte (sexual) phase of a fern's life cycle and for the development of a new fern plant which starts growth from a single cell, not from a pre-formed seed.

Within the wet tropics, the forest provides a variety of habitats in which spores can grow to produce gametophytes and in which new plants can develop. The variety of possible habitats has resulted in the evolution of ferns with different growth-habits, especially epiphytes which can grow on the branches of trees. Few epiphytes can withstand prolonged drought, and it is only in the tropical rainforest that they have full scope for development and evolutionary diversification. Though ferns existed before flowering plants (angiosperms) appeared on the earth, it is evident that many of them are the result of relatively recent evolutionary change. Present-day ferns are adapted to the environment provided by the dominant angiosperm trees of the forest. In Peninsular Malaysia, they are everywhere evident but nowhere dominant except in thickets where forest has been cleared.

The forest provides various kinds of habitats for ferns, but all are dependent on the existence of the trees which together control the humidity of the air within the forest and the amount of light which reaches ground level. The presence of streams and of steep hillsides involve local variations of habitat for ferns, also change of altitude above sea-level which involves change of temperature-range within the forest. The basic fact to be understood is that habitats for ferns are dependent on the existence of the forest. Of all plants in the forest community, terrestrial ferns are among the most vulnerable to changes which involve decreased humidity and increased light, though there are also many small angiosperms which are similarly vulnerable (Chapter 6).

7.2. FERN HABITATS

The temperature of the surface of the ground, and of the earth below the surface, varies according to the degree of exposure to direct sunlight. The ground surface in undisturbed forest is protected from direct sunlight and from lateral movement of air. At any one place temperatures typically vary not more than 1–2°C throughout the year. Above ground level the daily range of temperature increases gradually up to the crowns of the tallest trees where there is a rapid transition to open-air conditions. In the same way humidity of the air near the surface of the ground in the forest changes little, day and night, because there is little movement of air. Terrestrial ferns of the forest are adapted to these conditions. The few species of climbing ferns start growing under the same conditions as terrestrial ferns but, as they climb up the trunks of trees, they enter conditions of greater light and less humidity; in some of them this change is accompanied by a change of frond-form (Fig. 7.1).

Epiphytes growing on the trunks of trees or on branches of small trees have a protected environment varying with height. The majority of epiphytes live on the branches of the crowns of the larger trees where there is greater intensity of light and lower humidity in the daytime. But, as compared with epiphytes on isolated trees, they are in some measure protected from the drying effects of strong winds, and they have beneath them, night and day, the more humid air of the interior of the forest. Isolated trees, except in the cloud zone on mountains, do not maintain the quantity or variety of epiphytes found on trees growing together in the forest.

Some species of epiphytic ferns are pioneers, starting life on bare branches of trees. These ferns, in various ways, can accumulate humus and make the establishment of other species possible. Trees beside small rivers in the forest have the greatest load of epiphytes. Under such conditions there is strong light down to ground level and also high humudity, with deposit of dew every night. Terrestrial ferns also find a special habitat on river banks where they are subject to periodic flooding. Some are adapted to cling to rocks, others to establishment on the earth banks of streams (Fig. 7.2).

Some ferns are specially adapted to grow on large rocks in the forest and some will grow either on rocks or tree-trunks (Fig. 7.3). But rocks, apart from limestone, are not abundant in the Malaysian forest except near streams or waterfalls. Limestone provides a special habitat because the rock dries rather quickly during short rainless periods. There are some ferns which grow always on limestone. Some of them are perhaps adapted rather to the accentuated dry periods than to the calcium content of the rocks. No experimental studies have been made.

The edge of the forest provides partially shaded sites for the establishment of ferns. Several widely distributed species of ferns which would not become established in fully exposed places can grow in such partial shelter. This position is also is suited to several kinds of ferns which will form thickets, and some which can climb up the trees on the forest margin owing to specialized growth of fronds, attaining sometimes a height of 15 m or more. Thicket-forming ferns may protect the edges of the forest but may, owing to their density, delay the establishment of new trees (Fig. 7.4). Such ferns die back as the shade of forest spreads to cover them, but this may be a slow process.

7.3. ALTITUDINAL ZONATION

Lowland dry land forest in Peninsular Malaysia is dominated by members of the family Dipterocarpaceae (Chapter 3). With increased altitude above sea-level and corresponding reduction of temperature, forest on mountains changes in composition but, except for exposed ridges or summits where there is leaching of the soil, it affords similar habitats for ferns, though there may be greater humidity at night where heavy cloud formation regularly occurs, affording optimum conditions for very delicate ferns (Figs. 7.5, 7.6).

Fig. 7.1. *Teratophyllum aculeatum*. Young plant climbing on the base of a tree-trunk; its fronds at a high level are much larger and the pinnae unlobed.

As with the trees, species of ferns are also different on mountains from those in lowland forest. The development of such ferns appears to be primarily controlled by minimum night temperature which in some way controls the germination of spores and the development of the sexual phase of a fern's life-cycle. A striking example of this may be seen on ascending the hill on the island of Penang (Pulau Pinang). When one attains an altitude of about 620 m one suddenly finds by the roadside an abundance of young plants of *Dipteris conjugata*. It is evident that spores of this fern will not germinate at the higher temperature of lower altitudes. A remarkable fact is that plants of *D. conjugata* also grow on sea cliffs in

Fig. 7.2. Gametophytes and young plants of ferns on the vertical face of a rock by a stream in deeply shaded forest.

Singapore and at least one locality on the coast of Johor, but young plants do not occur. I take this as confirmation that at some time in the Pleistocene the temperature on those cliffs (then not near the sea) was like that at 620 m on Penang Hill today (see Chapter 2). It is also evidence that *Dipteris* plants can flourish at a higher temperature and that the control of establishment of new plants is through temperature-adaption of the sexual phase. Some ferns can tolerate a wider range of temperature than others, but few species which grow near sea-level can be found above 1200 m altitude.

7.4. THE FERN FLORA

Ferns have been more intensively collected in Peninsular Malaysia than in most comparable areas in the wet tropics. The local fern flora is thus on the whole well documented, though many species need

Fig. 7.3. Looking up at a branch of a tree covered with ferns of the genus *Drynaria*.

more study and there are many questions concerning their natural classification which are not yet clear. The classification of ferns is less easy to understand than that of angiosperms, and it is only in recent years that its main features have become more clearly defined. Biologically, ferns provide interesting subjects for experiments because in many ways their organization is simpler than that of angiosperms.

The total number of species of ferns now known to be native in Peninsular Malaysia is a little more than five hundred. About the same number of species occur in the whole of continental Africa; in England, similar in area, there are 54 species. The neighbouring island of Sumatra, considerably larger in area, has a larger number of species of ferns (including most of those occuring in Peninsular Malaysia) partly reflecting the larger land area but also the higher mountains, providing habitats for a greater range of species adapted to such conditions.

Fig. 7.4. Thicket of light-demanding ferns of the family Gleicheniaceae on the edge of forest.

Within the floristic region of Malesia, some genera are more diversified in the eastern or western parts, and some species occur only in the east or in the west. For its size, Peninsular Malaysia has about an average proportion of species and the great majority of the genera.

The Malesian fern flora is related to that of mainland Southeast Asia. There is a distinct but rather gradual change in floristic composition of the fern flora in peninsular Thailand, reflecting the transition between the ever-wet climate of low latitudes and the more seasonal climate further north. There are sheltered areas, especially by rivers, which permit Malesian species to penetrate northwards, and many species of Peninsular Malaysia have some northward distribution. On the other hand, some species which have their main area of dispersal in mainland Asia extend southwards only to northern parts of Peninsular Malaysia. A few mainland species, missing the ever-wet latitudes, have a further distribution in the seasonal climate of east Java and the Lesser Sunda Islands.

As noted above, most Malesian ferns represent recent evolutionary developments. So far as present knowledge goes, some species have a very restricted range, indicating a probable recent origin. More field study of such ferns by trained observers is needed; we have still too little knowledge of the extent of the occurrence of many species. Of the 500 now known in Peninsular Malaysia, about 50 are recorded from a single collection or from collections in one small area; nine of these species are apparently at the southern limit of their ranges, which lie mainly further north. Twenty-five species are known only in

Fig. 7.5. Trunk of small tree in mountain forest bearing small filmy ferns and also a larger epiphyte of the Polypodium family.

Peninsular Malaysia, but this number may be reduced through further field study in Sumatra and Borneo. Of the 25, eleven are known to exist in only one small area and are certainly not common elsewhere.

As regards adaptation to particular habitats, the following is an approximate statement:

terrestrial in forest	210	species
epiphytes	165	
rock plants in forest (including stream-banks)	40	
climbing ferns in the forest	15	
ferns on edge of forest (Fig. 7.7)	25	
ferns of open places	55	

Fig. 7.6. *Lomagramma perakensis*, a climbing fern in mountain forest; young plants start to grow on wet rocks by streams.

7.5. TAXONOMIC PROBLEMS

Most ferns in Malesia have no well-defined counterparts among fossils. As a consequence, an understanding of the interrelationships among them can only be derived from comparative study of existing species. Attempts in the 19th century to define genera were based mainly on rather crude

Fig. 7.7. *Blechnum orientale*, a light-demanding fern on a steep bank on the edge of forest.

observations of the structure and position of sori. In many cases the resulting groups were not natural, related species often being assigned to different genera. A few botanists made significant observations of the nature of scales, hairs, spores and other structures, but these were largely ignored in the summary account of ferns by Diels (1899–1900). The first subsequent attempt to arrive at a more natural classification of a major group, using new characters, was by Christensen who in 1911 subdivided the large unnatural group named *Dryopteris* in his *Index Filicum* (1906). The *Index* was mainly based on the genera of Diels, and in compiling it Christensen had come to realize the confusion it contained.

Christensen confined his work on *Dryopteris* to American species. Subsequently he turned his attention to those of Asia and Malesia, and in this was followed by others, so that the main natural groups have now become established. Observations of chromosomes, initiated by I. Manton in 1950, soon contributed a great deal of new evidence. But within the main groups there remains much work to be done, as some of them are complex, and there is still uncertainty as to their inter-relationships. One of the large complexes comprises species related to the north temperate species *Thelypteris palustris*. These form a family of almost one thousand species, mainly in the wet tropics and including plants of very diverse habit. The only way to find a natural classification within such a large group is to re-examine all species, making records of structures not previously observed (Holttum, 1959–81). By thus examining all species in the Old World, I arrived at a new subdivision of the *Thelypteris* complex into more than 20 groups to which I assigned generic status, but two of the largest of these are still not clearly separated. The significant fact is that, only by studying all species over a much wider area, could I gain a good understanding of the interrelationships between those native in Peninsular Malaysia. Incidentally, this study also shows how the five species of thelypteroid ferns in Europe are related to those in the rest of the world; each belongs to a separate genus when seen in this wider perspective.

Fig. 7.8. *Asplenium nidus*, a common epiphyte; the basket formed by its fronds catches falling leaves from trees which decay and form a compost for the fern.

Another major group, the constituents of which are still unclear, is that which includes the genus *Tectaria*. This needs the same treatment, and here it will be necessary to examine the New World species also to arrive at a good conclusion. I have begun to survey those species which certainly belong to *Tectaria* and have already found that several widely-distributed Malesian species have received more than one name, due to inadequacies of original descriptions. Out of sixteen species described in my book on the ferns (1955), six have names earlier than those I adopted. Since the book was published another six species have been found in Peninsular Malaysia, all of them known at present from single collections or from small areas.

The group of genera which includes *Diplazium* (33 species in the Peninsula) is even more confused than the *Tectaria* group. Few Malesian species have been adequately described and some, especially in eastern Malesia, are still undescribed and unnamed. A clear division of these into natural genera depends on a comparative study of those in mainland Asia with those of Malesia. In Peninsular Malaysia seven species of *Diplazium* are now only known from small areas, but some of them probably occur more widely. There is need for more critical field study.

In north temperate regions much cytotaxonomic study of ferns has been accomplished during the past thirty years, especially in the genera *Dryopteris (s.str)* and *Asplenium*. *Dryopteris* is mainly a temperate-zone genus, with a few species on mountains in the tropics, but *Asplenium* is mainly tropical with about 700 known species (Fig. 7.8). The few species of *Asplenium* in Europe which have been intensively studied need to be seen, like those of the *Thelypteris* family, against the great wealth of the tropical species, but tropical *Asplenium* has not yet been divided into the natural sub-groups which need to be established before the techniques of cytotaxonomy can be fully exploited. Through the study of chromosomes of individual species, through the production of artificial hybrids and through the discovery of natural hybrids, much information about the interrelationships between species in Europe has been obtained. It has been proved that some species which are polyploid have arisen through hybridization between known diploids. There is a very great potential for such studies in Peninsular Malaysia but they have hardly been attempted, though a few local species have been used in experimental work mainly involving species from other areas. Natural fern hybrids may be important in Peninsular Malaysia. In the genus *Bolbitis*, triploid plants which must be of hybrid origin occur in abundance alongside the Tahan river in Pahang and help to consolidate its banks.

REFERENCES

Christensen, C. (1906). *Index Filicium*. Copenhagen: H. Hagerup.
Diels, L. (1899–1900). In Englar, A. and Prantl, K. (eds.) *Die Naturlichen Pflanzenfamilien* 1(4).
Holttum, R.E. (1955). *Ferns of Malaya*. Government Printing Office, Singapore. Second edition, with new appendix (1968).
Holttum, R.E. (1959–81). Pteridophyta. In Steenis, C. G. C. J. van and Holttum, R. E. (eds.). *Flora Malesiana, Series II, Pteridophyta*. Vol.1, parts 1–5.

Completed January 1986

CHAPTER 8

Higher Fungi

E. J. H. Corner, C.B.E., F.R.S.

Emeritus Professor of Tropical Botany, University of Cambridge, U.K.

CONTENTS

8.1. INTRODUCTION

Fungi do not figure in the usual brief of wildlife or habitat conservation. No voice has been raised on their behalf. They, however, do the dirty work by clearing up, very largely, the mess of dead vegetation. The combination of putrescible plant remains with a host of fungal scavengers, especially the higher fungi, stokes the power plant of the forest. Over the years fungi can render a trunk into clouds of spores which settle as manna for animalcules at the beginning of food-chains and into friable rubble which is tilth for the soil and material for root-absorption. The more substantial the vegetation, the greater is the variety of fungi needed for its degradation. The high and varied forest of Peninsular Malaysia contains an inestimably rich assemblage of higher fungi essential for its continuance. From end to end of the Peninsula hyphae permeate the humus, dead wood and bark, fruit-husks, seed-coats, living and dead roots, and extend in these dead parts into the canopy. Their presence is, however, not obvious until they fruit, which they do seasonally, as over all the world.

8.2. SEASONS OF FRUITING

Despite the general uniformity of the perhumid climate (Chapter 1), the lowland fungi of Peninsular Malaysia fruit seasonally in March to May and August to October or November. These seasons apparently correlate with the return of wet weather after relatively dry periods of several weeks. Prolonged vegetative growth during the rainy weather in most cases does not lead to fruiting. That growth must be checked by the dry spell and, then, on the return of the rain, the hyphae change by some means, unknown, from diffuse foraging to compact growth as fruit-bodies. Suddenly, at these times, there appear on the forest-floor countless toadstools, puffballs, clubs, cups and brackets, in explanation of what has been occurring. The expanded fruit-bodies seldom last more than a few days or a week and species and genera fruit in succession. Unless one is on the spot for the few critical days, many fungi will be overlooked (Fig. 8.1).

Fig. 8.1. Fruit bodies of *Micropsalliota laeta* Heinem. on the forest floor; Angsi Forest Reserve, Negeri Sembilan.

First come small fruit-bodies of such as *Marasmius* and *Mycena*, growing on fallen leaves and twigs — first to be soaked with rain. Then begin the terricolous agarics such as *Amanita, Boletus, Russula*, and *Lactarius*, to be followed by other genera and by the more slowly growing, tough and woody, fruit-bodies of polypores and stereoid fungi. Details and variations, which I have often confirmed, are given in other works (Corner, 1935a, 1950a, 1966, 1972, 1930b; Heinemann, 1980; Maas Geesteranus, 1970).

There are some exceptions. Certain species of ephemeral habitats or those with short-lived mycelium respond to slight local variations in the weather, such as freak storms in a dry period. In addition, the perennial brackets of woody polypores and stereoid fungi are present at all times of year but they, too, respond seasonally by adding, twice a year, a new layer of tubes or of hymenium. Thus, although the forest is never devoid of some fruit-bodies, for most of the time there are few signs of its fungus wealth.

8.3. HABITS OF GROWTH

8.3.1. Mycorrhiza

Repeated visits to a given part of the forest show that many higher fungi appear on the ground at about the same place, season after season. This constancy suggests that their mycelia perennate and that some may be mycorrhizal. By analogy with temperate experience, one would expect species of *Amanita, Boletus, Russula* and *Cortinarius*, for instance, to be mycorrhizal. However, the great mixture of trees and the very great entanglement of their roots has prevented direct means of ascertaining this in Peninsular forests. Partial or complete rings of fruit-bodies may be found, but it has been impossible to decide on their focus. A likely example may be found on the sea coast, where the clavaria-like fruit-bodies of *Thelephora ramarioides* always appear in season growing in hundreds along the lines of the roots of *Casuarina equisetifolia* (Reid, 1958; Corner, 1968a). Brief inspection of the plantations at the Forest Research Institute, Kepong, has suggested that a few common species of *Russula* may be mycorrhizal with dipterocarps. Where Fagaceae, notoriously mycorrhizal, abound, there can be found many agarics of mycorrhizal genera. Hypogeal fungi, basidiomycete and ascomycete, may also be mycorrhizal. Such as have been collected in the Peninsula were described by Corner and Hawker (1953), but their host-connections, if any, remain unknown. Several of these fungi are but partly hypogeal and only their tops reveal their presence. In general, to find them, it is necessary to scrape away gently the loose humus in likely places.

8.3.2. Parasites

Compared with the host of saprophytic higher fungi, parasites are few. They are mainly facultative, lignicolous, and polyporoid. Thus the root- and collar- parasites of rubber plantations, *Rigidoprus microporus* (or *Fomes lignosus*) and *Ganoderma philippii* (*G. pseudoferreum*, Corner 1931), are saprophytic and parasitic in the forest. This is the habit of several brown polypores of *Phellinus*, such as *P. noxius* (Corner 1932b), *P. rimosus* and *P. pectinatus*. Formerly, when high forest stood unlogged in central Pahang, so many canopy trees were infested with *Ganoderma mirabile* (*G. fuscopallens*) and *Phellinus setulosus* that it seemed as if their demise must have been accelerated by these parasites (Fig. 8.2). Both species attacked dipterocarps, the *Koompassia* species (kempas and tualang) and various unidentified canopy trees. They produced, on the trunks and main branches up to 40 m above ground, rarely below 3 m, enormous perennial brackets up to 80 cm wide, commensurate in size with their hosts, and so firm that one could climb up on them. Presumably the fungus had entered through broken limbs and descended in the heartwood until this tapered into the buttresses. Another polypore (*Fomes tricolor*, erroneously referred to *Pyrofomes*), with much smaller brackets, attacks dipterocarps of the genus *Shorea* in the same lofty manner. Yet, after repeated search on Fraser's Hill (and Kinabalu, Sabah), I have never discovered any fungus that rots the heartwood of the montane oak *Trigonobalanus verticillata*.

Characteristically the first lofty trunk of this tree dies, moulders away, and crumbles right to the base leaving a hollow stool which coppice-shoots then enlarge as they, too, die in this mysterious manner. Old trees have several trunks around a large hollow in which several people can stand, but I have never seen

Fig. 8.2. *Phellinus setulosus* (Lloyd) Imaz. parasitic on a dipterocarp tree in lowland forest, Tembeling, Pahang.

a wholly dead specimen. How old are these trees, and do they ever die? A species of *Fistulina* rots the massive aerial roots of strangling figs. One or more species of *Veluticeps* seems to prefer the hard, dry wood of living *Tristania* (Myrtaceae), lowland and montane. On the ridge of Gunung Panti, Johor, I found that most small trees of *Timonius flavescens* (Rubiaceae) were attacked by *Phellinus pectinatus*, the ageing brackets of which, on trunks and branches, became fantastically lobed (Corner, 1978). By contrast, species of *Phylloporia* may attack herbaceous plants and develop small brown brackets from green stems, petioles and midribs; such a one was not infrequent on an acanthaceous herb on Langkawi island.

8.3.3. Rhizomorphs

As a preliminary to fruiting, many higher fungi aggregate their hyphae into white or yellowish threads or strings which grow from the humus or rotten wood to develop fruit-bodies at the end. Sometimes these rhizomorphs fan out again in contact with a suitable surface and resume vegetative growth. Rather coarse rhizomorphs are characteristic of many Gasteromycetes. Other rhizomorphs are more extensive and assist the spread of the fungus. Such are usually grey, brown, or black because of a tough coating of agglutinated hyphae. In very humid forest, such as fresh-water swamp-forest or thickets of *Saraca* (Leguminoseae) by forest streams, bushes may be entangled with these small threads, tough enough to impede progress and copious enough to enshroud the bushes with fallen leaves. Known as horse-hair blights, they belong chiefly to species of *Marasmius* and its allies. Their fruit-bodies develop in the saturated air directly from these threads and, if abundant (by no means always the case), the bush may be spangled with little toadstools. Stouter horse-hairs, 2–4mm thick, more like boot-laces, belong to various polypores.

8.3.4. Radicicolous fungi

The fruit-bodies of various higher fungi often seem to have rooting stems, although these are not roots of downgrowth but rhizomorphs of upgrowth which form the fruit-body at ground-level. The question is, from what have they grown. In some cases, such as *Panus* (Corner, 1981), it may be from a sclerotium concealed in the humus, or from an insect larva as with *Cordyceps*, or from a termite-nest as with the termite agaric *Termitomyces*, but generally it seems to be from dead or living roots. To trace the rhizomorphs through the humus into the depths of the soil is usually impossible. In digging one must also be cautious not to disturb coral snakes or other animal life: when I first extracted a termite agaric, my fingertip was sliced by a soldier!

Radicicolous fungi indicate the quantity of root-material, living or dead, to be exploited. They commonly produce large or very large fruit-bodies. The habit is distinctive of some species of *Amauroderma* and *Ganoderma*; the common polypore *A. rugosum* has the pileus up to 30 cm wide and the stem as long both above and below ground. For toadstools, there are the tropical form of *Oudemansiella radicata* (Corner, 1934), *Tricholoma crassum* (pileus to 24 cm wide, stem to 40 cm high and 6 cm thick, with up to 70 fruit-bodies in a clump from the one rooting base), *Paxillus pahangensis* (Corner, 1970b), *Panus giganteus* (pileus to 30 cm wide, Corner, 1981), and the smaller but numerous fruit-bodies of *Phaeocollubia* (Horak, 1976). For hydnoid fungi there are *Climacodon efflorescens* and *Hydnum* 'species 2' (Maas Geesteranus, 1970). There is, also, the curious ascomycete *Dendrosphaera* with yellow-brown and clavaria-like fruit-bodies arising from a slender rhizomorph that penetrates the soil so deeply and finely that its origin has never been determined (Boedijn, 1935a).

These large fruit-bodies contrast with the very small ones found in many epiphyllous agarics, even in such genera as *Russula* and *Boletus*. In these genera the simplification and diminution of the fruit-body can be traced as the mycelium is adapted to more restricted and ephemeral habitats. It is possible that the deeply subterranean habit may have been the primitive state for higher fungi and that which favoured the conversion of the agaric into the gasteromycete, the discomycete into the truffle, and the onygenaceous *Dendrosphaera* into the false truffle *Elaphomyces*.

Some lignicolous fungi may appear to be terricolous because their fruit-bodies grow from sclerotia in the humus; for instance, the well known chendawan susu rimau (*Polyporus*, or *Lignosus*, *sacer-rhinoceros* complex) and *Panus* spp. Actually, these sclerotia form in the soft mushy wood of fallen trunks from which they subside or drop into the humus as the trunk breaks up. The large fruit-bodies of these kinds also display the more primitive form: in the Peninsular forests, one can trace the simplification of *Lignosus* into *Microporus*.

8.4. SPECIAL HABITATS

The inference is that the richness of the Peninsular fungus flora is due largely to the abundance and variety of trees of the family Fagaceae *(Quercus, Lithocarpus, Castanopsis, Trigonobalanus)*, but that other plants from aroids and palms to canopy trees and lianes add their quota. In fact, the extremely rich angiosperm flora requires an equally rich, or richer, fungus flora to break down its constituents. No one fungus can decompose everything; all are severely limited chemically to certain substrates, and in this sense no angiosperm is a single substrate. Lamina, petiole, midrib, veins, twigs, branches, trunks, sap-wood and heart-wood, bark, roots, and fruit-husk have different fungi in different stages of their decay, and all these substrates will vary with the species, genus, and family of angiosperm. Canopy trees may often be identified from the fungi on the floor of the forest. Such specificity, well known in the parasitic rust and smut fungi, extends throughout the saprophytic world. Indeed, many fungi grow on wood only in its final stage of crumbling decay. A point not to be overlooked is the interaction of one kind of fungus with another. Higher fungi seldom grow intermixed; even on a dead trunk, one species may preponderate. It appears that the mycelia are antagonistic (antibiotic) and that, if more than one kind prefers the same substrate, it is a matter of first come, first served. Not a few tropical fungi have been tested for antibiotic activity against bacteria but little or nothing is known about fungus-fungus interactions, or the effect of fungal antibiotics on other soil-organisms.

Apart from this mixed array in the lowland forest, there are several special habitats.

8.4.1. Casaurina equisetifolia

Under these trees on sandy shores there occur, besides the apparently mycorrhizal *Thelephora, Inocybe casuarinae* (Horak, 1980a), an ally of *Naucoria, Tulostoma* sp., *Bovista* sp. and *Xylaria johorensis* (Morgan-Jones and Lim, 1968). In the swards where these coastal forests have been cleared, *Porania oedipus* may be common on cow-dung (Lim, 1968; Jong and Rogers, 1969).

8.4.2. Mangrove

Many polypores, small agarics and resupinate fungi occur on the dead wood and bark of mangrove trees, but the saline conditions are unsuitable to most fungi, and the detritus is carried away at high tide.

Yet, on Singapore island, where it may still occur at Kranji, as well as on the mainland, a toadstool of some size grew in troops from the saline mud in the upper reaches of the mangrove; it is an ally of *Tricholoma*, evidently undescribed.

8.4.3. Palms

The dead trunks and leaves of *Oncosperma* develop numerous fungi, but not, apparently, those of *Arenga*, *Livistona*, *Nypa*, or *Pholidocarpus*. The introduced betel *Areca*, *Cocos* coconut and oil-palm *Elaeis* gather more or less omnivorous fungi. The earthy mounds under bertam *Eugeissona* often supply particular fungi, as species of *Entoloma* (Horak, 1980b). *Cymatoderma dendritica* is characteristic of dead rattans (Reid, 1958b, 1965).

8.4.4. Canopy fungi

Broken and dying branches admit lignicolous fungi. Rotting bits of these branches with the fruit-bodies of polypores and resupinates that have grown on them in the canopy are often found on the forest-floor, where they may also thrive. A few polypores, as already mentioned, seem to be limited to these lofty situations. A few other species and some stereoid fungi grow on dead bark still attached to trunk and limbs; an example is an undescribed species of *Vararia* which is normally a resupiante on fallen wood and leaves. It is difficult to study these fungi except where felling is in progress, but this flora shows how the forest is decayed from above downwards.

8.4.5. Tree-ferns

Primitive fungi might be expected on tree-ferns, but their dead fronds and trunks yield only a few reduced species of basidiomycete and ascomycete. I have not associated any fungus with their roots.

8.4.6. Bryophyta

Numerous higher fungi with suitably small fruit-bodies grow on bryophytes in temperate regions; there appear to be very few in Peninsular Malaysia, e.g., *Leptoglossum seticolum* (Corner, 1966) and *Nectria egens* (Corner, 1935b). I have searched *Sphagnum* in lowland and mountain without reward. The soft, green, and bracket-like tufts of *Trentepohlia*, which grow along the edges of leaves in the undergrowth, commonly have little *Helotium*-like apothecia on their undersides. These are considered to be the fungus partner of a loosely constructed lichen (*Coenogonium*), but the apothecia lack algal cells and the hyphae appear to be parasitic on the filaments of the *Trentepohlia*, very much as those of the terricolous *Clavaria fossicola* of forest shade and *C. helicoides* of the open ground are parasitic on unicellular algae (Corner, 1950, 1970a).

8.4.7. Dung

Cropophilous fungi have been little studied in Peninsular Malaysia, except for *Poronia oedipus*. In the forest most dung soon disappears, only that of larger ungulates (tapir, elephant) is readily found in large piles. Elephant dung from Johor provided me with many small species of *Coprinus*, with white to deep crimson pileus, Ascobolaceae, *Sordaria*, and a rather massively volvate species of *Conocybe* (Watling, 1979).

8.4.8. Termite nests

The well known syndrome of *Termitomyces, Phaedropeziza*, and *Xylaria longipes*, all of which develop fruit-bodies from ascending rhizomorphs, occurs frequently throughout the Peninsula (see Chapter 14). The latest summary is that by Heim (1977). The Peninsular species of *Termitomyces* have yet to be identified for they do not always conform with Heim's descriptions. There are also other fungi to be found on the old nests of black termites. As wood-eaters, termites are the chief competitors of lignicolous fungi. With soft wood termites often win, though by no means always, and this competition calls for investigation.

8.4.9. Limestone hills

Besides the swampy forest round their bases, limestone hills have on their precipitous flanks many deep pockets of humus where unusual fungi can be found. This habitat deserves study, but the seasons of fruiting appear to be very uncertain.

8.4.10. Open places

While it is not clear what natural open places there may have been originally in the Peninsular Malaysian lowlands, cultivation has brought them about. *Clavaria helicoides* on bare laterite soil in the full open is an extreme example. In open grassland there are several agarics such as species of *Lepiota, Entoloma, Hygrophorus*, and the ordinary field mushroom (*Agaricus campester*; Heinemann, 1980), along with some puffballs (*Bovista, Calvatia*). They do not occur in the forest. Similarly on dead wood in the open there are many species which endure, or prefer, the heat, insolation, and incessant, if incipient, desiccation and are seldom found in the forest. Such are *Lentinus squarrosulus* (*L. subnudus*), *Panus lecomtei, Pleurotus djamor, Oudemansiella canarii, Lepiota zeylanica, Trametes scabrosa, T. sanguinea*, the complex of *Daedalea elegans* and *D. flavida* (referred to *Lenzites*), and various heterobasidomycetes. Several of these, perhaps all, are natural elements of the canopy flora.

8.4.11. Mountain forest

The lowland fungus flora extends up to 1000–1500 m altitude. At higher elevations, a montane element undoubtedly enters, but has not been studied in any detail in the Peninsula. The difficulty is to discover the seasons, which may vary from mountain to mountain. To this flora belong, apparently, such gasteromycetes as *Aseroe* and *Calostoma* (Lim, 1969), the edible shiitake (*Lentinula edodes*), which occurs at Cameron Highlands, and the strange *Trichocoma paradoxa* of Onygenaceae (Boedijn, 1935b; Kominami *et al.*, 1952) which used to be common on Fraser's Hill where we called it 'cigarette-ends' from the ashy little tips of the short fruit-bodies projecting in clusters along fallen trunks. To these must be added the deep blue 'Corticium' and the pinkish orange to red discs of *Aleurodiscus* which catch the eye. Whether the striking fungi which occur on Kinabalu about this altitude grow also in Peninsular Malaysia is not known, e.g., *Gomphus grandis* (Corner, 1969), *Hydnum erinaceum* (Maas Geesteranus, 1970), *Bondarzewia*, *Corneromyces* (Ginns, 1976), and the discomycete *Wynnea*. The mountain conifers *Agathis*, *Dacrydium* and *Podocarpus* seem devoid of any particular fungus flora. Whether this is true also for the myrtaceous trees *Baeckia* and *Leptospermum*, to which several agarics and boleti are ascribed in New Zealand, is not known. When I was in thickets of these plants on Mt Tahan in September, 1937, there was no fungus season. Among other Myrtaceae, mountain *Eugenia*, however, has many lignicolous fungi, and *Tristania* may carry *Veluticeps*.

8.5. EDIBILITY

Spores, hyphae and fruit-bodies are eaten by a great variety of invertebrates. The biomass of fungi in the forest is unknown, though undoubtedly considerable. Most fleshy fruit-bodies become infested with maggots before or soon after they expand and thus produce fungus-flies and beetles, to be eaten or decayed in their turn. Snails browse on many. Various grubs and snails rasp the hymenium off stereoid fungi. Beetles lay eggs in polypores and gasteromycetes. Dung-flies and carrion-flies relish the stinking slime which holds the spores of phalloids. Some staphylinid beetles are so minute and slender that they can enter the tubes of polypores, about 0.1 mm wide, and eat the hymenium. Tortoises, rats, squirrels, pig, deer and macaques, if not also tapir, rhinoceros and elephant, eat many of the larger toadstools, relishing always the termite agaric which may commonly be found merely as a cluster of bitten-off stems. Marks of squirrel teeth can be seen round the growing margin of woody polypores, and this kind of vegetable 'cheese' was enjoyed by the pig-tailed macaques that I kept. These monkeys were adept mycophagists, distinguishing edible from inedible or poisonous by taste; if mistaken, the stomach rejected. They ate many kinds, far beyond my knowledge of possibilities, refused many that surprised me, often vomited with no inconvenience, and were never poisoned. So far as the names of these fungi have been obtained, the facts have been recorded (Corner and Bas, 1962; Maas Geesteranus, 1970; Corner, 1972; Heinemann, 1980).

The true mushroom, gathered by Europeans, is doubtfully native. *Lentinula edodes*, esteemed by the people of Ranau, Sabah, is little known in Peninsular Malaysia, except as dried culinary specimens, though an attempt was made to cultivate it by the Japanese. Young fruit-bodies of *Lentinus sajor-caju* and *L. squarrosulus*, before they toughen into leather, are used in soups. The edible rice-straw mushroom (*Volvaria*), sometimes cultivated, is wild in the forest where it is lignicolous. The large parasol mushroom *Chlorophyllum molybdites (Lepiota morgani)*, which occurs on lawns and in grassland, is poisonous. So, too, is the mountain *Pleurotus decipiens* (Corner, 1981). Actually, in the tropics, these fleshy fungi are either so short-lived or become fly-blown so quickly that there is seldom enough for a repast.

It is to be noted that, for the identification of certain groups of agarics, taste must be ascertained, whether bitter, peppery, or mild. A nibble from the edge of a pileus should always be spat out because, if not actually poisonous, it may be emetic, and the peppery species of *Russula* and *Lactarius* can irritate the throat.

8.6. LUMINESCENCE

A few species of *Mycena, Dictyopanus,* and *Pleurotus* have luminous fruit-bodies or spores (Corner, 1950b, 1954, 1981). They were little studied in Peninsular Malaysia until the advent of Y. Haneda during the Japanese occupation. The luminescence, like the colours of the fruit-body, seems to have no function. The chemistry of the process has been reviewed in the work of Johnson and Haneda (1966).

8.7. IDENTIFICATION

The handicap to mycological research in Peninsular Malaysia, as in most tropical countries, is two-fold: there is a lack of any ready means of identifying the known, and many have never been described. A list of fungi recorded in the Peninsula was published by Chipp (1921), but the descriptions to which he referred were too often inadequate or erroneous. This has been the trouble with all the old descriptions which included, of course, all the common species. They lack the field-characters of the living fungi, for most of the descriptions were based on dried specimens. Fungi that in fact are white have been described as yellow, brown, grey or black! I have yet to find a good description of the living fruit-bodies of the common polypore *Amauroderma rugosum*. In the references to this chapter I have listed the modern systematic works which have appeared after 1921, so far as I know.

The initial task of identification is difficult because, unlike the flowering plants, most genera and many species are cosmopolitan or pantropical. The systematist must ransack the world flora in books and herbaria, for neither of which is there sufficiency in Peninsular Malaysia. Thus it takes many years to evaluate a single genus. As authorities become willing, so I have enlisted their help with my own collections. Perhaps one hundredth of the Peninsular fungus flora has now been more or less adequately presented. Species described from Europe, North America, South America, Africa, New Zealand, Australia, Tibet and Japan occur in Singapore.

There is certainly a Malesian fungus flora roughly co-extensive with the phanerogamic. Near at hand, Sumatra, Java and Borneo have much the same fungi. The Malesian fungus flora extends in much specific or varietal identity to Japan where there is a growing literature on the higher fungi. The Malaysian mycologist must consult the Japanese flora. New Guinea, with the southern beeches *Nothofagus*, introduces a new element. The Solomon Islands, without dipterocarps or oaks and chestnuts (Fagaceae), seems to lack several genera, such as *Amanita, Russula,* and *Boletus*. Yet, *Climacodon efflorescens* occurs in the Solomons. *Sarcodon conchyliatum* was first found, once only, on Bukit Timah in Singapore and it is now recorded from the Solomons without intermediate locality. *Phellodon maliense* is known from Bukit Timah, Singapore, South Australia and New Zealand (Maas Geesteranus, 1970). Several examples of this apparently disjunct distribution are given for *Entoloma* by Horak (1980b); they reflect the lack of collecting or the lack of occasion. The north temperate *Entoloma rhodopolium* has been found on Kinabalu. I once saw what I was sure was this species on Bukit Timah, but it was the day before I left for England. I failed to make notes or specimens and never saw it again in Peninsular Malaysia. This is the point. The mycologist must try not to miss any opportunity. Many higher fungi are rare, or fruit but seldom. Their very diversity implies a host of lost or lingering ancestors. A russule that I found on Bukit Timah in 1929, on my second visit to this little hill, I have not seen again; but I have notes and a specimen. After the Gardens Jungle in Singapore, Bukit Timah is the best collected locality in the region, yet I never went there without a discovery (Fig. 8.3).

Another difficulty in identification arises from the variability of many species. The field mycologist, aware of differences in size and colour, recognizes varieties but the so-styled type-specimen has too often

Fig. 8.3. *Agaricus brunneolus* (Lang) Pat. in Bukit Timah forest, Singapore.

been described from a dried specimen without field-notes; the type cannot be recognized. In contrast, systematists working with dried material and unacquainted with the living will not understand the variation and tend to describe varieties as new species, in spite of the existence of intermediates. As examples of very variable and common Peninsular fungi, I cite *Hygrophorus firmus* (Corner, 1936), *Stereopsis hiscens* (Reid, 1965), *Panus fulvus, Pleurotus djamor* and *P. eugrammus* (Corner, 1981). The Malaysian mycologist must be aware that he is ever breaking new ground.

Singapore has been the most fully explored. Fraser's Hill and Tembeling may be second. Cameron Highlands, Penang (P. Pinang), and Gunung Panti, Johor, may be poor thirds. Maxwell Hill (Bukit Larut), Perak, has been scratched. Mt Tahan is almost unknown. A beginning has recently been made with Langkawi (Kuthubutheen, 1981).

The following figures give the inventories of several genera that have been studied in some detail. The total number of species known from Peninsular Malaysia (including Singapore) is listed after the genus and, in brackets, the number of those species which are supposed to be endemic. Neither figure is final because more species will be found and many thought to be endemic will as surely be discovered in neighbouring countries.—*Amanita* 27 (22), *Lentinus* 5 (1), *Panus* 7 (1), *Pleurotus* 6 (2), *Cantharellus* 9 (6).

Entoloma 55 (34), *Astrosporina* 9 (6), *Inocybe* 3 (1), *Phaeocollybia* 3 (1), *Agaricus* 18 (8), *Micropsalliota* 16 (14).

Boletus 84 (54), *Heimiella* 2 (0), *Strobilomyces* 4 (1).—*Cymatoderma* 3 (0), *Podoscypha* 7 (1), *Stereopsis* 5 (1), *Thelephora* 13 (6).—*Clavulina* 8 (4), *Clavulinopsis* 11 (2), *Ramaria* 7 (0).

8.8. EVOLUTIONARY SIGNIFICANCE

The fungi of the primaeval forest of Peninsular Malaysia have as much to contribute to evolutionary thought as the flowering plants. Many outstanding and exceptional species do not fit present classification with its temperate bias; others enlarge the concept of genera. As for families, few are phyletically sound. With a flora so rich, the Malaysian mycologist has unrivalled opportunity, provided that the forest in all its aspects is not exterminated. The cleaned woodlands of Europe have lost and will go on losing many species. The primaeval forest of the Peninsula is replete with the richest plant-debris which has provided for millennia the richest habitat for fungi, but extermination is under way.

Conservation from root to canopy and sea-coast to mountain-top must ever be in the mycologist's thoughts. I give a few examples of remarkable fungi: *Aporpium* (David and Jaquenoud, 1972), *Araeocoryne* (Corner, 1950a), *Boletus longipes* and *B. nigropurpureus* (Corner, 1972), *Clavulina gigartinoides* (Corner, 1950a, 1970a), *Cyclomyces greenii*, *Dichantharellus* (Corner, 1966, 1974), *Dichopleuropus* (Reid, 1965), *Meiorganum* and *Merulius versicolor* (Corner, 1971), the orange-ferruginous *Pisolithus* in central parts of Peninsular Malaysia, *Strobilomyces mirandus* (Corner, 1972), *Thelephora alta*, *T. fragilis*, and *T. magnifica* (Corner, 1968a), *Tubilicrinis* (Julich, 1979), and *Tulostoma exasperatum*. Perhaps the most extraordinary higher fungi, structurally and biologically, are the phalloid gasteromycetes. Fortunately there is the illustrated account of the Indonesian genera (Boedijn, 1932), for most occur in the Peninsula, though I have not seen *Simblum*. Colour photography will be a help because the intricate fruit-bodies are tedious to draw. It is necessary, of course, to search for the unopened 'eggs' in which the fruit-bodies are constructed, and thus the Malaysian mycologist may help to unravel the coordination of hyphae in the most complex of fungal achievements. The problem to be borne in mind is, 'Given a hypha, make a phalloid'. Pyrenomycetes abound in all forms. With modern, if exotic, works on *Xylaria, Hypoxylon* and *Daldinia* it should be possible to identify the Peninsular Malaysian species. A perplexing fungus, which must have an intriguing life-cycle, grows with powdery (conidial) yellowish branches from the seeds of *Strychnos*; comparison should be made with the seminicolous fungi listed under the excluded species of *Pterula* (Corner, 1950a). By contrast, discomycetes seem poorly represented though, by analogy with Kinabalu, Sabah, unusual kinds may be discovered in the mountains. The red, yellow, and white cups and discs of *Cookeina* and *Phillipsia* are well known (Cash and Corner, 1958), but this vulgarity should not obscure the fact that no one has looked for their ascogonia.

8.9. COLLECTING

Good collections are paramount. Indifferent collecting in the past has led to the inadequate descriptions and to the lack of systematic interest. It is never possible to study all the living material as it is collected. Too many fungi appear all at once and decompose too rapidly. Collections must be preserved with labels which refer to the separate field-notes, and patience must be exercised because it will take many years to record the fungi of, even, a few acres of forest. Standard practice is to make dried specimens lightly pressed but not squeezed, and dried in a few hours, either in full sunshine or next to a fire: quick drying is essential. The specimens can then be kept in packets in plastic bags with the mouths twisted and tied up, but plastic bags are not recommended for collecting since excessive moisture wets the specimens and with dirt destroys them. Flat tins are recommended, small and medium size, and a basket with rain-proof lid for the larger specimens. Woody fungi can be wrapped in big leaves and put in the basket, never a vasculum which overheats and jolts. Care should be taken to prevent one fungus from sporing on to another. Sandy or gravelly specimens should be avoided for they may be impossible to study later under the microscope.Collecting and drying is not sufficient. Every collection needs a full description of the living fungus, such as size, form, colour, texture, smell and taste. Flowering plants are often collected with such simple notes as 'tree 20 m, flowers green', but such economy is too superficial for mycology. Daily collections must be limited, so that the field-notes can be written up the same evening. The task is great but the result enduring. It is well to practice with a good description of a living fungus as an example. Spore-prints are also desirable, even with Aphyllophorales, with a note on the fresh colour, because dried prints fade over the years. It is often difficult to find spores in dried polypores, for which reason on expeditions I used to take microscope-slides, on which to collect the spore-print. This precaution enabled me to know if a fertile specimen had been obtained. The slide was wrapped carefully and put with the dried collection.

But even copious field-notes are not enough. For microscopic study at a later date, living material should be preserved in alcohol-formalin, especially developmental stages. Polythene collecting tubes have now made this addition simple. Neglect of this adjunct has led to the almost universal failure to understand the agaric genus *Trogia* (Corner, 1970b). Woody polypores make no exception; for the fleshy, alcohol-formalin material is invaluable.

To begin with, the mycologist will be too busy to find time for photography and painting. These can be taken up later when familiarity provides some leisure and enables the selection of good specimens. By themselves such illustrations cannot substitute for the full description of the living specimen, without which no modern account is satisfactory.

REFERENCES

Boedijn, K. B. (1932). The Phallineae of the Netherlands East Indies. *Bull. Jard. Bot. Buitenz.*, ser. III, 12, 71–103.

Boedijn, K. B. (1935a). The genus *Dendrosphaera* in the Netherlands Indies. *Bull. Jard. Bot. Buitenz.*, ser. III, 13, 472–7.

Boedijn, K. B. (1935b). On the morphology and cytology of *Trichocoma paradoxa*. *Annls. Jard. Bot. Buitenz.* 44, 243-56.

Bouidin, J. and Lanquetin, Paula (1976). *Scytinostroma albocinctum* and *S. phaeosarcum* sp. nov. (Basidiomycetes, Lachnocladiaceae). *Kew Bull.* 31, 621–8.

Cash, Edith and Corner, E. J. H. (1958). Malayan and Sumatran Discomycetes. *Trans. Br. Mycol. Soc.* 41, 1–10.

Chipp, T. F. (1921). A list of the Fungi of the Malay Peninsula. *Gdns. Bull. Straits Settlements* 2, 311–418.

Corner, E. J. H. (1931). The identity of the fungus causing the Wet Root Rot of rubber trees in Malaya. *J. Rubber Res. Inst. Malaya* 3, 1–4.

Corner, E. J. H. (1932a). The fruitbody of *Polystictus xanthopus* Fr. *Ann. Bot. Lond.* 46, 71–111.

Corner, E. J. H. (1932b). The identification of the Brown Root fungus. *Gdns. Bull. Straits Settlements* 5, 317–50.

Corner, E. J. H. (1934). An evolutionary study in agarics: *Collybia apolosarca* and the veils. *Trans. Br. Mycol. Soc.* 19, 39–88.

Corner, E. J. H. (1935a). The seasonal fruiting of agarics in Malaya. *Gdns. Bull. Straits Settlements* 9, 79–88.

Corner, E. J. H. (1935b). A *Nectria* parasitic on a liverwort. *Gdns. Bull. Straits Settlements* 8, 135–44.

Corner, E. J. H. (1936). *Hygrophorus* with dimorphous basidiospores. *Trans. Br. Mycol. Soc.* 20, 157–84.

Corner, E. J. H. (1950a). A monograph of Clavaria and allied genera. *Ann. Bot. Lond., Mem.* 1, 1–740.

Corner, E. J. H. (1950b). Descriptions of two luminous tropical agarics (*Dictyopanus* and *Mycena*). *Mycologia* 42, 423–31.

Corner, E. J. H. (1954). Further descriptions of luminous agarics. *Trans. Br. Mycol. Soc.* 37, 256–71.

Corner, E. J. H. (1966). A monograph of Cantharelloid Fungi. *Ann. Bot. Lond., Mem.* 2, 1–255.

Corner, E. J. H. (1968a). A monograph of *Thelephora* (Basidiomycetes). *Beih. Nova Hedwigia* 27, 1–110.

Corner, E. J. H. (1968b). Mycology in the tropics — *Apologia pro monographia sua secunda*. *New Phytol.* 67, 219–28.

Corner, E. J. H. (1969). Notes on Cantharelloid fungi. *Nova Hedwigia* 18, 783–818.

Corner, E. J. H. (1970a). Supplement to "A Monograph of Clavaria and Allied Genera". *Beih. Nova Hedwigia* 33, 1–299.

Corner, E. J. H. (1970b). *Phylloporus Quél* and *Paxillus* Fr. in Malaya and Borneo. *Nova Hedwigia* 20, 793–822.

Corner, E. J. H. (1971). Merulioid fungi in Malaysia. *Gdns. Bull. Singapore* 25, 355–81.

Corner, E. J. H. (1972). *Boletus in Malaysia*. Government Printing Office, Singapore.

Corner, E. J. H. (1974). *Boletus* and *Phylloporus* in Malaysia: further notes and descriptions. *Gdns. Bull. Singapore* 27, 1–16.

Corner, E. J. H. (1976). Further notes on Cantharelloid fungi and *Thelephora*. *Nova Hedwigia* 27, 325–42.

Corner, E. J. H. (1978). The freshwater swamp-forest of South Johore and Singapore. *Gdns. Bull. Singapore* Suppl. 1, 1–266.

Corner, E. J. H. (1980a). *Boletus longipes* Mass., a critical Malaysian species. *Gdns. Bull. Singapore* 33, 290–6.

Corner, E. J. H. (1980b). *Entoloma* (Fr.) Kummer in the Malay Peninsula. *Gdns. Bull. Singapore* 33, 297–300.

Corner, E. J. H. (1981). The agaric genera *Lentinus*, *Panus*, and *Pleurotus*. *Beih. Nova Hedwigia* 69, 1–169.

Corner, E. J. H. and Bas, C. (1962). The genus *Amanita* in Singapore and Malaya. *Persoonia* 2, 241–304.

Corner, E. J. H. and Hawker, Lilian E. (1953). Hypogeous fungi from Malaya. *Trans. Br. Mycol. Soc.* 36, 125–37.

David, A. and Jaquenoud, M. (1977). Tremellales with tubular hymenophore found in Singapore. *Gdns. Bull. Singapors* 29, 151–3.

Ginns, J. H. (1971). The genus *Merulius*. V. Taxa proposed by Bresadola, Bourdot and Galzin, Hennings, Rick, and others. *Mycologia* 63, 800–18.

Ginns, J. H. (1976a). *Merulius* s.s. and s.l., taxonomic disposition and identification of species. *Can. J. Bot.* 54, 100–67.

Ginns, J. H. (1976b). *Corneromyces kinabalui* gen. nov., sp. nov. (Aphyllophorales, Coniophoraceae). *Mycologia* 68, 970–5.

Ginns, J. H. (1978). *Leucogyrophana* (Aphyllophorales): identification of species. *Can. J. Bor.* 56, 1953–73.

Heim, R. (1977). *Termites et champignons*. Paris, Société Nouvelle des Editions Boubée.

Heinemann, P. (1980). Les genres *Agaricus* et *Micropsalliota* en Malaisie et en Indonésie. *Bull. Jard. Bot. Nat. Belg.* 50, 3–68.

Horak, E. (1976). Further additions towards a monograph of *Phaeocollybia*. *Annls. Mycol.*, ser. II, 29, 27–70.

Horak, E. (1977). The genus *Melanotus* Pst. *Persoonia* 9, 305–27.

Horak, E. (1979). *Astrosporina* (Agaricales) in Indomalaya and Australasia. *Persoonia* 10, 157–205.

Horak, E. (1980a). *Inocybe* (Agaricales) in Indomalaya and Australasia. *Persoonia* 11, 1–37.

Horak, E. (1980b). *Entoloma* (Agaricales) in Indomalaya and Asutralasia. *Beih. Nova Hedwigia* 65, 1–352.

Johnson, F. H. and Haneda, Y. (1966). *Bioluminescence in progress*. Princeton University Press, Princeton.

Jong, S. C. and Rogers, J. D. (1969). *Porania oedipus* in culture. *Mycologia* 61, 853–62.

Jüulich, W. (1979). Studies in resupinate basidiomycetes, VI. *Persoonia* 10, 325–36.

Kominami, K., Kobayasi, Y. and Tubaki, K.(1952). Is *Trichocoma paradoxa* conspecific with *Penicillium luteum*? *Nagaoa* 1, 16–23.

Kuthubutheen, A. J. (1981). Notes on the macrofungi of Langkawi. *Malay. Nat. J.* 34, 123–30.

Lim, Gloria (1963). A species of *Porania* new to Malaya. *Mycologia* 60, 664–65.

Lim, Gloria (1969). *Calostoma sarasini* from Malaya. *Gdns. Bull. Singapore* 25, 109–10.

Lim, Gloria (1972). Some common large fungi in Malaysia and Singapore. *Malay. Nat. J.* 25, 84–9.

Maas Geesteranus, R. A. (1970). Hydnaceous fungi of the Eastern Old World. *Verh. K. Ned. Akad. Wet. Natuurk.*, ser. II, 60, n. 3, 1–176.

Morgan-Jones, G. and Lim, Glora (1968). *Xylaria johorensis* sp. nov. from Malaysia. *Trans. Br. Mycol. Soc.* 51, 165–7. Reid, D. A. (1958a). A new species of *Thelephora* from Malaya. *Kew Bull.* 13, 227–30.

Reid, D. A. (1958b). The genus *Cymatoderma* Jungh. (*Cladoderris*). *Kew Bull.* 13, 519–30.

Reid, D.A. (1965). A monograph of stipitate stereoid fungi. *Beih. Nova Hedwigia* 18, 1–382.

Watling, R. (1979). Observations on the Bolbitiaceae, XVII. Volvate species of Conocybe. *Annls. Mycol.*, ser II, Beih. VIII, 401–15.

Wolfe, C.B., Jr. (1979). *Austroboletus* and *Tylopilus* subgen. *Porphyrellus*, with emphasis on North American taxa. *Biblioth. Mycol.* 69, 1–136.

Completed July 1985

CHAPTER 9

Forest Tree Biology

Francis S.P. Ng

Forest Research Institute Malaysia, Kepong, 52109 Kuala Lumpur, Malaysia

CONTENTS

9.1. TAXONOMIC AND SPATIAL DIVERSITY

The indigenous trees of Peninsular Malaysia are now estimated to number 2650 species, distributed in 510 genera and 99 families (see Appendix). These trees are associated with each other in communities, but the composition of forest communities in the humid tropics is imprecise.

The mangrove community, which was the first to be documented in detail (Watson, 1928), is the simplest and best defined, with only about 35 component species (Table 9.1). Even so, many of its species may also be found sporadically, or even regularly, on sandy or rocky shores especially where streams enter the sea.

The frontline of trees along sandy and rocky coasts is made up of about 50 characteristic species (Table 9.2) and, provided we limit our definition to the narrow strip facing the sea, this community is well-defined. Nevertheless, about 50–100 other species from inland areas may also intrude onto the coastline.

Inland forest communities are richer in species and correspondingly more imprecise in composition. The limestone flora, for example, has about 331 species of trees (Chin, 1973) of which only about 50 species are considered exclusive to limestone (cf. Chapter 6).

Freshwater swamp forests are usually differentiated into 'peat' and 'alluvial' types, depending on whether the habitat accumulates peat or not. The situation is complicated by the existence of seasonal

TABLE 9.1. Trees of the mangrove forest community.

Avicennia alba	*Ceriops tagal*	*Oncosperma tigillarium*
Avicennia lanata	*Cycas rumphii*	*Podocarpus polystachyus*
Avicennia marina	*Excoecaria agallocha*	*Rhizophora apiculata*
Avicennia officinalis	*Heritiera littoralis*	*Rhizophora mucronata*
Bruguiera cylindrica	*Hibiscus tiliaceus*	*Sonneratia alba*
Bruguiera gymnorrhiza	*Intsia bijuga*	*Sonneratia caseolaris*
Bruguiera hainesii	*Kandelia candel*	*Sonneratia griffithii*
Bruguiera parviflora	*Lumnitzera littorea*	*Sonneratia ovata*
Bruguiera sexangula	*Lumnitzora racemosa*	*Thespesia populnea*
Brownlowia argentata	*Myrsine umbellulata*	*Xylocarpus granatum*
Brownlowia tersa	*Merope angulata*	*Xylocarpus moluccensis*
Cerbera odollam	*Nypa fruticans*	

TABLE 9.2. Characteristic trees and large shrubs of the sandy and rocky coastline. Species exclusive to this habitat are indicated with asterisks; the others prefer this habitat but may also occur in mangroves and elsewhere.

Adenanthera pavonina	*Gomphia serrata*	*Peltophorum pterocarpum*
Albizia retusa	*Guettarda speciosa*	*Pemphis acidula*
Atalantia monophylla	*Gyrocarpus americanus*	*Pericopsis mooniana*
Barringtonia asiatica	*Heritiera littoralis*	*Pisonia gradis*
Calophyllum inophyllum	*Hernandia nymphaeifolia*	*Planchonella firma*
Casuarina equisetifolia	*Hibiscus tiliaceus*	*Planchonella linggensis*
Cerbera manghas	*Manilkara kauki*	*Planchonella obovata*
Chaetocarpus castanocarpus	*Memecylon edule*	*Podocarpus polystachyus*
Colona serratifolia	*Messerschmidia argentea*	*Pongamia pinnata*
Cordia subcordata	*Mischocarpus sundaicus*	*Scaevola sericea*
Cycas rumphii	*Myristica guatteriifolia*	*Scyphiphora hydrophyllacea*
Diospyros ferrea	*Myrsine umbellulata*	*Sophora tomentosa*
Dodonaea viscosa	*Neolitsea zeylanica*	*Sterculia foetida*
Erythrina orientalis	*Ochrosia oppositifolia*	*Suregada multiflora*
Erythroxylum cuneatum	*Olea brachiata*	*Terminalia catappa*
Excoecaria agallocha	*Oncosperma tigillarium*	*Thespesia populnea*
Eugenia grandis	*Pandanus dubius*	*Ximenia americana*
Glochidion littorale	*Pandanus odoratissimus*	

swamps that are wet or dry, depending on the weather. In the more permanent swamps, the *Tree Flora of Malaya*, Vols. 1–3 (Whitmore, 1972, 1973; Ng, 1978) and the *Foresters' Manual of Dipterocarps* (Symington, 1943) have recorded 155 tree species of which only 33 are exclusive to such swamps.

The bulk of the dry land (non-swamp) forests are zoned roughly according to elevation (Chapter 2). The most extensive of the original inland forest types (now the most depleted) is Lowland Dipterocarp forest, beginning behind the coastline and rising to about 300 m elevation. It contains the largest number of species, with members of the Dipterocarpaceae most prominent. This forest type is almost impossible to define by composition, for it varies continuously from place to place, so much so that Symington (1943) wrote: "The dominants...are segregated and grouped so that they form an almost unlimited number of different plant communities. Obscure local climatic or edaphic variations may influence the distribution of species but it is more probable that the main determining factors are chance and opportunity".

The difficulty of characterizing forest types by floristic composition has led to the development of somewhat arbitrary definitions according to the prominence of one or at most a few 'indicator' species

(Chapter 2). Hence, the so-called Hill Dipterocarp forest, at about 300–800 m elevation, is usually distinguished by the prominence of a single indicator, the dipterocarp *Shorea curtisii*, and the Upper Dipterocarp (at about 800–1200 m) by the presence of *Shorea platyclados* and a few other indicator species. The 'Montane Oak' and 'Montane Ericaceous' zones are defined by physiognomy rather than species. Montane Oak forest was considered by Wyatt-Smith (1963) to be relatively stunted, to 20–25 m tall, while Montane Ericaceous forest is even more stunted, to about 10 m tall. On Gunung Jerai, the dominant family above the Upper Dipterocarp zone was actually found to be Myrtaceae, rather than Fagaceae (Kochummen, 1982). For riverine fringe vegetation, Corner (1940) has defined 'Saraca', 'Neram' and 'Rasau' types by their most prominent indicator species, *Saraca thaipingensis, Dipterocarpus oblongifolius* and *Pandanus helicopus*, respectively (Chapter 16).

In general, inland forest communities in the Peninsula, and possibly all inland tropical rain forests worldwide, are rather open ended in composition, with a small number of exclusive indicator species and a large and often undefinable number of other species which are relatively insensitive to site or have sensitivities too subtle to detect. The phenomenon has fascinated ecologists for a long time, evoking many attempts at quantification. Already, by 1927, Foxworthy had placed the number of tree species in the Peninsula at 2500, (*cf.* the present estimate of 2650) which, he noted, was "perhaps more than are recorded from all of British India and Burma". Foxworthy also noted, "It is not unusual to find, in the Malay Peninsula, single acres of forest which carry more than 100 species of trees".

If one acre (0.4 ha) can contain 100 species, what would two, three, four or more acres contain? This sort of spatial relationship has been worked out by Wyatt-Smith (1966) for two areas of forest at Bukit Lagong (550 m elevation) and Sungei Menyala (about sea level), corresponding to the Hill Dipterocarp and Lowland Dipterocarp forest types, respectively. All trees of 4 in (*c.* 10 cm) diameter and above at breast height (137 cm) were determined. The two studies gave virtually identical results. As one enlarges the area of study, one encounters species not met with before, though at a diminishing rate; but there is apparently no upper limit. This was further demonstrated by Poore (1964) in an enumeration of trees over 28 cm diameter in 22.3 ha of lowland forest in Jengka. The species area curve (Fig. 9.1) shows no sign of flattening, even at large areas.

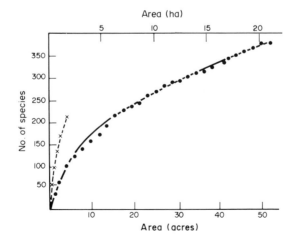

Fig. 9.1. Species/area curves: x--x--x curve for trees over 10 cm diam., lowland forest in Sungai Menyala (also applicable to hill forest, Bukit Lagong); •—•—• curve for trees over 28 cm diam., lowland forest in Jengka. (Adapted from Poore, 1964).

What this means for the conservation of tropical species is that any conserved forest that is smaller in size than the original community area must exclude a proportion of the species of that original community. The greater the reduction in area, the greater the loss of species (Wilcox, 1980). There is therefore no really 'safe' way to reduce the area of any complex tropical forest community without jeopardizing the existence of some of the component species.

A high proportion of the heterogeneity in species-rich inland forests is due to the occurrence of the so-called 'rare' species. In Poore's (1964, 1968) study area, no fewer than 38% of species were represented only by a single (over 28 cm diam.) individual each, of a total of 337 species enumerated (Fig. 9.2). It should therefore come as no surprise that the degree of endemism is also high. Ng and Low (1982) have estimated that 27% of the tree species of the Peninsula are endemic. Hence 27% may be taken as a minimum measure of the floristic difference between the Peninsula and the neighbouring lands of the Sunda Shelf.

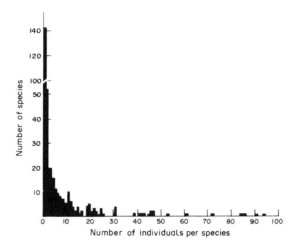

Fig. 9.2. Distribution of species in relation to their frequency status (number of mature individuals) in 23 ha of lowland forest. (After Poore, 1968).

9.2. THE EVERGREEN CONDITION

Like tropical rain forests everywhere, the forests of Peninsular Malaysia are evergreen and overwhelmingly broad-leafed (dicotyledonous). Gymnosperms are represented by only about a dozen species of trees. The evergreen condition is traditionally and popularly considered to be the opposite of the deciduous condition, but this distinction is too simplistic for the humid tropics. There are basically three ways in which a tree can stay evergreen, and all three are well represented here. The first is by continuous growth, whereby shoot growth and new leaf development proceeds without interruption throughout the vegetative life of the plant so long as it remains in good health, or until terminated by flowering. In the humid tropics, plants can develop to large size through continuous growth, e.g., the timber species *Shorea platyclados* and *Dryobalanops aromatica*. Continuous growth is also the habit of practically all palms, most climbers, most pioneer trees, most shrubs and of course all herbs. The rate of

continuous growth, whereby shoot growth and new leaf development proceeds without interruption throughout the vegetative life of the plant so long as it remains in good health, or until terminated by flowering. In the humid tropics, plants can develop to large size through continuous growth, e.g., the timber species *Shorea platyclados* and *Dryobalanops aromatica*. Continuous growth is also the habit of practically all palms, most climbers, most pioneer trees, most shrubs and of course all herbs. The rate of continuous leaf production has been measured for several species: the papaya *Carica papaya*, an introduced fruit tree, produces leaves at a steady rate of 13–15 per month; *Macaranga tanarius*, 7–8 per month; *Dillenia grandifolia*, 2–3 per month; *Shorea platyclados* and *Shorea ovalis*, about 1–1.5 per month (Ng, 1979). These rates were measured for juvenile trees that could be reached with a stepladder. Tall trees are difficult to measure directly but we can assume leaf production to be continuous if the trees have young expanding leaves at all times, as opposed to dormant buds or dead ends.
for a given species or individual tree.

This 'conifer' type of evergreen behaviour, when expressed in dicots, is mainly (but not exclusively) a phenomenon of understorey trees, i.e., trees that belong permanently in the understorey and juvenile individuals of canopy species. In a study on flush production in juvenile trees, Ng (1979) obtained a rate of 10 flushes in 24 months for *Harpullia confusa*, each flush consisting of 6–16 leaves, and 13 flushes in 29 months for *Dacryodes rostrata*, each flush consisting of 3–12 leaves. Mature trees flush at longer intervals than juveniles. We know this because it so happens that the sapling of *Dacryodes rostrata* referred to above was the offspring of a mother tree in Bukit Lanjan (Hill Transect, tree L') which was kept under observation from March 1972 to March 1974. This tree was recorded flushing new leaves in May 1972, October 1972, March 1973, September 1973, March 1974 and October 1974, i.e. at 5–7 month intervals compared to two-month intervals in the sapling. Medway's observations (1972) on a mixed population of canopy trees in hill forest at Ulu Gombak and Ng's observations (1981) on dipterocarps in the Kepong arboretum show that two flushes a year may be normal among mature trees.

The third way of maintaining an evergreen appearance has been termed 'leaf-exchange' by Longman and Jenik (1974). It involves flushing a new set of leaves at the same time that the previous set is shed. The coincidence of leaf flushing and leaf shedding is not necessarily exact. There can be periods when both sets briefly coexist or when the two sets are separated by a brief period of bare crown. This is the behaviour of most trees in the upper canopy of tropical rain forest. It is astonishing that such a dominant phenomenon did not get a name until 1974, after observations by Longman and Jenik in tropical Africa. Previously, such trees were sometimes referred to as deciduous, sometimes as evergreen. For example many of the 'deciduous' species studied by Holttum in his pioneering work on Malayan tree phenology (1930, 1940) were actually of the leaf-exchanging type.

Obligate deciduous trees are those that invariably have a prolonged bare period between leaf-shedding and flushing. Such trees are rather uncommon in the humid tropics. The best-known example in the Peninsula is *Bombax valetonii*, which is bare for several months before flushing.

Having outlined the four categories: evergrowing, 'coniferous', leaf-exchanging and (obligate) deciduous, it is necessary to emphasize that these are not fixed and static. One can merge into another depending on climatic conditions, developmental stage of the plant and degree of shoot specialization. If an English oak is grown in the humid tropics, as Holttum (1954) observed in Singapore, one flush of leaves is exchanged for another in approximately 5 monthly intervals with no intervening bare periods. Similarly, the rain tree *Samanea saman* (or *Pithecellobium saman*), is leaf-exchanging at 6 month intervals here (Ng, 1982), whereas in its native seasonal forest habitat in Costa Rica it flushes once a year and is deciduous for several months before each flush (Janzen, 1978). Even *Enterolobium cyclocarpum*, considered an obligate-deciduous species in Tropical America (Medina, 1983), behaves as a leaf-exchanging tree in the vicinity of Kepong, at annual intervals. Apparently, whether a tree is deciduous or leaf-exchanging depends very much upon the climatic conditions under which it grows. It can therefore be misleading to refer to a species as having a fixed growth habit.

Furthermore, a tree may have a juvenile habit quite different from its mature habit. The rubber tree *Hevea brasiliensis*, introduced from Brazil and now better known than any native tree, starts by accumulating one flush of leaves after another in fairly rapid succession so that the sapling carries, at any time, five or more consecutive sets. This 'coniferous' habit gives way, at maturity, to annual leaf-exchange which the rubber industry refers to as 'wintering' because the period from first discoloration of the leaves until complete refoliation takes about two months, with a brief period during which all trees are bare (Edgar, 1960). Among indigenous forest trees, similar changes in behaviour are known, e.g., the giant heavy hardwood tree chengal *Neobalanocarpus heimii*, in its very prolonged juvenile stage up to about 15 m tall, can accumulate up to seven consecutive flushes of leaves on its branches and form a very dense crown. Yet the mature tree at 30–40 m tall tends to be leaf-exchanging (Ng, 1983). On occasion, chengal has even been observed, in southern Thailand, to be deciduous in dry weather (Tem Smitinand, pers. comm.).

These examples show that the 'coniferous', leaf-exchanging and deciduous habits are alternatives influenced by developmental stage and climatic condition. Collectively, they may be regarded as three variants or phases of intermittent growth. The opposite of intermittent growth is continuous growth. The distinction between intermittent and continuous growth is probably more fundamental than that between evergreen and deciduous. Intermediate or mixed states are apparently rare, although it is possible that further examples may be found with more intensive study. The best known intermediate example is 'foxtailed' Caribbean pine *Pinus caribaea*. In its native habitat in Central America, Caribbean pine grows intermittently and each new flush is accompanied by the production of a tier of branches. When introduced to the humid tropics, most trees grow intermittently but, in a rather high proportion of trees, the apical bud may (in the absence of a strong climatic cue) fail to rest, and the result of such continuous growth is a long, unbranched apical shoot or 'foxtail'. Eventually such trees do develop branches, for fox-tailing is apparently a juvenile phenomenon. Other species that may be expected to behave similarly are those that, like the Caribbean pine, have been introduced from the seasonal tropics to the humid tropics. Thus, teak *Tectona grandis* grows intermittently in its native seasonal habitat but reveals an evergrowing tendency while juvenile, when planted in non-seasonal locations. The best known example of a mixed state occurs in *Shorea ovalis* (Ng, 1979), one of the more important of the red meranti timbers. This has an ever-growing leader shoot but intermittently growing branches. Such different behaviour between leaders and branches may be more frequent than at present realised.

9.3. TREE-BUILDING

Under extreme conditions, such as at high elevations or in mangroves, Peninsular Malaysian forests can be so stunted that nearly all trees, regardless of species, may be able to contribute to the single-layered canopy. On better soils and at lower elevations, the trees segregate by species so that only some, predominantly dipterocarps, contribute to the formation of the upper canopy. Others remain permanently at lower sub-canopy levels, known collectively as 'understorey'.

The tallest tree species in our region is not a dipterocarp but the legume *Koompassia excelsa*. Individual heights have been measured as 84 m (275 ft) tall, in Sarawak, and another as 81 m (265 ft) in the Peninsula (Foxworthy, 1927). At the other extreme, *Eurycoma longifolia* (Simaroubaceae) can barely reach 10 m (Kochummen, 1972). Both trees have the capability to make a trunk, but what enables one to exceed 80 m and what limits the other to 10 m?

The basic elements of tree-construction have been subjected to detailed analysis (Halle *et al.*, 1978) and as a result, we are now able to place all trees among 23 basic models, based principally on the manner of construction of the trunk and its branches. However, within each model, there are dwarfs and giants. We must conclude that the attainment of height cannot be a matter of tree architecture. In fact, there are many plants, notably in the genera *Phyllanthus* (Euphorbiaceae) and *Ardisia* (Myrsinaceae), which develop a single leading erect stem bearing lateral horizontal branches, thereby closely resembling trees in form but lacking the ability to grow to 3 m; some even stop at *c.* 30 cm. Insight may be found by analysis of the way in which forest giants develop. Even giants must begin as seedlings, but a mature tree of the upper canopy is not merely a seedling magnified a thousand times, any more than a skyscraper is a house magnified a thousand times. A big tree experiences a different environment, is subjected to different stresses and, consequently, has different structural and physiological requirements. However, whereas a skyscraper and a house can be designed and built separately, an upper canopy tree must have the inherent ability or genetic programme to adapt itself to a vastly different environment as it emerges from the understorey into the upper canopy. Halle and Ng (1981) have adopted the word 'metamorphosis' to express the physio-morphological changes involved as an upper canopy species makes such a transition. Species that cannot make the transition are presumably unable to adapt to more than a single environment, and thus some remain permanently in the understorey while others, the pioneer species, are confined to the open.

The juvenile form of a canopy species shows all the characteristics of an understorey species. The leaves are shade-tolerant and presumably long-lived. The branches are slender and diverge from the trunk at large angles, and as they elongate may even become horizontal or drooping. As the juvenile crown accumulates one layer of foliage after another, it becomes narrow-oblong in shape. As it approaches the upper canopy, the upper branches are produced at sharper angles from the trunk. Such branches tend to grow upwards rather than sideways and eventually behave as competing leaders, putting on conspicuous secondary thickening, while development of the original leader slows. In this way, a crown is produced with a rounded top, supported by large branches diverging from a transition zone on the trunk. The base of this zone marks the biological clear bole height of the trunk. Above this zone, branch breakage (if it occurs) leaves large knobbly scars. Below it, the branches of the juvenile phase are cleanly shed, leaving the majestic cylindrical columns that are so highly valued by the timber industry (Fig. 9.3).

Meanwhile the leaves too undergo drastic changes. They become sun-demanding and short lived, limited in distribution to a narrow zone around the periphery of the mature crown. When viewed from below, the branches and twigs are revealed clearly, to their extremities, because the foliage lacks depth (Fig. 9.4). The aversion of upper canopy leaves to shade is dramatically underscored by the phenomenon of crown-shyness (Ng, 1977) whereby crowns of adjacent trees avoid each other, leaving a halo of open space around each and every one (Fig. 9.5). Viewed directly from below, the upper canopy appears as a gigantic jigsaw puzzle, in which the crowns fit with no overlap. The phenomenon is all the more remarkable when compared to the behaviour of a 'wolf' species such as the coastal *Terminalia catappa* (Combretaceae), in which the crowns of adjacent trees will grow straight into each other.

Without the low-density foliage and crown-shy habit of the upper canopy, there could not be enough light passing through to sustain the abundance of plant life below. The species of the upper canopy do not monopolize the light when they reach the top, yet are able to maximize their use of diffused light in the understorey by temporarily adopting a conifer-like structure of deep narrow crowns, thin branches and persistent leaves.

9.4. PHENOLOGY

Phenology is the study of recurrent phenomena such as leaf-change, flowering and fruiting. In principle, it requires no more than the recording of dates and events. In practice, there are many

Fig. 9.3. A stand of *Dryobalanops aromatica* showing mature trees with long clear boles and acutely-angled branches below which are juvenile trees (poles) with oblong crowns formed by multiple tiers of horizontal and drooping branches.

Fig. 9.4. Crown of durian, *Durio zibethinus* viewed from below; branches are visible almost to the extremities because of the restriction of leaves to a thin layer at the periphery of the crown.

Fig. 9.5. Crown shyness in a stand of jelutong, *Dyera costulata*. Each crown is separated from those of its neighbours by a clear zone; however, from a fixed camera position, it is possible to show this only for one crown at a time.

complications, not the least of which is that different people may understand and record the same event in different ways.

A visiting Japanese botanist, Koriba (1958), was so amazed to see lack of synchronization between branches on the same tree as well as between trees of the same species, that he coined the terms 'branchwise manifold growth' for the first category of behaviour and 'stockwise manifold growth' for the second. Had Koriba stayed longer, he would have realised that loss of synchrony between branches can occur on any tree and that non-synchronization between individuals can occur in any species. Consequently, his categories can be simultaneously applied to all trees and species in the humid tropics, thereby defeating the purpose of such categorization.

In Japan when cherries, for example, are in bloom, all trees of the same species and all branches of each tree would be in synchrony, hence there is no need to consider separately the phenology of branches, trees, and species. In Malaysia, when we say, for instance, that the durian *Durio zibethinus* has two fruiting seasons a year, we mean that the species fruits twice a year. What individual durian trees do is quite a different matter. Some fruit in both the seasons, some fruit annually in one or the other season, some fruit irregularly. Furthermore, it is quite possible for one branch to be out of phase with another. We are constantly reminded in the humid tropics that a tree may appear like a colony of seemingly autonomous shoots.

At the unit shoot level, vegetative growth is either continuous or intermittent. Flowering (and fruiting) is also either continuous or intermittent. A continuously growing shoot may flower continuously, e.g., coconut *Cocos nucifera*, intermittently, e.g., *Fagraea fragrans*, or terminally at the end of growth of the shoot concerned, e.g., *Corypha elata* which dies after reproduction, and *Dryobalanops aromatica* which continues to live by branching below the inflorescence. However the shoots of most trees, perhaps as much as 90% of our tree flora, grow and flower intermittently. Why more trees have not adopted the ever-growing and ever-flowering habits, when the climate is generally favourable, remains a mystery.

In whole-tree phenology, we need to consider the necessity for quantification, since a crown is usually made up of many shoots. Should a crown be recorded as active if only a minority of its shoots are active? In a study carried out in the Dipterocarp Arboretum at Kepong, between 1972 and 1977, a total of 120 trees representing 67 species (listed in Ng, 1981) were monitored monthly and each event was recorded by proportion of crown involved, as follows:· 0, >0 -1/3, $>1/3$ -2/3, $>2/3$. The results are given in Figure 9.6. The most surprising finding is that it does not matter which level of quantification is used. All levels, >0, $>1/3$ and $>2/3$ reveal the same rhythm of activity in the population. We interpret this to mean that when an external stimulus to flush or flower is perceived in a community of trees, some trees may respond fully, some partially, and some only a little, but all these responses are valid indicators of the stimulus. Small responses, involving a minority are not random false starts which might obscure the rhythm of seasonality. On the contrary, our evidence shows that the inclusion of small responses improves the sensitivity of the phenologic record.

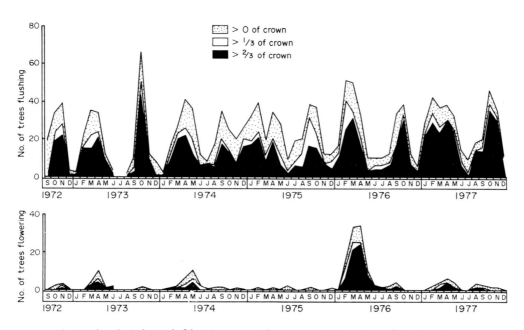

Fig. 9.6. Phenological record of the Dipterocarp Arboretum at the Forest Research Institute, Kepong.

A tree is capable of more than mechanical all-or-nothing growth responses. It can regulate and, more importantly, determine the position of response. Such selectivity enables a tree to (a) grow into available territory, (b) hold back from territory that is already occupied, (c) repair localized damage in its own crown and (d) shed overshadowed parts.

The concept of a tree as a territorial organism (Ng, 1978) helps us to understand many of the peculiarities of tree behaviour. A tree signals its occupation of territory by casting a foliage-shadow. This foliage-shadow, and the ability to react to such a shadow, govern the spatial relationship between adjacent trees. With the exception of 'wolf' trees one tree will not grow into the foliage-shadow of another. This explains crown-shyness, which is the result of crowns stopping short of each other. It explains why forest-grown trees are so often asymmetric compared to open-grown trees. Such asymmetry is the result of preferential growth on those parts of the crown that are not inhibited by the foliage-shadow of another crown. It also explains why tropical rain forests are evergreen: a deciduous tree, having no foliage-shadow during its bare period, is vulnerable and can lose territory to actively growing trees, lianes or epiphytes. Under such competitive pressure, it is in the interest of trees to keep up their guard and stay evergreen. Conversely, if a lower branch gets overshadowed by an upper branch and can no longer function effectively, the lower branch would stop growing, weaken, and be shed or rendered vulnerable to decay. If part of a crown is damaged, there is often an immediate response by way of new shoot growth, to fill up the gap with new foliage (Ng, 1980).

The more dramatic manifestations of territorial responsiveness, such as crown-shyness and 'branchwise manifold growth', are well developed in the humid tropics, probably because foliage is always present to guide and mutually adjust the growth of every shoot and tree. In deciduous woodlands, the absence of foliage through the winter 'fools' all viable buds into flushing in spring as if there were no limitations to territory. Such 'mistake' flushing not only causes adjacent crowns to intrude into each other, it also gives a new lease of life each year to the moribund and long-persistent lower branches that characterize deciduous woodlands.

The next level for phenologic consideration is the species. To obtain an idea of the response profile for a species, we need to study a fairly sizable number of its representatives. In one such study, Yap (1982) monitored populations of 18 understorey species, comprising 5 to 48 adult trees per species, for 40 months, in Pasoh Forest Reserve. Yap was interested in flowering behaviour. His observations are summarized in Figure 9.7, from which we learn that, instead of all-or-nothing responses, the percentage activity can vary enormously between one flowering pulse and the next, as in *Baccaurea parviflora* among which 21% of individuals flowered in 1974, 2% in 1975 and 62% in 1976. Seldom did a flowering pulse involve 100% of the mature individuals of a species. Most times, the response was below 50%. This illustrates one of the biggest paradoxes in reproductive biology, that humid tropical species, occurring in low population densities, are the most poorly synchronized in flowering, whereas species in limiting climates, occurring in high population densities, are the most highly synchronized. If there are advantages in synchronous flowering and cross-pollination, one would expect species with low-density populations to be best synchronized, to compensate for their spatial disadvantage.

Finally, at the community level, phenology is the summation of the individual activities of the component trees of the community. The pattern obtained depends on the particular composition of the communities being studied. In practice, observers are forced by various limitations to define their communities arbitrarily.

The community studied by Ng (1981) was totally artificial, being composed of dipterocarp species from all over the country, raised together in an arboretum. An unobstructed view of every crown was possible. There were two annual peaks of flushing, in April and October, and one annual peak of flowering, in April (Fig. 9.6). The flowering peak of April 1976 was exceptionally high, and corresponded to a 'general' flowering throughout the Peninsula.

The community studied by Yap (1982) was in natural forest, defined by choosing 18 understorey species, no dipterocarps among them, and locating their representatives by ground search. All other species were ignored. Indeed, it is virtually impossible to study upper canopy phenology from the ground in natural forest, because of obstruction by understorey trees. Yap's data (Fig. 9.7) revealed an annual flowering peak between February and May, with the peak reaching an exceptional level in March 1976, coinciding with 'general' flowering throughout the Peninsula.

An earlier seven-year study by Medway (1972), in Ulu Gombak, was made from a platform constructed on a tall tree overlooking a forested valley. This enabled 61 upper canopy trees, representing 45 species, to be studied. Eighteen of these trees (13 species), were dipterocarps. Medway detected an annual flowering peak, occurring between February and July, and two flushing peaks, in March–May and October–December.

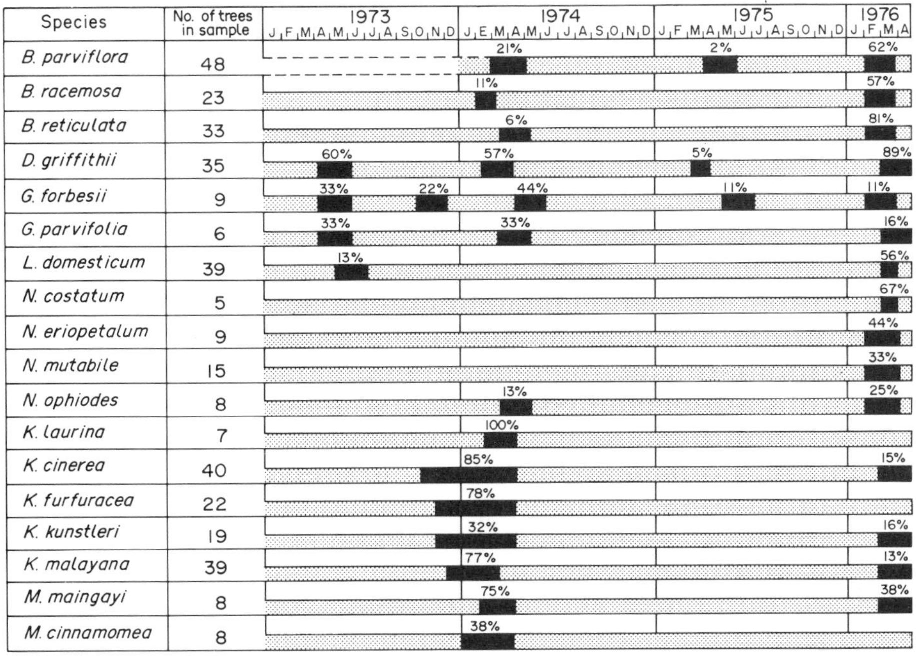

Species	No. of trees in sample	1973	1974	1975	1976
B. parviflora	48		21%	2%	62%
B. racemosa	23		11%		57%
B. reticulata	33		6%		81%
D. griffithii	35	60%	57%	5%	89%
G. forbesii	9	33% 22%	44%	11%	11%
G. parvifolia	6	33%	33%		16%
L. domesticum	39	13%			56%
N. costatum	5				67%
N. eriopetalum	9				44%
N. mutabile	15				33%
N. ophiodes	8		13%		25%
K. laurina	7		100%		
K. cinerea	40		85%		15%
K. furfuracea	22		78%		
K. kunstleri	19		32%		16%
K. malayana	39		77%		13%
M. maingayi	8		75%		38%
M. cinnamomea	8		38%		

Fig. 9.7. Flowering periods and percentages of trees involved, in forest populations of *Baccaurea, Durio, Garcinia, Lansium, Nephelium, Knema* and *Myristica*. (After Yap, 1982).

Overall, it has therefore become clear that a single annual flowering peak occurs, in the early half of each year. This applies to dipterocarps and non-dipterocarps, upper canopy as well as lower canopy. The peak varies greatly in intensity from year to year. In most years, the peak is so low that it can only be detected by careful analysis of flowering records. At infrequent intervals, high peaks occur. These high peaks have long attracted attention under the name of 'general' or 'gregarious' flowering.

The literature on gregarious flowering goes back as far as Ridley (1901) who declared that, "as a rule, the Shoreas and Hopeas flower only once in 6 years". Foxworthy (1932) corrected Ridley by pointing out that most species (of dipterocarps) flower to some extent each year, in some part of the Peninsula, but occasional years show exceptionally high levels of activity. There is also no magic in the figure of six years; gregarious flowering has occurred at intervals of anything between one and six years (Ng, 1977). Nor is gregarious flowering confined to dipterocarps and the upper canopy. The phenomenon is relatively general.

The localization of the single annual flowering peak to the early half of the year indicates that the climatic stimulus for such flowering must be an annually recurrent event. This stimulus must obviously occur before the flowering and it must be of variable intensity in order to account for the occasional high peaks and the more usual low peaks. Ng (1977) has pointed out that the only climatic event that fits the bill is the annual change from cloudy weather in November–December to sunny weather in January–February.

The rainfall pattern has two peaks each year in most parts of the Peninsula, in about May and October, and minima in January–February and again in June–August (Chapter 1). This semi-annual change ties in with the semi-annual pattern of flushing detected by Medway (1972) and Ng (1981). It also tallies with the semi-annual pattern of litter fall detected by Ogawa (1978) in Pasoh. Litter fall reflects leaf flushing, because most trees are of the leaf-exchanging type in which flushing is synchronous with the shedding of old leaves.

The only study differing from this overall pattern is that carried out at Bukit Lanjan from 1972 to early 1976, just missing the general flowering of 1976. The data was analysed by Putz (1979) who found no seasonality in flowering and flushing. Perhaps the particular composition of species at Bukit Lanjan is aseasonal but it is also possible that human error had a role, because the Bukit Lanjan observations were carried out by five different people taking turns, whereas all the others were single-observer studies.

Following the flowering comes the fruiting. If there is an annual peak in flowering, should there be an annual peak in fruit ripening? Medway (1972) did detect such a peak, in September–November. Such a fruiting peak following a flowering peak, in a mixed-species community, must imply that most species have fruit maturation periods of somewhat similar duration. This has been shown to be the case in a study by Ng and Loh (1974) on 89 species of Malaysian trees. The results, in Table 9.3, show that the fruit maturation period is heavily clumped (75% of species) within the third to sixth (2 – <7) months after flowering.

TABLE 9.3. Frequency distribution of species according to minimum period recorded for fruit maturation, in months.

Class interval	<1	1<2	2<3	3<4	4<5	5<6	6<7	7<8	8<9	9<10	10<11	11<12
No. of species (Total 89)	1	3	17	19	18	13	4	7	1	1	1	1

In the above discussion, flowering and flushing have been treated as separate events. In fact, they are closely connected morphologically and physiologically, for both are expressions of shoot development. A flowering peak is always associated with a flushing peak. In this respect there is no difference between temperate forest, dry tropical forest and humid tropical forest. The unexpected twist, in humid tropical forest, is that flushing may occur without flowering. In our case, where flushing peaks develop twice a year, flowering is associated with the first peak, but not with the second. It would appear that a separate stimulus for flowering must act together with the stimulus for flushing in order for a flowering peak to develop.

9.5. POLLINATION ECOLOGY

Animals and wind are the main carriers of pollen. In temperate regions, wind-borne pollen is responsible for the annual bouts of hay fever that many people suffer. In the humid tropics, hay fever is practically unknown; the Malaysian atmosphere is practically free of pollen. Consequently, it is of some interest to enquire how a wind-pollinated species would behave when introduced to our environment, and whether any of our indigenous species are wind-pollinated.

The only wind-pollinated species that has been introduced on a large scale is Caribbean pine, from central America. It grew well enough to be considered a promising species for plantation establishment. However, it would not seed, except where planted on mountain ridges and coastal areas. This is significant when one considers the distribution of the local conifers, viz., *Agathis, Dacrydium* and *Podocarpus*, which are concentrated on mountain ridges, coastal areas, and (one species of *Podocarpus*) on the tops of steep limestone hills. These are the only habitats that are assured of air movement, day and night.

In contrast, the air in inland forests is relatively still. But the stillness of the air is not the only factor working against wind-pollination. The extreme heterogeneity of the forest and the low level of seasonality must also be considered. For a wind-pollinated species, every other plant in the vicinity belonging to another species depresses the probability of pollination success, by not contributing the same kind of pollen to the air, as well as by intercepting pollen that is of no use to it. Furthermore, if the area occupied by the wind-pollinated species is too small, pollen is easily blown out of the stand and lost. The typical wind-pollinated species of temperate regions occur gregariously and cover considerable areas of ground. Questions regarding critical levels of air movement, critical pollen density, and critical area of ground, are important in wind-pollination. In addition, wind-pollination requires a high degree of synchronization in the production of pollen and pollen-receptors.

After considering the difficulties of wind-pollination in our environment, it appears all the more remarkable that about a dozen conifer species have managed to persist, albeit only in special environments.

The very factors that lower the effectiveness of wind as a pollination agent in the humid tropics, open the way for animals which can move around obstacles, seek out flowers by sight or scent, and match species correctly. The vital difference is that animals discriminate while wind does not. As the degree of forest heterogeneity increases, so does the need for discrimination where transfer of pollen is concerned.

Of the animals, insects are the most commonly encountered flower-visitors. Bats are important too, but are seldom noticed because they are active only in the late evening and at night. Some birds are heavily dependent on nectar in the diet (Chapter 13.4.2). Among insects, the most indefatigable flower-visitors are the many species of bees; some specialize in nectar, but the sweat bees, as a group, are most highly conspicuous as pollen collectors.

During the past few years the pollinating activities of two groups of animals, the thrips and the bats, have been worked out in some detail. Their stories will be summarized here in order to illustrate a number of important principles.

Among the least conspicuous of insects, the thrips, minute creatures with extremely poor powers of flight, have recently been shown to play an important role in the pollination of some major giants of the forest, of the genus *Shorea*. The story (Appanah and Chan, 1981) is that thrips breed and feed on the floral tissues, getting dusted with pollen in the process. The quantity of pollen carried is very small, averaging only 2.5 grains per animal, but this is compensated by the enormous number of thrips produced during a flowering event. The weakness of flight of the thrips is compensated by the fact that they are easily carried along even by slight air movement.

The *Shorea* flowers open in the evening. In the morning, the imbricate corollas, weighted in the centre by a short tuft of stamens, spiral very gently to the ground. Any air movement would carry them away. Why have corollas evolved that are so well adapted for wind-dispersal? It so happens that these corollas carry lots of thrips on them, and the thrips get dispersed in the process. In the evening, the thrips leave the corollas on the ground, to fly weakly up to the canopy, apparently attracted by the scent of newly opened *Shorea* flowers. During the upward flight, the thrips may again be subjected to wind drift. In this way, a tiny insect with weak flight and poor pollen-carrying capacity, is enabled to cover more ground than it would otherwise be able. The randomness of wind drift is compensated by the ability of the thrip to make the necessary fine adjustments to locate a flower should it land on the wrong spot. This wind-thrip combination neatly fills the ecological gap between wind- and insect-pollination.

The role of bats in pollination has been described in detail by Start and Marshall (1976). The bats involved are those species adapted to a diet of nectar and pollen. While on the flowers, the bats probe rapidly for nectar and occasionally lick pollen directly from the stamens, but most pollen is obtained by grooming the fur which gets liberally dusted during flower visits. The bats flit from inflorescence to inflorescence, so that each can expect a number of visits from pollen-dusted bats each night. The bats are not limited to one particular species of plant. Their diet is based on a number of species that flower throughout the year, e.g., wild bananas, *Arenga, Duabanga, Sonneratia*. However, one species of bat, *Eonycteris spelaea*, is an important pollinator of such seasonally flowering species as the durian and petai *Parkia speciosa*. They turn to the ever-flowering species only when their preferred trees are out of season. In this way, the productivity of our most important orchard fruit (durian), is linked to that of our most popular jungle fruit (petai) as well as to the ostensibly rather 'useless' *Sonneratia* of the mangroves. The *Sonneratia*, in fact, plays a vital role in sustaining the bats when durian and petai are not in flower.

9.6. DISPERSAL

The subject of plant dispersal throughout the world has been treated by Ridley (1930) in a massive 744-page work. For Malaysian trees, I have constructed Table 9.4 to provide a simpler basis for discussion. The descriptive terminology and classification of fruits dates back to the murky beginnings of plant classification in Europe. As an unavoidable consequence of its long history, ideas and terminology became fixed before an adequate number of fruit types, especially tropical ones, could be studied. We now find that some tropical fruits do not fit into the conventional categories. The following treatment of fruit types and dispersal mechanisms is not yet exhaustive, but will, perhaps, be adequate to serve as an introduction to the subject.

TABLE 9.4. Classification of Peninsular Malaysian forest tree fruits.

Principal characteristics (i) (ii) (iii)	Type name	Examples	Subsidiary characteristics
1. Dehiscent Unicarpellate Dry	FOLLICLE	*Firmiana, Pterocymbium, Scaphium*	Single-seeded; seed without pulp; ovary wall split very early and expanding to form a wing, the split developing on the inner side.
(**Note:** Among dehiscent unicarpellate fruits, a distinction is made between those that split only on one side, theoretically the 'inner' side, and those	FOLLICLE	Connaraceae	Single-seeded; seed with pulp; fruit wall split on inner side.
that split on both sides.	—	*Amesiodendron*	Single-seeded, seed without pulp; fruit wall split on both sides.
The former is a follicle, the latter a legume but in practice, the term 'legume' has become restricted to the fruits of Leguminosae.	FOLLICLE	*Alstonia, Dyera*	Multi-seeded; seed coat produced into a wing; fruit wall split on inner side.
This leaves the fruit types of *Amesiodendron*, Myristicaceae and *Anaxagorea* without a name.)	FOLLICLE	*Sterculia*	Multi-seeded; seeds with pulp; fruit wall split on inner side.
	LEGUME	*Acacia, Adenanthera, Intsia, Saraca*	Multi-seeded; seeds without pulp; fruit wall split on both sides.
2. Dehiscent Unicarpellate Fleshy	—	Myristicaceae	Single-seeded; seeds with pulp; fruit wall split on both sides.
(**Note:** See note under the previous section.)	—	*Anaxagorea*	Multi-seeded; seeds slimy but without pulp, shot out by the drying and contraction of the valves after dehiscence; fruit wall split on both sides into two valves.

(Table 9.4 continued)

	Principal characteristics (i) (ii) (iii)			Type name	Examples	Subsidiary characteristics
3.	Dehiscent	Multicarpellate	Dry	CAPSULE	*Durio, Sloanea*	Seeds with pulp.
	(**Note:** Multicarpellate fruits are nearly always multi-seeded, hence the distinction between uni- and multi-seeded conditions is not relevant here. A distinction may be made between septicidal fruits which split along the margins of the carpels and loculicidal fruits which split along the midline of the carpels; this is analogous to splits along the inner and outer sides, respectively, of a unicarpellate fruit. In this table, no distinction is made between septicidal and loculicidal fruits.)			CAPSULE	*Elateriospermum, Hevea*	Seeds without pulp; fruit splitting to scatter seeds explosively.
				CAPSULE	*Leptospermum, Rhododendron*	Seeds minute, shaken out by wind.
				CAPSULE	*Bombax, Ceiba*	Seeds embedded in cottony fibres, blow out by wind.
				CAPSULE	Bignoniaceae, *Casuarina, Lagerstroemia, Triomma*	Seed coat expanded into a wing.
				CAPSULE	*Commersonia, Neesia*	Seeds without pulp and without other dispersal mechanisms.
4.	Dehiscent	Multicarpellate	Fleshy	—	*Baccaurea*, some *Dillenia, Dysoxylum, Guioa, Pittosporum*	Seeds with pulp.
	(**Note:** See note under previous section.)					
5.	Indehiscent	Unicarpellate	Dry	SAMARA	*Heritiera, Koompassia*	Single-seeded; fruit-wall expanded into a wing.
	(**Note:** An achene differs from a nut only by being smaller.)			SAMARA	*Peltophorum, Pterocarpus*	As above but multi-seeded.
				NUT	*Parishia*	Single-seeded; sepals expanded into wings.
				NUT	*Melanorrhoaea, Swintonia*	Single-seeded; petals expanded into wings.
				NUT	*Engelhardtia*	Single-seeded; fruit-bract expanded into a wing.
				ACHENE	*Vernonia*	Single-seeded; fruit small, bearing a tuft of hairs.
				NUT	Fagaceae, *Helicia, Pongamia*	Single-seeded; fruit wall hard, smooth.
				SCHIZOCARP	*Desmodium*	Multi-seeded but fruit breaking up into single-seeded parts.
6.	Indehiscent	Unicarpellate	Fleshy	DRUPE	*Calophyllum, Mangifera, Prunus*	Single-seeded; inner layer of fruit-wall hard.
				—	*Cynometra, Dialium*	Single-seeded; inner layer of fruit-wall fleshy.
				BERRY	*Alphonsea, Tamarindus*	Multi-seeded; inner layer of fruit-wall fleshy.

(continued over page)

(Table 9.4 continued)

Principal characteristics (i) (ii) (iii)	Type name	Examples	Subsidiary characteristics
	—	*Dimocarpus, Litchi, Nephelium*	Single-seeded; ovary 2–3 carpellate but fruit effectively unicarpellate; fruit-wall peelable; seed surrounded by pulp.
7. Indehiscent Multicarpellate Dry	SCHIZOCARP	*Acer, Trigoniastrum*	Multi-seeded; fruit splitting into single-seeded parts, each with its wall expanded into a wing.
	NUT	Dipterocarpaceae	Single-seeded; sepals expanded into wings.
	SCHIZOCARP	*Ehretia*	Multi-seeded but splitting into single-seeded parts.
	—	*Hydnocarpus, Monocarpia*	Multi-seeded but not splitting; fruit wall hard, decaying over a long period before seeds are released.
8. Indehiscent Multicarpellate Fleshy	DRUPE	*Alangium, Gironniera*	Single-seeded; inner layer of fruit-wall hard.
	—	*Canarium, Elaeocarpus, Spondias*	As above, but multi-seeded, the seeds enclosed within a single stone.
	BERRY	*Ilex*	As above, but the seeds enclosed individually in separate stones.
	BERRY	Ebenaceae, Sapotaceae, *Garcinia*	Multi-seeded; fruit-wall fleshy throughout.
	—	*Lansium*, some *Baccaurea*	Multi-seeded; fruit wall peelable; seeds surrounded by pulp.
	—	*Styrax*	Single-seeded; fruit wall fleshy.
	—	*Artocarpus, Ficus*	Multi-seeded; perianth or receptacle forming the fleshy jacket.
9. Indehiscent fleshy outer layer, dehiscent dry inner layer; multi-carpellate	—	*Phyllanthus emblica, P. pectinata*	Multi-seeded; outer layer fleshy, indehiscent; inner layer stony, exploding to scatter seeds.

The framework for fruit classification is based on three principal sets of characteristics, namely (i) dehiscent *versus* indehiscent, (ii) unicarpellate *versus* multicarpellate and (iii) dry *versus* fleshy. The first and third characteristics are intimately related to seed dispersal, while the second relates to the carpel theory of ovary and fruit development, which is admittedly of little relevance to dispersal. Nevertheless, it would hardly be possible to discuss dispersal without reference to fruit classification, hence a brief description of the principle characteristics is given here. A dehiscent fruit is defined as one which opens to expose the seeds. Some fruits break up into seed-containing segments; these are technically considered to be indehiscent because the seeds themselves are not exposed.

A carpel is the female reproductive unit, consisting, in its most basic form, of a pollen-receptive area (stigma) connected by a neck (style), to a swollen hollow bulb (ovary), which contains one or more ovules in its chamber (cell or locule). Some fruits develop from a single carpel, e.g., those of the Leguminosae and Myristicaceae. Others are considered to have developed from two or more carpels joined together, in which case the number of styles or stigmatic lobes (seen before they are shed or obscured) would indicate the number of carpels. The number of cells in the ovary is also a good indicator except that in some plants, an internal partition develops in the middle of each carpel so that the number of cells is double the number of carpels.

A fleshy fruit is one with the fruit wall, or any layer of the fruit wall, of whatever origin (ovary, perianth or receptacle) being fleshy or pulpy. Fleshy tissue associated with the seed (aril or sarcotesta) is not counted. In contrast, a dry fruit has a dry wall. However, the distinction between a dry and a fleshy fruit wall is not always clear. The fruit wall of the durian, for example, is intermediate, at least at the time of fruit maturity; drying occurs after the fruit has split open. The same situation occurs with the fruits of Connaraceae, Myristicaceae, *Nephelium* and *Sterculia*. In *Mangifera* and *Spondias*, the fruits of the wild types are quite fibrous compared to those in cultivation. The categorization of fruits as dry or fleshy is therefore, in some cases, a little arbitrary.

These three sets of principal characteristics give rise to eight combinations. Various other characteristics are then used selectively to give a finer breakdown of fruit types, e.g., for unicarpellate dehiscent fruits, a distinction is made between those that split on one side and those that split on both sides of the fruit. Such 'subsidiary' characteristics are listed in Table 9.4, but some of the fruit types are without names because they do not fit within the established nomenclatural system. For example, a 'nut' is technically an indehiscent, dry, one-seeded, hard-shelled fruit, as found in Fagaceae and Dipterocarpaceae. *Hydnocarpus* qualifies as a nut in all characteristics except that it is multi-seeded; we could call it a multi-seeded nut. The fruits of Myristicaceae and *Anaxagorea* are technically 'legumes', a legume being defined as a dehiscent unicarpellate fruit opening on both sides of the fruit, but the term has in practice been applied only to the fruits of the Leguminosae. It would upset everybody if we called a nutmeg a legume. To make things even more confusing, many tropical species of the Leguminosae have fruits that are technically not legumes at all. Instead, they may be 'samaras' (*Koompassia, Peltophorum, Pterocarpus*), 'nuts' (*Pongamia*) or 'berries' (*Tamarindus*). Some have fruits that do not fit any named category, e.g., *Cynometra, Dialium*.

A revision of fruit names is needed, but this would not be the appropriate place for new proposals. From the viewpoint of plant dispersal, the classification scheme in Table 9.4 is adequate for highlighting the main dispersal features.

In contrast to its almost negligible role in pollen dispersal, wind is important for seed dispersal. The range of structures that enable seeds or fruits to float, glide, drift or scatter in the wind is so extensive that the whole effort looks like an exaggerated attempt to catch what little air movement there is in our forests. The attempt is not futile because the target is the ground, in contrast to wind-dispersal of pollen for which the target is the minute and elusive stigma. The most remarkable winged structures are those of several genera of the Sterculiaceae, namely *Firmiana, Pterocymbium* and *Scaphium*. These develop as a result of the ovary wall splitting on its inner side quite early in development and expanding into a papery, boat-shaped wing. The developing seed is as 'naked' as that of a gymnosperm. Corner (1976) has remarked, "If such fruits were found immature and fossilized, theorists would pen abundance of speculation on the primitive angiosperm."

From the review by Tamari and Jacalne (1984), it is apparent that the vast majority of winged dipterocarp fruits falls within 20 to 40 m of the mother tree, but dispersal distances of up to 800 m (half a mile) have been noted; such a distance would not have been possible without winged structures. Hence the fact that most winged fruits fall close to the parent tree, anomalous as it may seem, does not invalidate the usefulness of the wing as a dispersal mechanism for the species. In the case of seeds that are winged and light, e.g., *Dyera* and *Oroxylum*, it can actually be quite difficult to find the seeds after they are released from the tree.

Adaptations for dispersal by animals are equally diverse. To attract an animal, the fruit usually offers a reward in the form of edible tissue, or at least something that appears edible. If edibility is located in the fruit wall, we have either a 'drupe' or a 'berry'. Both are indehiscent and fleshy. A drupe is single-seeded and fleshy, but with the innermost layer of the fruit wall hardened to protect the seed; a berry is multi-seeded, with the seeds individually hardened, and the innermost layer of the fruit wall remaining soft. The Peninsular Malaysian flora includes typical drupes (*Calophyllum, Mangifera*) and berries (Ebenaceae, Sapotaceae). But we also have anomalous 'single-seeded berries' (*Cynometra*) and 'multi-seeded drupes' (*Canarium, Elaeocarpus, Spondias*).

The greatest tropical innovation of all is, perhaps, the transfer of edibility from the fruit wall to special parts of the seed coat known as sarcotesta or aril. In conjunction with this transfer, the fruit wall splits when ripe, in order to present the sarcotesta or aril to animals. Such fruits, by virtue of dehiscence, can be accommodated within the definition of a 'capsule', provided that the fruit wall is dry at maturity. If the dehiscent fruit wall is fleshy, it may be termed a 'fleshy capsule', as seen in some *Baccaurea*, some *Dillenia*, *Dysoxylum, Guoia* and *Pittosporum*. In such cases, the fruit wall often develops a colour which contrasts with that of the pulp around the seed. Such eye-catching combinations are thought to be attractive to certain animals. The pulp around the seed is obviously the edible reward, and the fruit wall is fleshy only to the extent necessary for the development of the bright hues of purple, red, pink, orange, yellow, etc., which are difficult to develop in dry tissues. The fleshy capsule, in association with the development of contrasting colours between the fruit wall and the seed pulp, may be considered the second greatest innovation in tropical fruits.

Having evolved, independently in many different families, the transfer of edibility from the fruit wall to the outer layer of the seed, it would seem perverse to enclose such a seed in an indehiscent fruit. Yet, this has happened in Euphorbiaceae (some *Baccaurea*), Meliaceae (*Lansium*) and Sapindaceae (*Nephelium*). It takes a smart mammal to peel off the skin of such fruits to get at the pulp. Yet, if the skin of such fruits is not peeled off, the seeds would quickly perish. The germination of such species has become absolutely dependent on the animals that feed on their fruits.

Dispersal by water calls for the development of corky bouyant tissues in the fruit. There are no special names for water-dispersed fruits, because there are no special morphological adaptations involved. Many bouyant fruits begin as 'proper', i.e., single-seeded, drupes, e.g., *Calophyllum inophyllum*, or as 'multi-seeded drupes', e.g., *Cerbera odollam* and *Cordia subcordata*. Such fruits actually ripen as drupes but the fleshy tissue subsequently dries out to become corky. There are also instances of corkiness developing in a persistent basal calyx of certain fruits such as *Shorea sumatrana* and *Vatica stapfiana*. An additional class is needed to accommodate the remarkable fruits of *Phyllanthus emblica* and *P. pectinata*. These are 'multi-seeded drupes' at first but, after the fleshy outer layer has been eaten or otherwise removed, the stone dries and splits explosively. Such species are apparently dispersed firstly by animals eating the fruits, and secondly by the explosive scattering of the seeds as the stones dry out.

9.7. GERMINATION

Many studies on germination are focused, not on the germination process itself, but on dormancy, i.e., the period of 'rest' prior to germination. This period is important for humans because we want to store, transport, buy, sell, or plant seeds at will. Seeds that do not stay dormant during such transactions will suffer spoilage.

The keeping properties of seeds are often foreshadowed by their behaviour in nature, e.g., the seeds of *Intsia palembanica* (Leguminosae) and *Parkia javanica* (Leguminosae) lie dormant for years and can easily be obtained by sieving the soil under a mature tree. Such seeds can be depended upon to retain viability for years under artificial storage.

In contrast, seeds that germinate immediately in nature, such as those of most dipterocarps, suffer rapid spoilage if prevented from germinating under test conditions (Yap, 1981). Six months is about the maximum storage period for a minimum germination standard of 50%. Most Peninsular Malaysian species cannot last even three months under storage, and are termed 'recalcitrant'.

Seeds deposited on the forest floor thus lie there for varying periods of time and germinate or die at various rates according to the nature of the species. The soil of the forest floor may be regarded as a seed bank from which withdrawals are made through germination or seed death. The characteristics of this seed bank may be used to define the regenerative capacity of the community from its own stored seed reserves. Unfortunately, direct studies are not possible because of uncertainties over the age of seeds already there, as well as the impossibility of arranging for a representative sample of species to seed at the same time and place. We can get around this problem by building up a model from accumulated *ex situ* experiments in which fresh seeds are planted in soil and kept warm, shaded and regularly watered, i.e., conditions normally favourable for germination. Such a model has been built for Malaysian woody plants (Ng, 1980).

The model, represented graphically in Figure 9.8, started with 335 species. By week 6, only 50% of the species were still represented by viable seeds in the soil. By week 12, the number of species represented had been halved again, and so on, halving for every doubling of time lapsed. This rapid rate of attrition makes it clear that the soil seed bank is virtually useless as a source of seeds for the re-establishment of the original forest after the loss of mature vegetation.

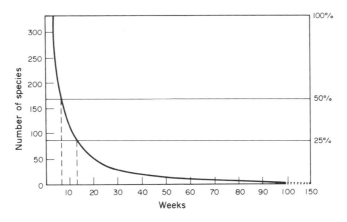

Fig. 9.8. Rate of loss of species from hypothetical soil seed bank under conditions normally suitable for germination (after Ng, 1980b). The curve approximates the equation $xy = 300$ if y is expressed in % and x in weeks.

The reversed-'J' curve in Figure 9.8 may be visualized as the response of the forest as a community to two opposing selection forces. One favours rapid germination and pulls the response curve towards zero on the time axis, while the opposite force favours delayed germination and pulls towards infinite time. The balance is overwhelmingly in favour of rapid germination. For this situation to have evolved, the climate must have been consistently warm and moist for an immense period of evolutionary time.

Nevertheless, the presence of long-lived dormant seeds in the soil cannot be overlooked. Data suggest that about 5% of our species have seeds able to persist for a year or more under moist, warm conditions.

Not surprisingly, hard-seeded legumes are well represented, e.g., *Adenanthera parvonina, Dialium maingayi, Intsia palembanica, Parkia javanica, Sindora echinocalyx*. This raises the question of whether the Leguminosae may have evolved elsewhere, in a more seasonal climate, before arriving in Peninsular Malaysia.

Somewhat more surprising are the large seeds of several species of *Anisophyllea* (Rhizophoraceae), *Barringtonia* (Lecythidaceae) and *Scorodacarpus* (Olacaceae). These lie around on the forest floor for many months or years without germinating, in a fully imbibed condition. Unlike legumes, the seed tissue of these species is fleshy and rather high in moisture content.

For some pioneer species such as *Vitex pinnata* and *Sapium baccatum*, there is evidence (Aminuddin and Ng, 1982) that seeds can lie dormant for months, fully imbibed, under a forest canopy, ready to germinate when the canopy is removed, probably in response to the change in light conditions.

9.8. APPENDIX

Taxonomic composition of the tree flora of Peninsular Malaysia (excluding arborescent monocots and cycads).

Family	No. of tree-genera	No. of species in tree genera	No. of endemic species	No. of tree species
Aceraceae	1	1	0	1
Actinidiaceae	1	10	4	10
Alangiaceae	1	7	1	7
*Anacardiaceae	16	77	11	76
Annonaceae	25	130	54	110
Apocynaceae	10	36	5	27
Aquifoliaceae	1	19	10	15
Araliaceae	10	47	27	22
Araucariaceae	1	2	1	2
Bignoniaceae	7	10	0	10
Bombacaceae	5	23	4	23
*Boraginaceae	3	7	?	6
Burseraceae	6	37	8	37
Capparidaceae	1	3	0	3
Casuarinaceae	1	1	0	1
Celastraceae	11	33	4	32
Clethraceae	1	2	2	1
Combretaceae	2	10	0	10
Compositae	1	8	2	1
Connaraceae	1	1	0	1
Convolvulaceae	1	18	4	1
Cornaceae	1	3	0	3
Crypteroniaceae	2	3	0	3
Cunoniaceae	1	1	0	1
Daphniphyllaceae	1	2	0	2
Datiscaceae	1	1	0	1
Dilleniaceae	1	10	1	9
Dipterocarpaceae	10	157	28	157
Ebenaceae	1	70	28	70
*Elaeocarpaceae	2	27	?	23
Epacridaceae	1	1	0	1
Ericaceae	5	37	15	9
Erythroxylaceae	1	2	1	2
Euphorbiaceae	57	344	101	286

* Estimated Figures

Family	No. of tree· genera	No. of species in tree genera	No. of endemic species	No. of tree species
Fagaceae	4	64	14	64
Flacourtiaceae	12	50	17	49
Gnetaceae	1	8	2	1
Guttiferae	4	120	40	120
Hamamelidaceae	6	6	1	6
Hernandiaceae	2	2	0	2
Hypericaceae	1	6	0	6
Icacinaceae	7	17	1	16
Illiciaceae	1	3	2	3
Juglandaceae	1	4	0	4
*Lauraceae	15	174	?	156
Lecythidaceae	4	19	4	19
Leeaceae	1	8	1	3
Leguminesae	27	127	30	103
Linaceae	2	3	1	3
Loganiaceae	2	18	1	12
Lythraceae	2	7	2	7
Magnoliaceae	5	20	5	20
Malvaceae	2	5	1	4
Melastomataceae	3	33	6	25
Meliaceae	14	91	6	91
*Monimiaceae	1	3	?	2
Moraceae	7	133	5	79
Myricaceae	1	1	0	1
Myristicaceae	4	53	13	53
*Myrsinaceae	3	72	?	24
Myrtaceae	9	209	80	204
Nyctaginaceae	1	3	0	2
Nyssaceae	1	1	0	1
Ochnaceae	3	4	0	4
Olacaceae	5	8	4	8
*Oleaceae	3	13	?	10
Opiliaceae	2	2	0	2
Oxalidaceae	2	6	3	6
Pentaphylacaceae	1	1	0	1
Pittosporaceae	1	2	0	2
Podocarpaceae	2	10	2	10
Polygalaceae	1	25	6	25
Proteaceae	2	12	5	12
*Rhamnaceae	2	3	?	2
Rhizophoraceae	8	23	5	23
Rosaceae	9	28	6	28
*Rubiaceae	24	237	?	71
Rutaceae	14	47	19	33
*Sabiaceae	1	8	?	7
Santalaceae	1	1	1	1
*Sapindaceae	17	46	?	46
Sapotaceae	11	76	18	76
Sarcospermataceae	1	2	0	2
*Saxifragaceae	1	12	?	10
Simaroubaceae	7	9	0	9
Sonneratiaceae	2	5	0	5
Staphyleaceae	2	4	0	4
Sterculiaceae	11	50	11	46
Styracaceae	1	5	2	5
Symplocaceae	1	21	3	21
Tetrameristaceae	1	1	0	1
Theaceae	8	37	21	37
Thymelaeaceae	2	8	1	8

* Estimated Figures

(continued over page)

Family	No. of tree-genera	No. of species in tree genera	No. of endemic species	No. of tree species
Tiliaceae	8	46	17	42
Trigoniaceae	1	1	0	1
Ulmaceae	3	10	0	10
*Urticaceae	2	3	?	3
Verbenaceae	8	45	14	25
Violaceae	1	9	0	7
	500	3220	>681	2646

* Estimated Figures

REFERENCES

Aminuddin, Mohamad and Ng, F. S. P. (1982). Influence of light on germination of *Pinus caribaea, Gmelina arborea, Sapium baccatum* and *Vitex pinnata. Malaysian Forester* 45, 62–8.

Apannah, S. and Chan. H. T. (1981). Thrips: the pollinators of some dipterocarps. *Malaysian Forester* 44, 234–52.

Chin, S. C. (1973). *The limestone flora of Malaya.* M.Sc. Thesis, University of Malaya, Kuala Lumpur.

Corner, E. J. H. (1940). *Wayside trees of Malaya.* Government Printer, Singapore.

Corner, E. J. H. (1976). *The seeds of dicotyledons.* Cambridge University Press.

Edgar, A. T. (1960). *Manual of rubber planting.* Incorporated Society of Planters, Kuala Lumpur.

Foxworthy, F. W. (1927). *Commercial timber trees of the Malay Peninsula.* Malayan Forest Record No. 3. Forest Department, Kuala Lumpur.

Foxworthy, F. W. (1932). *Dipterocarpaceae of the Malay Peninsula.* Malayan Forest Record No.10. Forest Department, Kuala Lumpur.

Halle, F. and Ng, F. S. P. (1981). Crown construction in mature dipterocarp trees. *Malay. Forester* 44, 222–33.

Halle, F., Oldeman, R. A. A. and Tomlinson, P. B. (1978). *Tropical trees and forests.* Springer Verlag.

Holttum, R. E. (1930). On periodic leaf-change and flowering of trees in Singapore. *Gardens' Bulletin Straits Settlements* 5, 173–206.

Holttum, R. E. (1940). Periodic leaf-change and flowering of trees in Singapore. *Gardens' Bulletin Straits Settlements* 11, 119–75.

Holttum, R. E. (1954). *Plant life in Malaya.* Longmans, London.

Janzen, D. H. (1978). Seeding patterns of tropical trees. In Tomlinson, P. B. and Zimmermann, M. H. (eds.), *Tropical trees as living systems.* Cambridge University Press.

Kochummen, K. M. (1972). Simaroubaceae. In T.C. Whitmore (ed.), *Tree Flora of Malaya* Vol. 1. Longman, Kuala Lumpur.

Kochummen, K. M. (1982). *Effects of Elevation on Vegetation on Gunung Jerai, Kedah.* Research Pamphlet No. 87. Forest Research Institute, Kepong.

Koriba, K. (1958). On the periodicity of tree-growth in the tropics, with reference to the mode of branching, the leaf-fall, and the formation of the resting bud. *Gardens' Bulletin Singapore* 17, 11–81.

Longman, K. A. and Jenik, J. (1974). *Tropical forest and its environment.* Longman, London.

Medina, E. (1983). Adaptations of tropical trees to moisture stress. In Golley, F. B. (ed.), *Tropical rain forest ecosystems.* Elsevier Scientific Publishing Co., Amsterdam.

Medway, Lord (1972). Phenology of a tropical rain forest in Malaya. *Biol. J. Linn. Soc. Lond.* 4, 117–46.

Ng, F. S. P. (1977a). Shyness in trees. *Nature Malaysiana* 2, 34–7.

Ng, F. S. P. (1977b). Gregarious flowering of dipterocarps in Kepong 1976. *Malay. Forester* 40, 126–37.

Ng, F. S. P. (ed.) (1978a). *Tree flora of Malaya,* Vol.3. Longman, Kuala Lumpur.

Ng, F. S. P. (1978b). Territoriality in trees. *Nature Malaysiana* 3, 12–7.

Ng, F. S. P. (1979). Growth rhythms in tropical juvenile trees. *Bull. Soc. Bot. France* 125, 139–49.

Ng, F. S. P. (1980a). The phenology of the yellow-flame tree, *Peltophorum pterocarpum. Malay. Nat. J.* 33, 201–8.

Ng, F. S. P. (1980b). Germination ecology of Malaysian woody plants. *Malay. Forester* 43, 406–37.

Ng, F. S. P. (1981). Vegetative and reproductive phenology of dipterocarps. *Malay. Forester* 44, 197–221.

Ng, F. S. P. (1982). Trees for towns. *Nature Malaysiana* 7, 4–15.

Ng, F. S. P. (1983). Ecological principles of tropical lowland rain forest conservation. In Sutton, S. L., Whitmore T. C. and Chadwick, A. C. (eds.), *Tropical rain forest: ecology and management*. Blackwell, Oxford.

Ng, F. S. P. and Loh, H.S. (1974). Flowering to fruiting periods of Malaysian trees. *Malay. Forester* 37, 127–32.

Ng, F. S. P. and Low, C. M. (1982). *Check list of endemic trees of the Malay Peninsula*. Research Pamphlet No. 88. Forest Research Institute, Kepong.

Ogawa, H. (1978). Litter production and carbon cycling in Pasoh Forest. *Malay. Nat. J.* 30, 367–73.

Poore, M. E. D. (1964). Integration in the plant community. *J. Ecol.* 52 (Suppl.), 213–26.

Poore, M. E. D. (1968). Studies in Malaysian rain forest I. The forest on the Triassic sediments in Jengka Forest Reserve. *J. Ecol.* 56, 143–96.

Putz, F.E. (1979). Aseasonality in Malaysian tree phenology. *Malay. Forester* 42, 1–24.

Ridley, H. N. (1901). The timbers of the Malay Peninsula. *Agricultural Bulletin, Straits Settlements and Federated Malay States* 1, 53.

Ridley, H. N. (1930). *The Dispersal of plants throughout the world*. L. Reeve and Co., U.K.

Start, A. N. and Marshall, A. G. (1976). Nectarivorous bats as pollinators of trees in West Malaysia. In Burley, J. and Styles B. T. (eds.), *Tropical trees: variation, breeding and conservation*. Academic Press.

Symington, C. F. (1943). *Foresters' Manual of Dipterocarps*. Malayan Forest Record No.16. Forest Department, Kuala Lumpur.

Tamari, C. and Jacalne, D. V. (1984). Fruit dispersal of dipterocarps. *Bull. Forestry and Forest Products Res. Inst.* (Japan) No.325, 127–40.

Watson, J. G. (1928). *Mangrove Forests of the Malay Peninsula*. Malayan Forest Record No. 6. Forest Department, Kuala Lumpur.

Whitmore, T. C. (ed.) (1972). *Tree Flora of Malaya* Vol. 1. Longman, Kuala Lumpur.

Whitmore, T. C. (ed.) (1973). *Tree Flora of Malaya* Vol. 2. Longman, Kuala Lumpur.

Wilcox, B. A. (1980). Insular ecology and conservation. In Soule M. E. and Wilcox B. A. (eds.) *Conservation biology*. Sinauer, Sunderland, Mass.

Wyatt-Smith, J. (1963). *Manual of Malayan Silviculture for Inland Forests*. Malayan Forest Record No. 23. Forest Department, Kuala Lumpur.

Wyatt-Smith, J. (1966). *Ecological Studies on Malayan Forests*. Research Pamphlet No. 52. Forest Research Institute, Kepong.

Yap, S. K. (1981). Collection, germination and storage of dipterocarp seeds. *Malay. Forester* 44, 281–300.

Yap, S. K. (1982). The phenology of some fruit tree species in a lowland dipterocarp forest. *Malay. Forester* 45, 21–35.

Completed July 1985

CHAPTER 10

Forest Management

Salleh Mohd Nor

Forest Research Institute Malaysia, Kepong, 52109 Kuala Lumpur, Malaysia

CONTENTS

10.1. INTRODUCTION

The earliest reference to forestry in Peninsular Malaysia was a report by a Major McNair in 1879 which described the principal timber trees and recommended the creation of a forest department. A Forest Department was duly established in the Straits Settlements in 1883. Early work concentrated on the gazettement of forest reserves in the Straits Settlements of Malacca and Penang. During the late nineteenth century, the high price of gutta percha (resin of *Palaquium gutta*) and problems with the supply of railway sleepers led ultimately to the establishment of a combined Forestry Department for the Straits Settlements and the Federated Malay States in 1901 (Wyatt-Smith, 1963). Peninsular Malaysia (Malaya prior to 1963) has therefore a long history of forest management, by tropical standards.

10.2. THE RESOURCE

As a consequence of a century of forest management, Peninsular Malaysia is in the position, fortunate by comparison with many other developing countries, that statistics such as forest area and forest types are fairly well documented (Table 10.1). Peninsular Malaysia undertook a Forest Resources Reconnaissance Survey (FRRS) in mid 1960s as part of a national programme for drawing up a Land Capability Classification. While reports for all States were not completed, the FRRS did produce a complete photo-type map of the peninsula resulting from photo-interpretation of 1966 aerial photographs. Forest typing was attempted, based on the forest type classification of Wyatt-Smith (1963).

TABLE 10.1 Change in forested area of Peninsular Malaysia since 1960 (compiled by the Forest Economics Unit, Kuala Lumpur).

Year	Forest Area (thousand ha)	Year	Forest Area (thousand ha)
1900	about 13,000	1971	7,875
1960	9,465	1972	7,583
1961	8,964	1973	7,450
1962	8,816	1974	7,319
1963	8,733	1975	7,290
1964	8,629	1976	7,199
1965	8,556	1977	6,968
1966	8,102	1978	6,839
1967	8,193	1979	6,443
1968	8,068	1980	6,360
1969	8,035	1981	6,293
1970	8,009	1982	6,212

The FRRS formed the basis of the first National Forest Inventory (NFI) carried out under the auspices of a United Nations Development Programme project in the early 1970s. This first NFI produced a comprehensive forest type and forest resources map, including estimates of volumes by species groups and by regions (Anon., 1973). It was then estimated that the total forest resource in Peninsular Malaysia amounted to about 0.9 billion m^3. Table 10.2 gives the breakdown by forest type and species group.

Besides these two major efforts at resource definition, the Forestry Department initiated a Disturbed Forest Inventory (DFI) and a second NFI in the early 1980s. The DFI attempted to focus on areas identified as disturbed or logged-over by the first NFI, while the second NFI attempted to carry out a second national inventory to determine changes in the forest resources. For various reasons, including shortage of funds, neither survey was completed. Figure 10.1 shows the best available plot of present forest resources.

Other than resource inventories, research plots have been established in virgin as well as logged forests to determine the development of individual trees as well as stands. Results from such experiments indicate growth rates for different species groups, varying from increases in diameter of 0.45 cm/yr/tree for Heavy Non-Dipterocarp Hardwoods to 0.66 cm/yr/tree for the meranti group of dipterocarps. Table 10.3 gives mean growth rates of some groups of species based on periodic measurements of logged-over forest (Tang and Wan Razali Mohd., 1981).

TABLE 10.2 Total computed gross volumes of trees with diameters > 18 inches (>48 cm), rounded to nearest
0.5 × 10⁶ m³; 1972 data from Anon. (1973), with discrepancies as in the original.

Forest Resource Type	Species Group	West Coast and South	East Coast and Central	Total
		100 000 m³	100 000 m³	100 000 m³
Primary Hill Forest	Mer.	480	980	1 460
	ODip	300	540	840
	Mar.	315	820	1 140
	Pot.	185	340	525
	Res.	430	895	1 325
	SUB TOTAL	1 720	3 590	5 310
Recently Harvested Hill Forest	Mer.	60	245	305
	ODip.	30	220	250
	Mar.	80	305	385
	Pot.	25	100	125
	Res.	110	310	420
	SUB TOTAL	295	1 185	1 475
Disturbed Forest and Forest Harvested prior to 1966	Mer.	115	165	275
	ODip.	70	90	160
	Mar.	145	220	370
	Pot.	70	75	140
	Res.	215	225	440
	SUB TOTAL	615	785	1 405
Poor Edaphic and Upperhill Forests	Mer.	35	40	75
	ODip.	25	30	50
	Mar.	45	55	100
	Pot.	40	55	95
	Res.	85	110	190
	SUB TOTAL	220	275	500
Swamp Forests and partly Harvested (types Wi, LWi, DWi)	Mer.	15	10	30
	ODip .	—	—	—
	Mar.	80	80	160
	Pot.	15	25	40
	Res.	50	35	85
	SUB TOTAL	170	140	310
TOTAL (all Types)	Mer.	710	1 445	2 150
	ODip.	420	875	1 295
	Mar.	670	1 485	2 150
	Pot.	340	595	935
	Res.	890	1 545	2 435
	GRAND TOTAL	3 020	5 975	9 000

Mer.:   Meranti
ODip.:   Other Dipterocarp (non-Meranti)
Mar.:   Non-Dipterocarp fully marketable in 1972
Pot.:   Non-Dipterocarp partially or potentially marketable
Res.:   Non-Dipterocarp residual

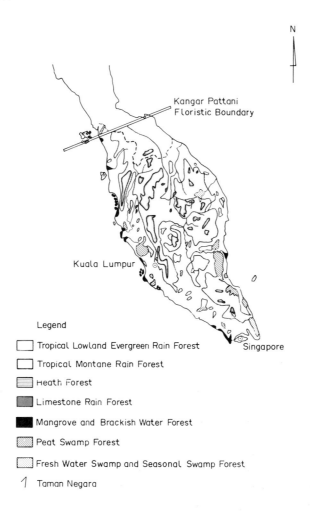

N

Kangar Pattani
Floristic Boundary

Kuala Lumpur

Singapore

Legend

☐ Tropical Lowland Evergreen Rain Forest

☐ Tropical Montane Rain Forest

▤ Heath Forest

▨ Limestone Rain Forest

■ Mangrove and Brackish Water Forest

▨ Peat Swamp Forest

▦ Fresh Water Swamp and Seasonal Swamp Forest

1 Taman Negara

Fig. 10.1 Tropical rain forests in Peninsular Malaysia.

Accumulated statistics show that the forests of Peninsular Malaysia have been fast dwindling over the last two decades. Taking the total land area of Peninsular Malaysia as 13.2 million ha, forested land has diminished from 72% in 1960 to 47% in 1982 (Table 10.1). Logging of the residue has continued apace, and 1972 statistics (Table 10.2) are now undoubtedly much altered. Such a marked decline in forest resources, especially of natural forests, is a common phenomenon over all the tropics and has been closely documented by the FAO.

Pleas by conservationists and international agencies are, however, beginning to take effect. It is hoped that, with improved education and greater awareness for environmental and conservational needs, the remaining natural forests may be better managed than before. Somehow or other, forest management has to adapt itself to the task of meeting the needs of an increasing population for a host of forest-based commodities, from a declining area of forest.

TABLE 10.3 Mean annual diameter increment, in cm/year/tree, by species groups; from Tang and Wan Razali Mohd. (1981).

Species Group	Year					
	1974–75	1975–76	1976–77	1977–78	1978–79	Mean
Dipterocarp (Meranti)	0.72	0.51	0.71	0.71	0.56	0.66
Dipterocarp (Non-Meranti)	0.71	0.63	0.65	0.64	0.58	0.64
Non-Dipterocarp Light Hardwood	0.54	0.38	0.62	0.50	0.42	0.49
Medium Hardwood	0.50	0.37	0.51	0.46	0.41	0.45
Heavy Hardwood	0.43	0.37	0.57	0.47	0.40	0.45
Mean	0.52	0.39	0.57	0.49	0.43	0.48

10.3. HISTORICAL PERSPECTIVE

Forests have always been looked upon as resources to be exploited. They are clearly visible, and during the past few decades they have become readily marketable and easily harvested compared with other resources. The problem of the high heterogeneity in timber properties, resulting from the large number of species, has been partially overcome by grouping timbers according to similarities in anatomy, density, colour, etc. In Peninsular Malaysia, the development of the Malayan Grading Rules (MGR) spurred the acceptance of Malaysian timbers in the export market, which in turn led to greater logging intensities and increased export of logs and sawn timber. This trend continued and developed as the world economy prospered into the early and mid-1970s.

The 1970s also saw an expansion of land development activities in Peninsular Malaysia, which resulted in the large-scale opening up and conversion of forest lands to agriculture or tree-crop plantation. The land-take for development in one decade is shown in Table 10.4. The Federal Land Development Authority (FELDA) is the main agency for land conversion, and all its projects involve conversion of forest areas. Although specifications require that the areas must first be logged, shortage of time as well as logistic and marketing problems often have not allowed for a complete harvest, and generally there has been abundant wastage of timber resources.

The conflict in land use and multifarious demands on land led to the Land Capability Classification (LCC) programme of the early 1970s. The land capability classification was biased in favour of agriculture. As a result, forestry was relegated to a residual land use status. All good soils and soils on low and flat terrain were designated as agricultural soils. Forestry was pushed to soils and terrain for which no other land use could be proposed. It was the LCC that determined that forestry be located on hilly terrain, generally above 18° slope. This decision had far reaching impacts on forest management in the country, as was discovered soon after. Forest Reserves which had been designated in early days, prior to Malayan independence (1957), were located mostly on flat, fertile lands. With the decision of the LCC, States (which have complete jurisdiction on land matters) began to focus attention on these forests as potential agricultural development areas, using the LCC as justification. As a result, with minor exceptions, most of the reserved forests on such 'agricultural' soils were soon degazetted and converted to agricultural or plantation crops.

TABLE 10.4 Development land, in ha, by State, 1971–1980.

State	FELDA	FELCRA	RISDA	Regional authorities	State agencies†	Private sector	Total land developed
Selangor*	342	—	—	—	10,189	8,278	18,810
Johor	81,645	12,782	2,308	16,307	18,735	11,038	142,815
Melaka	2,087	1,621	—	—	—	3,373	7,081
N. Sembilan	62,710	6,735	2,807	—	5,567	12,878	90,697
Pahang	164,869	8,143	10,434	18,255	47,033	13,491	262,225
Perak	17,133	9,332	7,252	—	13,530	7,206	54,453
P. Pinang	—	—	—	—	—	—	—
Kedah	6,879	4,377	1,324	—	10,942	184	23,706
Kelantan	10,693	1,473	7,338	5,663	11,048	379	36,494
Perlis	3,187	—	—	—	—	—	3,187
Terengganu	22,732	6,247	—	15,623	41,492	4,599	90,693
TOTAL	372,277	50,710	31,464	55,848	158,536	61,327	730,162

† Includes land schemes undertaken by State Economic Development Corporations, State Agricultural Development
  Corporations, State Land Development Boards, State Agricultural Departments and District Officers.
* Includes the Federal Territory.

FELDA: Federal Land Development Authority
FELCRA: Federal Land Consolidation and Rehabilitation Authority
RISDA: Rubber Industry Smallholders Development Authority

Source: Anon. (1981) Fourth Malaysian Plan 1981–1985 Kuala Lumpur

The evident lack of security for forest reserves spurred attempts to find a long-term solution, through the political acceptance of the concept of a Permanent Forest Estate (PFE). It was hoped that with the establishment of a PFE on a State by State basis, the area and location of forests would be made permanent, in order to facilitate management on a long term basis. However, the establishment of the PFE has not been smooth sailing, and it has yet to be finalized. In spite of the LCC, the formal reservation even of hill forests that are of little or no interest to agriculture has turned out to be a prolonged procedure, compared with the comparative ease with which Forest Reserves are degazetted. The delay in gazettement of the PFE puts forest authorities in Peninsular Malaysia in a dilemma, as proper management can be practised only in gazetted reserves.

10.4. NATIONAL FOREST POLICY

As noted above (Section 10.3) land is under the jurisdiction of the State Governments and attempts at overall policy formulation made little progress until 1978, when a National Forest Policy for Peninsular

Malaysia was adopted by the National Forestry Council and endorsed by the National Land Council. This was a landmark in Malaysian forestry and represented a major step towards coordinated and uniform forestry management in Peninsular Malaysia.

The policy recognizes the vital role of forests in the national welfare and economy. It recognizes the need for the establishment of a Permanent Forest Estate, to be located strategically throughout the country. It calls for the reservation of protective forests, productive forests and amenity forests. The policy calls for rational forest management, a sound programme of forest development, efficient resource utilization, promotes industrial development, encourages research and development and fosters cooperation and publicity (Anon., 1978).

In an effort to coordinate standardized forestry practices in the various states of Peninsular Malaysia, the National Forestry Act was passed by Parliament in late 1984 (Anon., 1984). This Act specifies the rules and procedures for the effective implementation of the National Forest Policy, covering a whole spectrum of activities relating to the management of the permanent reserved forest.

10.5. LOGGING MANAGEMENT

The basic concern in the management of the natural forest in Peninsular Malaysia is to ensure that logging is conducted in a regulated manner, that all financial charges are collected and that all silvicultural activities are performed effectively. While this is the ultimate aim of management, exigencies of time often necessitate compromise and variations from the optimum.

The determination of logging area is based on area control, with a rotation of 50 years or a felling cycle of 30 years, depending upon the silvicultural system used. The annual coupe is determined for each State by the State Forestry Department, but logging permits are awarded by the approving authority, which is usually the State Executive Council, on the advice of the State Forestry Director.

Long term logging permits are given to certain 'timber complexes' or State corporations, which usually have processing capability. Forest logging areas are levied a premium, which is a forestry charge equivalent to stumpage. Forest areas not under long term agreement are generally awarded by a tender system, which replaces the premium and maximizes revenue to the state. The logger is also charged a royalty on the volume of logs removed, the rates being different for different species or species groups. Beside premium and royalty, a silvicultural 'cess' is also charged, based upon the volume of timber removed. This charge is to provide funds for silvicultural treatment after logging.

In actual logging control, the logging area is demarcated on the ground into blocks with clearly marked boundaries. The logger is free to put in access roads, feeder roads and skidding roads. However he has to comply with the road alignments and conditions as specified in his logging permit. Logging is controlled by size class and species, and differs with the silvicultural system used. Any tree or log which is considered by the Forestry Department to have been felled but not removed, is recorded during a 'closing report' evaluation, carried out after logging. This encourages the logger to remove as much as is legally required of him, as royalty and silvicultural cess are charged on anything recorded in the closing report. While the logging permit has specifications on roading, riverine protection strips, damage on residual trees, etc., control is mainly based on volume and species removed. Other conditions are seldom strictly enforced.

Logging of forests outside Forest Reserves is not strictly controlled and cutting limits are not enforced, on the premise that these areas are not for long term management but are destined for conversion sooner or later. Nevertheless, all forest charges mentioned earlier are levied. Such forest areas have been major sources of timber, but are now depleted. Logging will soon be limited wholly to Forest Reserves.

Forest Management plans used to be drawn up for some States in the past (Ismail, 1965), but have not become a regular practice. This has been understandable, due to the uncertainty of tenure in the 1960s and 1970s. Timber complexes, such as the Jengka Timber Complex in Pahang, draw up their own management plans which require the approval of the State Forestry Departments.

Now that the PFE has in principle been adopted and the area determined on a State by State basis, and the annual coupe by State has been agreed upon, it is imperative that management plans be drawn up for each State. Such plans should document the existing resource in each State and ensure that logging is based on the capacity of the resource to renew itself on a sustained yield basis. Some States may find it difficult to adopt this approach but with a management plan, the problem could be highlighted, attention focussed on the long term implications. In large forested States, working plans or logging plans should be drawn up. The current practice of maintaining an Annual Felling Plan record for each Forest Reserve, while enabling the monitoring of each Forest Reserve by compartment, does not reflect the overall linkages and dependencies of which managers should be constantly aware, but which a management plan and a working plan would consider and emphasize.

10.6. SILVICULTURE

Silviculture, which may be defined as the theory and practice of growing trees, forms an important component of forest management. The system of silviculture used is influenced by the nature of the trees and forest and by the nature of human demands upon the forest. Changing demands have to a large extent been responsible for major changes in the silviculture systems practised in Peninsular Malaysia.

10.6.1. Regeneration improvement fellings (1900–1940)

Before 1940, all timbers extracted were used locally. Foxworthy (1921) estimated that the annual total wood consumption in the Peninsula at that time was about 8 million cubic metres, of which 90% was used as fuel, roughly in the proportion of two-thirds as domestic firewood and one-third as industrial fuel to power steam engines in the mining, rubber, transport and manufacturing industries. The mining industry also consumed wood in the form of poles. For construction and for railway sleepers, only the naturally durable heavy hardwood species, principally chengal *Neobalanocarpus heimii* and merbau *Intsia palembanica*, were in demand. In addition, there was strong overseas demand for gutta percha, obtained by tapping the bark of the tree *Palaquium gutta*. Gutta percha was used industrially as a coating to insulate submarine telegraph wires. Gutta percha's brief fling with high technology came to an end with the development of plastics.

The demand situation for forest products during the period 1900–1940 can be summarized as follows:

(a) Strong overseas demand for gutta percha, produced almost entirely by one species of tree, *Palaquium gutta*.

(b) Small local demand for naturally durable heavy hardwoods for construction and for railway sleepers. Such demand was highly specific and limited almost entirely to two species, chengal and merbau.

(c) Moderate local demand for fuelwood and poles. Such demand was non-specific and could be met by cutting any species or mixture of species.

In order to increase the production of gutta percha from forests, the practice was adopted of gradually eliminating other vegetation by girdling and cutting operations, so as to free gutta percha trees and seedlings from competition. The gutta percha seedlings respond with improved rates of survival and growth. This practice was then extended to favour heavy hardwood species and species of meranti (*Shorea*), with the same encouraging results. In fact, not only did seedlings of the favoured species respond with improved survival and growth, but the density (stocking) of the favoured species was also observed to increase. This was due to the improvement of forest floor conditions for newly produced seeds of the favoured species to germinate and grow. In this way, the so-called Regeneration Improvement Felling (RIF) system of silviculture came into being, whereby the felling of unfavoured trees and other vegetation was carried out with the aim of improving the regeneration of the favoured species.

The unwanted trees (consisting mainly of what we now call 'lesser-known species') that were felled under RIF were, in some cases, consumed as fuelwood and poles, for which there was demand in the more densely populated west coast states and in tin-mining areas. Otherwise the Forest Department had to finance regeneration improvement fellings from its own budget. Consequently, despite marked improvements in the stocking of commercial species, RIF was never practised on an extensive scale.

10.6.2. The Malayan Uniform System (1949–1980)

World War II and the Japanese occupation in 1942–1945 resulted in extensive uncontrolled felling and exploitation of the high forests. However, assessment and evaluation of these forests after the war revealed that such disturbed areas often had adequate or even abundant regeneration especially of meranti (*Shorea*) species. The ability of meranti seedlings to survive and thrive after a single heavy felling of mature trees gave rise to the idea that wherever a good stocking of meranti seedlings was already present, the mature trees could be logged in one operation without impairing the ability of the forest to regenerate. Thus was born the Malayan Uniform System (MUS) in 1949 and the formulation of the Linear Sampling (LS) techniques (Ismail Haji Ali, 1966).

The Malayan Uniform System involved the removal of mature trees in one felling operation; any mature trees not felled were poison-girdled. The removal of the upper canopy ('canopy opening') allowed more light to penetrate, and so stimulate the growth of seedlings. Since the starting time for all seedlings in one working area of forest would be the same, in theory the next crop would mature at uniform age and size.

For all this to work, there had to be seedlings of the desired or acceptable species already present prior to logging. Linear sampling techniques were devised whereby estimates of seedling stocking were made along parallel lines in the forest. In principle, logging was not permitted unless an area had been checked in this way and found to bear an adequate stocking of seedlings. Linear sampling was also supposed to be carried out periodically after logging to monitor the growth of the seedlings through sapling, pole and young tree stages, in order that remedial treatments could be prescribed as necessary.

The adoption of the MUS was made possible not only because it could work silviculturally, but also because of other factors:-

(a) Introduction of the chain saw and the 'Sun Tai Wong' (=King of the Hills), 4-wheel-drive winch lorries, modified from war-surplus vehicles. From then on, logging became a mechanized business and it became possible to log a whole block of forest in one quick operation, i.e., within a year or so, to ensure the uniform seedling response desired under the MUS.

(b) Receptivity of the European market towards a wide range of Malaysian timbers especially dipterocarps (principally red, white and yellow meranti, keruings, kapur, mersawa, balau) for which there had been little or no demand before the war.

(c) Introduction of chemical preservatives, which improved the acceptability of timbers that were of low natural durability.

(d) Three decades of research in the botany and wood properties of dipterocarps and other species which provided the technical basis for the Malayan Grading Rules and paved the way for world-wide acceptance of Malaysian timbers.

Unfortunately, the Emergency period of 1948–1959 prevented the full scale implementation of the MUS. Logging was determined more by security factors rather than by linear sampling, and sampling often could not be carried out. Logging was often prolonged, and often no post-felling monitoring and treatment were carried out.

When it was adhered to in meranti-rich forests, the MUS worked like a charm. But no thought was given to meranti-poor lowland forests, to hill forests and other non-meranti forests. MUS cannot be applied if adequate regeneration is not present at time of felling and, once the axe is put to the tree, a whole series of interdependent operations are set in motion. Felling of the economic crop and poison girdling of uneconomic trees are not two independent activities but rather one single operation of canopy opening. Oftentimes, attention was focused to the meranti-rich forests, and the meranti-poor forests, which need more attention, were ignored.

10.6.3. The Selective Management System (Since 1977)

A form of the Selective Management System came into use on a small scale in Jengka, Pahang, in 1977. It was adopted in Terengganu in 1979 and subsequently by other states in the Peninsula. At the same time MUS was phased out or modified to such an extent that it bore little resemblance to the original formulation. Nevertheless a modified form of MUS is still said to be practised wherever the SMS has not been officially adopted.

The factors that led to the formulation of SMS were as follows:

(a) Regeneration based on seedlings was considered to take too long. Most timber species take 50–80 years to reach commercially harvestable size of 50 cm diameter or more.

(b) The easily accessible lowland forests were lost to agriculture. Forestry shifted to the hills where, because of the heterogeneous habitat, seedling regeneration was patchy and difficult to manage.

(c) Sodium arsenite, the cheapest and most effective of the tree poisons, was banned for agriculture and its use for poison-girdling unwanted trees came to an end. Under MUS, all big trees still left standing after commercial logging (mostly lesser-known species) had to be poisoned in favour of a uniform crop of seedlings. If interpreted strictly even 'advanced growth', i.e., trees intermediate in size between seedlings and fully mature trees, also had to be poisoned.

(d) Changing local and overseas demand for tropical hardwoods, increased costs of logging, improved processing and dwindling forest resources, resulted in a wider range of species and size classes being utilized. A number of species earlier considered non-commercial (and hence poison-girdled) were now marketable.

Under the SMS, emphasis turned from seedlings to advanced growth as the basis of the next harvest. In principle, a logger would be permitted to cut trees above a certain minimum diameter limit, say 45 cm. Of the advanced growth left behind, some trees would be quite advanced, say 30–45 cm diameter. If these could grow at, say, 1 cm diameter per annum, they would be ready for harvesting in 25–30 years. By that time there would be another back-up batch of advanced growth.

For the SMS to work, two important conditions must apply. Firstly the forest must carry an adequate stocking of currently commercial trees of all sizes. Only then will it be possible to fix a minimum felling size in such a way that the logging operation is commercially viable above that limit, while enough advanced growth of the correct sizes and species are saved below that limit to guarantee the commercial viability of the next cut. Clearly, the fixing of the minimum felling size is the most important decision under the SMS. A lower size limit would allow more to be taken from the forest this time, but the residue would take longer to grow to commercial size.

Secondly, the advanced growth must be able to grow and develop at the necessary rate after the removal of the larger trees. This is not something that can be taken for granted, because many of the advanced-growth trees will have already been suppressed for many years by the proximity of the large trees. Many will also have sustained damage to their crown, bark, or roots during the felling and extraction of the large trees.

In principle, however, SMS should confer the following advantages:
(a) Reduction in cutting cycle, from 50–80 years to 25–30 years.
(b) Reduction in silvicultural costs.
(c) Less damage to the environment.
(d) Less wastage of timber (from poisoning of unwanted trees).

10.7. CONCLUSION

Management and silviculture in Peninsular Malaysia can best be described as being in a state of transition. Under such conditions, caution is prudent and logical. As the success of SMS depends on advanced growth, or 'residual trees', logging takes on a new role and responsibility, and truly becomes the first silvicultural operation. Its function will be to ensure adequate and appropriate opening up of the canopy to allow for growth and development of residuals. While this is also true under MUS (but with emphasis on seedlings), there is the added responsibility of minimizing logging damage to the residuals which will become the next crop. The adoption of good logging practices and greater supervision become critically important and new regulations may need to be instituted. The number, species and condition of trees left after logging form the basic ingredients to a successful SMS.

This responsibility should not be left entirely to the forest manager but should also be borne by the logging industry, if not the whole timber industry as well. Loggers should be called upon to assist in the implementation of the SMS. The key role that they play needs to be communicated and emphasized. While the large timber complexes may have been informed, most other loggers in the country have not been made aware of their responsibility in ensuring the continued well-being of the forest through proper execution of the SMS. The logging industry, and indeed the whole timber industry, must take account of the complexity of managing the natural forests, the aims and objectives of management, and the importance of the extractor's role in safeguarding the regenerative capacity of the forests for future generations.

Forest managers, on the other hand, must ensure that silvicultural operations, whether by Departmental staff or by contract, are actually carried out. Natural forest silviculture can be very subjective and open to individual interpretation. This is particularly true in the implementation stage on the ground, where control and monitoring of work is difficult, due to the nature of the environment, slow response to treatment and problems of accessibility.

With the impending projected shortfall in forest resources supply in the early 1990s, it is inevitable that the pressure on the natural forests to supply more logs as quickly as possible will be greater. This will be particularly true in areas close to industries, such as the west coast states of the Peninsula. Indeed this pressure is already felt by some of these states, such as Selangor. Choice of species, while still remaining important, will become a secondary issue, as will size of logs. We are confident that the industry can and will adapt to changing resource supplies. This is already taking place. Branchwood could become a supplementary resource. Thinning may become necessary before the complete felling cycle.

Environmental concern and the conservation lobby will grow stronger. Silviculture today must direct itself to the resource demands of the future.

Related to all this is the needling question of whether silviculture is worth the cost at all; whether it is just as well to leave the forests by themselves after logging; a do-nothing silviculture? Emphasis and effort could then be directed at controlling logging and extraction, to minimize damage to residual stands and to ensure adequate residual stocking. There is merit in this notion, but data are still inadequate to reach a definite conclusion. Existing growth figures after logging are not encouraging, as even a 1 cm dbh annual growth for 30 cm dbh trees is difficult to attain. This means that a 30 cm dbh tree can hardly be expected to attain 60 cm dbh after 30 years, assuming the growth rate is maintained.

While the circumstances in Peninsular Malaysia have been discussed in these pages, we hope that our experiences can be used as guidance for our neighbours, who have the same aspirations and dreams of managing the natural forest. Under the conditions that prevail generally in the region, the problems, and the discouraging statistics, one has to look closely at the merits and demerits of natural forest silviculture. This dilemma is aptly reflected by Leslie's (1977) statement, "The best reasons for not completely abandoning natural management of moist tropical forests lie in the insurance it provides against the distinct possibility that decisions based on it being an uneconomic proposition could be mistaken."

REFERENCES

Anon., (1973). A National Forest Inventory of West Malaysia, 1970–1972. *FO: DP/MAL/72/009 Technical Report 5*, UNDP/FAO, Rome.

Anon., (1978). *Dasar Perhutanan Negara (National Forest Policy)*. Forestry Department, Kuala Lumpur.

Anon., (1984). *Laws of Malaysia Act 313*. National Forestry Act 1984.

Foxworthy, F. W. (1921). Commercial Woods of the Malay Peninsula. *Malay. Forest Record* 1.

Ismail Haji Ali (1964). The impact of land development on forest reservation and management in Selangor. *Malay. Forester*. 28, 264–70.

Leslie, A. (1977). Where contradictory theory and practice co-exist. *Unasylva* 29(115), 2–17.

Tang, H. T. and Wan Razali Mohd. (1981). *Report on growth and yield studies in inland mixed indigenous forests in Peninsular Malaysia*. Unpubl. report compiled for FAO. Dept. of Forests, Peninsular Malaysia.

Wyatt-Smith, J. (1963). Manual of Malayan Silviculture for Inland Forest. *Malay. Forest Record* 23. Vols. 1 and 2.

Completed September 1985

CHAPTER 11

Mammals: Genetic Diversity and Evolution

H. S. Yong

Department of Zoology, University of Malaya, 59100 Kuala Lumpur, Malaysia

CONTENTS

11.1. INTRODUCTION

The mammal fauna of Peninsular Malaysia is diverse in size, habits and taxonomic units (see Chapter 12). This introductory chapter is concerned with diversity due to genetic factors, which arises as a result of the processes of mutation (including both genic and chromosomal changes). Measures of genetic diversity may be based on the number of families in an order, the number of genera in a family, the number of species in a genus, and genic differentiation among populations/races/subspecies/species.

11.2. DIVERSITY AT THE HIGHER TAXONOMIC LEVEL

Diversity in taxonomic units is primarily the result of cladogenetic evolution (i.e., a phylogenetic lineage splitting into two or more independently evolving lineages), which results in adaptation to a greater variety of 'niches' (or ways of life).

The 11 orders of mammals in Peninsular Malaysia are not individually represented by the same number of families, genera, or species (Table 11.1). It is evident that the bats (Chiroptera), represented by 8 families, form the most diverse order, followed successively by the carnivores with 5 families, the rodents and artiodactyls or even-toed ungulates each with 4, the insectivores and primates (3), perissodactyls or odd-toed ungulates (2) and the remaining four orders with a single family each.

TABLE 11.1. The taxonomic diversity of mainland Peninsular Malaysian mammals.

Order	No. of families	No. of genera	No. of species
Insectivora	3	6	9
Dermoptera	1	1	1
Chiroptera	8	32	83
Tupaioidea	1	2	3
Primates	3	4	10
Pholidota	1	1	1
Rodentia	4	29	54
Carnivora	5	19	29
Proboscidea	1	1	1
Perissodactyla	2	3	3
Artiodactyla	4	6	9
Total (11)	33	104	203

Figures are derived from Medway (1983), updated

At the family level, only five families are each represented by ten or more genera, while the remainder are represented by less than five genera each (Tables 11.1 and 11.2). The more diverse are the bat families Pteropodidae (fruit bats) and Vespertilionidae (common bats), the rodent families Sciuridae (squirrels) and Muridae (rats and mice), and the carnivore family Viverridae (civets and mongooses). At the generic level, only six genera contain five or more species. These are the horseshoe bats *Rhinolophus* (12 species) and *Hipposideros* (11 species), the common bats *Myotis* (7 species) and *Pipistrellus* (6 species), the true rats *Rattus* (6 species), and the diurnal squirrels *Callosciurus* (5 species). Families represented by 10 or more genera comprise 15.6% (5/32), and genera represented by 5 or more species comprise 5.8% of the total (6/104).

11.3. INTERSPECIFIC DIVERSITY

As in other outbreeding sexual organisms, a species of mammal may be defined as an array of Mendelian populations, the individuals of which share a common gene pool and are capable of interbreeding to produce viable and fertile offspring. Arrays which are reproductively isolated constitute different species. In nature, the mechanisms which ensure reproductive isolation may be pre-mating (e.g., mechanical, behavioural, ecological isolation) or post-mating (e.g., gametic or zygotic mortality, hybrid inviability or sterility).

Natural populations contain a large amount of variability in qualitative and quantitative characters. The variations can either be genetically or environmentally induced, or can be the product of a combination of these two factors. Quantitative characters, such as stature, are generally affected by both

TABLE 11.2. Summary of the more diverse orders of Peninsular Malaysian mammals.

Order	Family	No. of genera	No. of species
Chiroptera	Pteropodidae	11	15
	Emballonuridae	2	4
	Nycteridae	1	1
	Megadermatidae	1	2
	Rhinolophidae	1	12
	Hipposideridae	2	13
	Vespertilionidae	11	32
	Molossidae	3	4
Rodentia	Sciuridae	14	25
	Rhizomyidae	1	2
	Muridae	11	24
	Hystricidae	3	3
Carnivora	Canidae	1	1
	Ursidae	1	1
	Mustelidae	4	5
	Viverridae	10	15
	Felidae	3	7

genetic and environmental factors, while genetic factors alone almost always determine the variation in qualitative characters such as gene–enzyme systems and other biochemical markers. Such genetic variations result from genic variation at the DNA level.

Interspecific biochemical genetic diversity has been amply demonstrated in the murid rodents of Peninsular Malaysia. In rats of the genus *Rattus*, the albumin of the Polynesian rat *R. exulans* exhibits unique starch-gel electrophoretic mobility (Yong, 1972a). The albumins of the house rat *R.r. diardii*, the Malaysian wood rat *R. tiomanicus* and the ricefield rat *R. argentiventer* possess equal migration rate in starch gel but are not identical, as shown by immunoelectrophoresis (Yong and Dhaliwal, 1980). Interspecific differences are also found in various gene-enzyme systems (Chan, 1977). Similar results have been obtained in species of other murid genera, e.g., *Leopoldamys, Niviventer* and *Maxomys* (Chan, Dhaliwal and Yong, 1978, 1979; Yong, 1972), but it appears that the leaf-monkeys *Presbytis* spp. do not exhibit species-characteristic biochemical markers (Yong *et al.*, 1982).

The degree of genetic diversity serves as a measure of the affinity of the animals under consideration. Closely related species show a high coefficient of similarity and a small genetic distance. Conversely, a small coefficient of similarity or a large genetic distance indicates that the species are distantly related.

Interspecific diversity is also manifested in the chromosome complement. The diploid number may or may not be constant within a genus (e.g., Yong, 1969; Yong *et al.*, 1971, 1973. 1975, 1976). Likewise, the chromosome complement may in some cases be distinctive, but in others indistinguishable between related species. Examples illustrating chromosomal diversity in Peninsular Malaysian mammals are summarized in Table 11.3.

TABLE 11.3. Chromosome complement of some Peninsular Malaysian mammals.

Genus	Species	Diploid number	Pairs of Autosomes			Sex Chromosomes	
			m + sm	sa	a	X	Y
1. **Bats**							
Cynopterus	C. brachyotis	34	11	2	3	sa	a
	C. horsfieldi	34	11	2	3	sa	a
Macroglossus	M. minimus	34	12	2	2	m	a
	M. sobrinus	34	12	2	2	m	a
Tylonycteris	T. pachypus	46	4	2	16	a	m
	T. robustula	32	11	2	2	a	m
2. **Squirrels**							
Callosciurus	C. caniceps	40	7	8	4	sm	a
	C. nigrovittatus	40	8	8	3	sm	a
	C. notatus	40	7	8	4	sm	a
	C. prevostii	40	7	8	4	m	a
3. **Rats**							
Rattus	R.r. diardii	42	7	2	11	A	A
	R. tiomanicus	42	7	4	9	A	A
	R. argentiventer	42	7	2	11	S	A
	R. norvegicus	42	7	4	9	A	A
	R. exulans	42	7	2	11	A	A
	R. annandalei	42	6	3	11	A	A
Sundamys	S. muelleri	42	6	2	12	S	A
Berylmys	B. bowersii	40	8	5	6	A	A
Leopoldamys	L. edwardsi	42	3	4	13	A	A
	L. sabanus	42	2	4	14	A	A
Niviventer	N. cremoriventer	46	3	1	18	A	A
	N. rapit	46	3	4	15	A	A
	N. bukit	46	3	1	18	A	A
Maxomys	M. whiteheadi	36	8	9	0	M	A
	M. inas	42	10	10	0	M	A
	M. rajah	36	6	3	8	M	M
	M. surifer	52	4	2	19	M	M

11.4. INTRASPECIFIC DIVERSITY

Intraspecific genetic diversity is usually measured by the proportion of polymorphic loci and the average heterozygosity per locus. A locus is considered polymorphic if the frequency of the commonest allele is equal to or less than 0.99 or 0.95, depending on the definition used. The heterozygosity (h) at a locus is defined as $h = 1 - X_i^2$ where X_i is the frequency of the ith allele.

The best parameter for measuring genic variation is average heterozygosity (H) or gene diversity, which is the mean of the heterozygosities of all the loci studied. This is evident in the data for biochemical variation in mainland species of murid rodents (Table 11.4). For example, the house rat and the Polynesian rat which have equal proportion of polymorphic loci (two out of nine loci), have quite different gene diversity – 0.092 for the house rat and 0.031 for the Polynesian rat. The gene diversity seen in Peninsular Malaysian rats is similar to that in rodents from other parts of the world. The low level of gene diversity has been generally attributed to the relatively small effective population sizes.

TABLE 11.4. Intraspecific biochemical diversity in some mainland Peninsular Malaysian rats (after raw data of Chan, 1976).

Species	Proportion of polymorphic loci	Average heterozygosity
Rattus r. diardii	2/9	0.092
Rattus tiomanicus	1/9	0.059
Rattus argentiventer	0/9	0
Rattus exulans	2/9	0.031
Rattus norvegicus	1/9	0.061
Rattus annandalei	2/9	0.044
Sundamys muelleri	3/9	0.022
Berylmys bowersii	1/9	0.013
Niviventer cremoriventer	1/9	0.013
Niviventer rapit	0/9	0
Niviventer bukit	1/9	0.02
Maxomys rajah	1/9	0.032
Maxomys surifer	1/9	0.048
Maxomys whiteheadi	3/9	0.068
Maxomys inas	1/9	0.042
Leopoldamys sabanus	1/9	0.044
Leopoldamys edwardsi	0/9	0
Lenothrix canus	4/9	0.055

Heterozygosity may also vary considerably with locus. For instance, in the grey tree rat *Lenothrix canus* the heterozygosity for adenylate kinase is 0.32 while those for phosphogluconate dehydrogenase, malate dehydrogenase and haemoglobin are 0.058.

As in the case of interspecific diversity, intraspecific diversity is also manifested in the chromosome complement, so that the karyotype may vary from individual to individual. In the house rat, the longest autosome is represented by either a subacrocentric or an acrocentric element (Yong, 1972b). This gives rise to three possible combinations of the longest pair of autosomes, viz. homomorphic acrocentric (A/A), homomorphic subacrocentric (S/S), and heteromorphic (A/S comprising one acrocentric and one subacrocentric element). The frequency of the three karyotypic classes may differ between populations.

In addition to the morphological difference caused by pericentric inversion in autosome pairs No.1 and No.13, the karyotype may also differ numerically in the house rat, with diploid chromosome number ranging from 41 to 45. Several mechanisms are responsible for numbers other than the standard 42. Additional chromosomes are accessory or B-chromosomes (Yong, 1972b; Yong and Dhaliwal, 1972), and one B-chromosome is found in the majority of rats with an increased chromosome number. Reduction in number from 42 may be due to the loss of one chromosome, or to the fusion of two acrocentric chromosomes to form a biarmed one. Three female rats have so far been recorded to have a 41, XO chromosome complement (Yong, 1971a), and only a single rat has been reported to be characterized by fusion (Yong, 1971b).

Numerical variability in the chromosome complement has also been found in the house shrew *Suncus murinus*, in this case ranging from 35 to 40. The mechanism responsible for this diversity is the fusion of non-homologous acrocentric chromosomes (Yong, 1971c, 1972c, 1973; Sam *et al.*, 1979). In addition, the Y-chromosome of the house shrew is variable in size and morphology (Yong, 1974).

Another interesting intraspecific chromosome variation is the numerical difference between male and female individuals of the barking deer or muntjak *Muntiacus muntjak*. The male possesses 9 chromosomes while the female has 8 chromosomes.

11.5. EVOLUTION OF DIVERSITY

Evolution is the process whereby changes in the gene pool are affected. Due to mutation and natural selection, the genetic make-up of a population or species will change gradually, i.e., will evolve. The genetic change may or may not result in the immediate appearance of new species or higher categories. Nonetheless, over a very long period of time, successive generations may change sufficiently in genetic composition to be regarded as a different species. Likewise, if populations are isolated for a sufficiently long period, they may acquire enough genetic differences to become separate species. Geographical isolation of populations, coupled with the gradual accumulation of genetic changes (mutations), is an important mechanism of cladogenic speciation. Other processes, such as chromosomal changes and interspecific hybridization, also play a role in speciation.

In the flat-headed bats, of the genus *Tylonycteris*, the two species have very different chromosomal number: 2n = 46 in the lesser flat-headed bat *T. pachypus*, and 2n = 32 in the greater flat-headed bat *T. robustula*. The difference is due to Robertsonian translocations (Yong *et al.*, 1971). It is presumed that this chromosomal change serves as an effective isolating mechanism.

Similar variations in chromosome number have also been reported for the species pair of Whitehead's rat *Maxomys whiteheadi* and Malayan mountain spiny rat *Maxomys inas* with 2n = 36 and 42 respectively, and the species pair of brown spiny rat *Maxomys rajah* and red spiny rat *Maxomys surifer* with 2n = 36 and 52 respectively. The numerical differences in these two cases, however, cannot be readily explained by simple mechanisms.

There are many clear-cut examples of chromosomal evolution in Peninsular Malaysian mammals. In the mouse-deer, both the lesser mouse-deer *Tragulus javanicus* and the large mouse-deer *Tragulus napu* have 32 chromosomes (Yong, 1973b, and 1973c). They differ only in the morphology of a pair of autosomes, which is the result of pericentric inversion (c.f. Table 11.3). In many other cases chromosomal changes have not resulted in speciation. Two good examples are found in the house rat and the house shrew (Section 11.4).Unlike chromosomal changes, the amount of genic changes can be readily quantified. This is usually expressed as genetic identity and genetic distance. For example, the genetic identity between the species pair of Whitehead's rat and Malayan mountain spiny rat is 0.671 while the genetic distance is 0.399 (calculated from data of Chan, 1976) but the exact degree of chromosomal change cannot be easily ascertained. Similarly, the species pair of brown and red spiny rats, which possess distinct karyotypes, have a genetic identity of 0.111 and a genetic distance of 2.198. These values indicate the genetic relationship or the extent of divergence between the taxa compared.

In general, closely related taxa have a small genetic distance while a large genetic distance indicates a greater degree of differentiation. In the above examples, the genetic distance between the brown and red spiny rats indicates that these two species probably belong to different genera. It has been shown that these two species are also markedly differentiated from the Whitehead's and Malayan mountain spiny rats (Chan *et al.*, 1978), and it is possible that the existing generic classification ought to be reviewed.

11.6. GENETIC RESOURCES

The genetic resources of Peninsular Malaysian mammals (and also other wildlife) are potentially of economic, recreational, aesthetic and scientific values (Yong, 1979). Many mammals are hunted for food, including the deer, mouse-deer, squirrels, bamboo rats, flying foxes, pigs and civet cats, and (illicitly) serow, pangolin and seladang. Other mammals continue to be of economic importance for other reasons. For instance, fruit bats play an important role in pollinating certain plant species, such as the durian pollinated by the cave fruit bat (*Eonycteris spelaea*). Bat guano has been utilized as fertilisers. There is a potential in utilizing these genetic resources in the development of game ranching, or for the development of new domesticated or semi-domesticated species. Some mammals also serve as good models in fields of biological and medical research. Malaysia will have a great deal to contribute to science and for the benefit of mankind if these resources are carefully and intelligently managed.

REFERENCES

Chan, K. L. (1976). *Biochemical systematics of Malayan rats (Rodentia : Muridae, genus Rattus Fischer)*. Ph.D. thesis, University of Malaya, Kuala Lumpur.

Chan, K. L. (1977). Enzyme polymorphism in Malayan rats of the subgenus *Rattus*. *Biochem. Syst. Ecol.* 5, 161–68.

Chan, K. L., Dhaliwal, S. S. and Yong, H. S. (1978). Protein variation and systematics in Malayan rats of the subgenus *Lenothrix* (Rodentia:Muridae, genus *Rattus* Fischer). *Comp. Biochem. Physiol.* 59B, 345–51.

Chan, K. L., Dhaliwal, S. S. and Yong, H. S. (1979). Protein variation and systematics of three subgenera of Malayan rats (Rodentia:Muridae, genus *Rattus* Fischer). *Comp. Biochem. Physiol.* 64B, 329–37.

Medway, Lord (1983). *The Wild Mammals of Malaya (Peninsular Malaysia) and Singapore*. 2nd edition, with corrections. Kuala Lumpur, Oxford University Press.

Sam, C. K., Yong, H. S. and Dhaliwal, S.S. (1979). The G- and C-bands in relation to Robertsonian polymorphism in the Malayan house shrew, *Suncus murinus* (Mammalia, Insectivora). *Caryologia* 32, 355–63.

Yong, H. S. (1969). Karyotypes of Malayan rats (Rodentia-Muridae, genus *Rattus* Fischer). *Chromosoma* 27, 245–67.

Yong, H. S. (1971a). Presumptive X-monosomy in Black rats from Malaya. *Nature* 232, 484–5.

Yong, H. S. (1971b). Centric fusion in the Malayan house rat, *Rattus rattus diardii* (Rodentia, Muridae). *Experientia* 27, 467–8.

Yong, H. S. (1971c). Chromosome polymorphism in the Malayan house shrew, *Suncus murinus* (Insectivora, Soricidae). *Experientia* 27, 589–91.

Yong, H. S. (1972a). Starch-gel electrophoretic patterns of the serum albumin of Malayan rats (Rodentia-Muridae, genus *Rattus* Fischer). *Malaysian J. Sci.* 1(A), 7–12.

Yong, H. S. (1972b). Population cytogenetics of the Malayan house rat, *Rattus rattus diardii*. *Chromosomes Today* 3, 223–7.

Yong, H. S. (1972c). Robertsonian translocations in the Malayan house shrew, *Suncus murinus* (Insectivora, Soricidae). *Experientia* 28, 585–6.

Yong, H. S. (1973a). Cytogenetics of the house shrew, *Suncus murinus* (Mammalia, Insectivora). *Genetics* 74, s 303.

Yong, H. S. (1973b). Complete Robertsonian fusion in the Malaysian lesser mouse-deer (*Tragulus javanicus*). *Experientia* 29, 366–7.

Yong, H.S. (1973c). The mouse-deer of Pulau Tioman. *Malay. Nat. J.* 26, 172–4.

Yong, H. S. (1974). Geographic variation in the sex chromosomes of the west Malaysian house shrew *Suncus murinus* (Insectivora, Soricidae). *Caryologia* 27, 65-71.

Yong, H. S. (1979). Genetic resources of terrestrial animals in Malaysia. *Malaysian Appl. Biol.* 8, 67–71.

Yong, H. S., Dhaliwal, S.S. and Teh, K.L. (1971). Somatic chromosomes of the flat-headed bats (*Tylonycteris* spp.). *Experientia* 27, 1353–55.

Yong H.S. and Dhaliwal, S. S. (1972). Supernumerary (B-) chromosomes in the Malayan house rat, *Rattus rattus diardii* (Rodentia, Muridae). *Chromosoma* 36, 256–62.

Yong, H. S., Dhaliwal, S. S., Lim, B. L., Teh, K. L. and Start, A. N. (1973). Uniformity in the karyotypes of the fruit bats *Cynopterus* (Mammalia:Chiroptera, Pteropidae). *Malaysian J. Sci.* 2(A) 19–23.

Yong, H. S., Dhaliwal, S. S., Lim, B. L., Muul. I. and Teh, K. L. (1975). Karyotypes of four species of *Callosciurus* (Mammalia, Rodentia) from Peninsular Malaysia. *Malaysian J. Sci.* 3(A), 1–5.

Yong, H. S. and Dhaliwal, S. S. (1976). Chromosomes of the fruitbat subfamily Macroglossinae from Peninsular Malaysia. *Cytologia* 41, 85–9.

Yong, H. S. and Dhaliwal, S. S. (1980). Immunoelectrophoretic serum-protein patterns of three commensal rats in Peninsular Malaysia. *Malaysian J.* Sci. 6(A), 5–7.

Yong, H. S., Dhaliwal, S. S. and Mak, J. W. (1982). Glucose phosphate isomerase and 6-phosphogluconate dehydrogenase polymorphism in Malaysian leaf monkeys. *Primates* 23, 312–5.

Completed September 1985

CHAPTER 12

Mammals: Distribution and Ecology

Earl of Cranbrook

Glemham House, Great Glemham, Saxmundham, Suffolk IP17 1LP, U.K.

CONTENTS

12.1. INTRODUCTION

In terms of the number of native wild species, the mammal fauna of Peninsular Malaysia shows the richness and diversity characteristic of tropical regions (Chapter 11). Denmark provides a temperate zone comparison, also a peninsula projecting from the Eurasian continental landmass and not much smaller in area. Respective figures for the numbers of all land mammals are as follows: species (PM) 203 / (D) 45, genera 104/32, families 32/13; and for selected groups: bats 83/12, rodents 54/14, carnivores 29/8. With the exception of the non-cavernicolous bats, which were difficult to collect before mist-nets and harp traps came into general use, by the start of World War II the mammal fauna of Peninsular Malaysia was adequately known, chiefly from the collections made by staff of the (then) Raffles Museum, Singapore, and the Federated Malay States museums at Taiping and Kuala Lumpur. This initial phase culminated in the publication of F.N.Chasen's (1940) *Handlist*. Taxonomic revision has since changed many of the names used by Chasen, but his list remains a key source for mammalogists of the region. After the war there followed a second phase, of ecologically oriented investigation using the cage trap as the principal tool. This was initiated at the Institute for Medical Research, Kuala Lumpur, by J.L.Harrison in the late 1940s and was later continued by others, notably by B.L.Lim and I.Muul (see bibliography). The underlying objective of I.M.R. research was to investigate the parasites of mammals, particularly in relation to zoonoses and 'arbor' (*arthropod-borne*) viruses. The by-product was a great augmentation of knowledge of the smaller mammals, with useful information on subjects such as distribution, reproductive periodicity, growth rates, life-spans, individual ranges, etc.

A third phase in mammalogical studies followed the opening of universities in Singapore and Peninsular Malaysia. Staff, students and overseas visitors have all contributed. University research has tended to follow individual special interests, covering a wide variety of subjects, including studies of single species and general topics such as growth rates, reproductive periodicity, or immunology and karyology (Chapter 11).

In its early years, what was then called the Game Department traditionally attracted officers interested chiefly in mammals, several of whom published their observations from time to time. Latterly, with the transformtion to a National Park and Wildlife service and the injection of substantial research funds, conservation-related studies of selected large mammal species have been undertaken (Chapter 17).

An introductory handbook of local mammals was produced by Harrison in 1966, and a more comprehensive field guide by Medway in 1969. The latter has been up-dated in subsequent editions; the third (Medway, 1983) contains a general bibliography of 280 entries. Tweedie's (1978) volume on mammals, in the Malaysian Nature Handbooks series, gives general coverage of the fauna of Peninsular Malaysia and the Bornean states of East Malaysia.

12.2. FOREST-ADAPTED MAMMALS

Steven (1968) tabulated the preferred habitats of Peninsular Malaysian mammals other than bats, concluding that 78% are obligatory forest-dwellers, 12% forest-dwellers which enter and 'can subsist' in cultivated areas and 10% confined to cultivated or urban habitat. Thus, in contrast to tropical Africa, where many mammals live in the open plains or savannah, the great richness of Malaysian mammals reflects a predominantly forest-adapted fauna. If 'forest' is defined in a very general sense, only nine (4.4%) of the 203 native wild mammals of Peninsular Malaysia have not been encountered in this habitat. These nine include three very rare bats of unknown habit, each recorded by a single specimen, and two common commensals of man, the house shrew *Suncus murinus* and house rat *Rattus diardii* (both of which have on occasion been trapped in secondary woodland and scrub). Among better known mammals, some that are normally associated with other habitats also freely enter adjoining forest: for instance, aquatic mammals, the rats of degraded scrub or grassland and the wild cattle.

12.3. ZOOGEOGRAPHY AND LOCAL DISTRIBUTION

The mammals making up this rich assemblage are related to and ultimately derived from the fauna of adjoining continental Asia. In zoogeographical terms, the Indian subcontinent, Southeast Asia and southern China together constitute the Oriental realm (Darlington, 1957). Only five Peninsular Malaysian forest mammal species (2.5% of the total forest fauna) appear to have world ranges that extend significantly beyond the limits of this realm. These are: two bats, *Miniopterus schreibersii* and *Myotis mystacinus*, a pigmy shrew *Suncus etruscus*, the leopard *Panthera pardus* and the Eurasian wild pig *Sus scrofa*. This list may, in fact, be still shorter since the taxonomy of the two bats is controversial, and close study may show that the Malaysian forms have local ranges and are not in truth representatives of widespread species.

Among the remaining forest mammals of Peninsular Malaysia, 107 species are also found over more or less extensive areas of the continental mainland of southern and eastern Asia. In addition to these, there are 76 species which do not occur (or not to any great extent) outside the lands of the Sunda shelf, i.e., Peninsular Malaysia, Sumatra, Borneo, Java and Bali (see Preface). In mammalogical terms, this area forms a coherent zoogeographical subregion which is inhabited by a distinctive assemblage of species, for which the term 'Sundaic' has been used.

A special class of Sundaic mammals consists of those endemic to Peninsular Malaysia, unknown outside its borders. The two forest-dwelling endemics are the Selangor pigmy flying squirrel *Petaurillus kinlochii* and Malayan mountain spiny rat *Maxomys inas*. The former is rare and little known, while the latter has a limited montane range (see below).

12.3.1. Latitudinal distributions : non-flying mammals

Most of the Sundaic species present in Peninsular Malaysia extend north of the political border, into the peninsular parts of Thailand and Burma. The transition zone between the Sundaic and the continental southeast Asian mammal fauna lies at about the Isthmus of Kera, lat. $10°30'$N (Chasen, 1940) as it does for birds (Chapter 13) and other vertebrate classes. The effective factors are probably complex. Climate is likely to be important, showing as it does a progressive transition from the perenially humid equatorial environment in the south to a seasonal, monsoon regime at higher tropical latitudes.

A few mammals have latitudinally restricted ranges within Peninsular Malaysia. Forest-adapted examples are the hoary bamboo rat *Rhizomys pruinosus* and the white-bellied rat *Niviventer bukit*, both of which extend extend from the northern border to about $4°$ N. Conversely, the bearded pig *Sus barbatus* has been found only in southern parts, from Johor north to about the same latitude. In none of these cases is there an identifiable topographical factor to account for the limited range.

Latitudinal variation in abundance has been recorded among two common squirrels, the grey-bellied, *Callosciurus caniceps* and plantain squirrel, *C. notatus*. Both frequent plantations and orchards as well as forest (especially disturbed or secondary forest). In such habitat, the grey-bellied squirrel is the more common in the north and the plantain squirrel in the south; relative representation in counts made in 1949–50 is shown in Figure 12.1.

Both these squirrels, and also Prevost's *C. prevostii*, the black-banded *C. nigrovittatus*, the common giant squirrel *Ratufa affinis*, the slender squirrel *Sundasciurus tenuis* and the three-striped ground squirrel *Lariscus insignis*, show clinal variation from north to south in the Peninsula, principally involving the pattern or shade of the pelage. The differences are sufficiently prominent in all species cited to have been recognized by the description of subspecies.

12.3.2. Other patterns

Two primates have restricted distributions that have been interpreted as evidence of geologically recent invasion from Sumatra. One, the silvered leaf monkey *Presbytis cristata*, is confined to a maritime

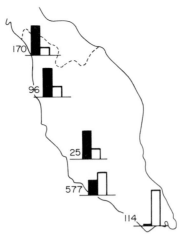

Fig. 12.1. The relative abundance of grey-bellied squirrels *Callosciurus caniceps* and plantain squirrels *Callosciurus notatus* counted in rural and suburban localities on the west coast in 1949–50. Derived from J.L.Harrison (1950, map 1), omitting data from Malacca town (100% plantain squirrel). Solid bar = grey-bellied, open = plantain squirrel; figures denote total sample size, all squirrels.

strip on the west coast, extending no more than 40 km inland. The other, the agile gibbon *Hylobates agilis*, occupies a sector of Perak and southern Kedah intervening between subspecifically distinct populations of lar gibbon *H. lar* (Chivers, 1980). The range of the agile gibbon appears to be limited by the physical barriers of the Perak and Muda rivers. The silvered leaf monkey in the Peninsula is largely confined to mangrove, riverine and freshwater swamp forest. Habitat preferences may be important in the control of its distribution in this part of the species' total range (Marsh and Wilson, 1981).

Another primate, the siamang *Hylobates syndactylus* (Fig. 12.2), is among a small assemblage of mammal species characteristic of forested hills and mountain slopes at moderate elevations (Table 12.1). In the Kerau Game Reserve, siamang occur in the lowlands and on Mt Benom up to at least 1800 m. Whereas lar gibbons become progressively scarcer with increasing altitude on the mountain, siamang numbers peak at 700–1000 m above sea level (Figure 12.3). Siamang (like the other *Hylobates* in the Peninsula) are territorial, and family groups remain within established home ranges. Caldecott (1980) found no evidence of seasonal shifts of territory on the hillside, and concluded that features of the siamang such as body size, thick fur and catholic diet are adaptations for the submontane habitat.

In the case of the marmoset rat *Hapalomys longicaudatus*, distribution has been linked with one principal factor: its dependence on a bamboo. This scansorial rat has been found in close association with the bamboo *Gigantochloa scortechinii*, which has a forest edge distribution on the flanks of the main range (Chapter 5), in Kelantan (at least) mostly at elevations above about 350 m (Medway, 1964). The marmoset rat finds shelter and nest sites chiefly, if not exclusively, in the internodal spaces of the bamboo, and in captivity does not thrive unless bamboo is provided in its diet (Fig. 12.4).

The lesser gymnure *Hylomys suillus* has occasionally been taken at moderate to low elevations in forested hills, but is included among the montane specialists listed in Table 12.1 because it is common only at upper elevations.

The full complement of specialized montane mammals can only be found in the central mountains of the Main Range. On the peripheral peaks, they are represented by a variable selection. Thus, seven out of nine occur on Mt Tahan, four on Benom, one on Tapis, Pahang, and one on Kedah peak (Gunung Jerai), but none has been found to the south on the mountains of Johor, nor on Lawit, Terengganu, the highest peak of the Eastern Range. This small assemblage of specialized non-flying montane mammals comprises seven species of continental affinity (some ranging to the Himalayas), one exclusively Sundaic and one endemic.

Fig. 12.2. An adult female siamang *Hylobates syndactylus*, feeding on the fruit of *Aglaia* sp. in the Ulu Gombak forest reserve. (Photograph by D. J. Chivers.)

TABLE 12.1. Peninsular Malaysian mammals (excluding bats) with altitudinally or topographically restricted ranges.

Species	Montane	Submontane
INSECTIVORA		
Long-tailed shrew, *Crocidura attenuata* 1, 5	+	
Lesser gymnure, *Hylomys suillus* 1, 4, 5	+	
Short-tailed mole, *Talpa micrura* 5	+	
PRIMATES		
Siamang, *Hylobates syndactylus*		+

(Table 12.1 continued)

Species	Montane	Submontane
RODENTIA		
Belly-banded squirrel, *Callosciurus flavimanus* 1, 2, 5	+	
Himalayan striped squirrel, *Tamiops macclellandii* 1, 2, 5	+	
Red-cheeked ground squirrel, *Dremomys rufigenis* 1, 5	+	
Bowers's rat, *Berylmys bowersii*		+
Long-tailed mountain rat, *Niviventer rapit* 4	+	
Malayan mountain spiny rat, *Maxomys inas* 1, 2, 3	+	
Edwards's rat, *Leopoldamys edwardsi* 1, 2, 5, 6	+	
Marmoset rat, *Hapalomys longicaudatus*		+

1. Present on Mt Tahan, Pahang (7174 ft = 2187 m)
2. Present on Mt Benom, Pahang (6913 ft = 2107 m)
3. Endemic to Peninsular Malaysian mountains

4. Also on mountains of Sumatra and Borneo
5. Widespread on continental eastern Asia
6. Present on Kedah Peak.

12.3.3. Bats

Although they have greater mobility, and in some cases are known to fly long distances, the bats exhibit comparable patterns of distribution. Confining discussion to those collected in reasonable numbers, there are four which are apparently restricted to northern parts of Peninsular Malaysia and three with a southerly distribution (Table 12.2). There is also a group of species among which none has been found south of the latitude of the termination of the Main Range. One of these, the grey fruit bat *Aethalops alecto* is truly montane in distribution; two others are characteristic of middle slopes ('submontane'), but the remainder frequent lowland habitats. Another fruit bat, the island flying fox *Pteropus hypomelanus*, roosts on islands but flights to the mainland to feed, as do some birds (Chapter 13).

TABLE 12.2. Bats with latitudinally restricted ranges in Peninsular Malaysia.

Species	LATITUDINAL LIMITS			Notes
	Not below *c*. 5° N	Not below *c*. 2° 30' N	Not above *c*. 3° 30' N	
Short-nosed fruit bat *Cynopterus sphinx*		X		
Grey-fruit bat *Aethalops alecto*		X		Montane
Dayak fruit bat *Dyacopterus spadiceus*			X	
Black-capped fruit bat *Chironax melanocephalus*		X		Submontane
Indian false vampire *Megaderma lyra*		X		
Peninsular horseshoe bat *Rhinolophus robinsoni*		X		
North Malayan horseshoe bat *R. malayanus*	X			
Acuminate horseshoe bat *R. acuminatus*	X			
Big-eared horseshoe bat *R. macrotis*		X		
Croslet horseshoe bat *R. coelophyllus*	X			
Lawas roundleaf horseshoe bat *Hipposideros sabanus*		X		
Boonsong's roundleaf horseshoe bat *H. lekaguli*	X			
Singapore roundleaf horseshoe bat *H. ridleyi*			X	
Cantor's roundleaf horseshoe bat *H. galeritus*		X		
Common roundleaf horseshoe bat *H. cervinus*			X	
Shield-faced bat *H. lylei*		X		
Great roundleaf horseshoe bat *H. armiger*		X		
Lesser large-footed bat *Myotis hasseltii*		X		
Round-eared tube-nosed bat *Murina cyclotis*		X		Submontane
Papillose bat *Kerivoula papillosa*		X		

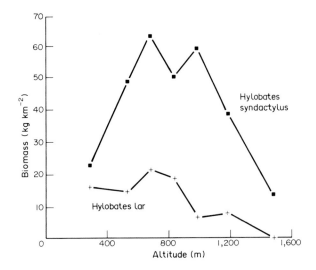

Fig. 12.3. Changes in estimated biomass of siamang *Hylobates syndactylus* and lar gibbons *Hylobates lar* with altitude on the east flank of Mt Benom, Pahang. Reproduced from Caldecott (1980, fig. 2).

12.4. HABITAT PARTITIONING

12.4.1. Altitudinal zonation

Serow *Capricornis sumatraensis* are specialists of steep terrain, from lowland limestone karst to mountain ridges. Other large ungulates (tapir *Tapirus indicus*, deer) and big carnivores (sun bear *Helarctos malayanus*, leopard), as their tracks show, range widely over all elevations. The true montane specialists are all small mammals and they constitute a tiny minority among the total Peninsular fauna.

The richest and most diverse fauna is that of the forest of the non-swampy alluvial lowlands. Certain mammals of this habitat, e.g., Prevost's squirrel, are rare above the hill-foot boundary. Others, e.g., the slender squirrel, appear to be equally common all elevations. Between these extremes there is great variation. Characteristically the number of species encountered falls away progressively with increasing altitude. The results of a transect following the northeastern flank of Mt Benom (Table 12.3) provide a quantified illustration of the general pattern. Statistical treatment of the results confirms the distinctness of the lowland and montane faunas, demonstrating a discontinuity around 3000 ft (915 m) elevation. But it is also clear that this broad division is complicated by differences in the distributions of member species of each fauna, which do not necessarily all share common altitudinal limits.

In the virgin habitat of Benom, trapping results showed no contact between the two rat faunas, with a gap of 600 ft (183 m) between the highest record (at 2900 ft = 884 m) of any member of the lowland assemblage (including red, *Maxomys surifer*, brown, *M. rajah* and Whitehead's spiny rat *M. whiteheadi*, and the long-tailed giant rat *Leopoldamys sabanus*) and the lowest record of montane congenerics at 3500 ft (=1070 m), the Malayan mountain spiny rat and Edwards's rat *Leopoldamys edwardsi*. No lowland rat species existed at densities comparable to those found among the highland rats, which became relatively more abundant at the highest elevations (Table 12.4). Comparison may be made with the contrasting altitude-related changes in density of gibbons and siamang plotted in Figure 12.3.

TABLE 12.3. Numbers of forest mammal species recorded in the Kerau Game Reserve, Pahang, (a) in the lowlands at Kuala Lompat, and (b) in successive contoured sectors on the northeastern flank of Mt. Benom. Data from Medway (1972a) and Payne (in Chivers, 1980).

Altitude range	Total species	Bats	Squirrels[1]	Rats	Others
(a) Kuala Lompat					
@ 200 ft (= 61 m) a.s.l.	66	18	13	8[2]	26
(b) NE Benom					
600–1000 ft (180–305 m)	62	17	14	7	24
–2000 ft (–610 m)	37	10	10	4	13
–3000 ft (–915 m)	28	6	4	4	14
–4000 ft (–1220 m)	21	5	5	3	8
–5000 ft (–1524 m)	15	2	4	2	7
–6900 ft (–2103 m)	10	1	3	2	4

[1] Including flying squirrels
[2] Records of two non-forest rats, *Rattus argentiventer* and *R. exulans,* have been excluded

TABLE 12.4. Abundance of small mammals on the eastern flank of Mt. Benom, shown by catches in conventional wire-mesh cage traps; the trapping rate is expressed as catches per 100 trap nights. Data from Medway (1972a, Table 3).

Median altitude (m)	244	457	762	1067	1372	1677	1966
Trap-nights	3380	1716	155	200	122	104	100
Species caught	16	5	4	3	2	2	2
Individuals	77	13	4	4	5	4	8
Trapping rate (all)	2.3	0.8	2.6	2.0	4.1	3.8	8.0
Trapping rate (rats)	1.5	0.6	2.6	2.0	4.1	3.8	8.0

It is likely that changes in the plant community affect the animals, and reasonable to conjecture that the altitudinal zonation of the forest vegetation (Chapter 2) is reflected in the mammals. Certainly, altitude alone is not a sufficient factor. In the disturbed montane habitat in the Main Range at Fraser's Hill the two rat species of genus *Leopoldamys* have been trapped at the same elevation, although not in the same place. Moreover, on outlying mountains, where the specialized montane rats are absent, the lowland rats have been trapped at high elevations, e.g., Whitehead's rat and the red spiny rat near the summit of Mt Lawit, and the long-tailed giant rat at 4500 ft (1370 m) in the Larut hills, Perak. In the Main Range, there is normally altitudinal separation between the plantain squirrel and its montane relative, the belly-banded squirrel *C. flavimanus*; but on outlying peaks where the belly-banded squirrel is absent, the plantain squirrel has been found up to summit elevations (e.g., Lawit). While there is some inconsistency in these observations, it appears that, in the absence of a similar-sized montane congener, uphill dispersal of some normally lowland rat or squirrel species may be unaffected by the botanical zonation.

Fig. 12.4. The marmoset rat *Hapalomys longicaudatus*, a scansorial rat obligatorily associated with bamboo. (Photograph by Jane Burton.)

12.4.2. Ecological segregation

The partitioning of the lowland forest environment by the diverse mammal fauna was examined in several papers by J.L.Harrison, whose final scheme (1966) envisaged divisions based on activity zone, active period and diet. The categorization recapitulated below can be compared with his earlier (1962) tabulation, which omitted the bats (Table 12.5). Inevitably there are elements of arbitrariness in any such classification. It will also be noted that aquatic mammals have not been included. Important points to be emphasized are the relative representation of nocturnal and diurnal species, the examples of specialization for different levels above the ground, and the diversity of diets.

(1) *Upper air community*: i.e., insectivorous bats which hunt above the tree canopy.

(2) *Canopy community*: confined to the crowns of trees, feeding predominantly on leaves, fruit or nectar, but including a few insectivores or mixed feeders:

(a) *Diurnal*: e.g., gibbons, leaf monkeys, giant squirrels *Ratufa* spp.

(b) *Nocturnal*: e.g., pentail treeshrew *Ptilocercus lowii*, slow loris *Nycticebus coucang*, colugo *Cynocephalus variegatus*, flying squirrels, arboreal rats and fruit bats.

(3) *Middle-zone flying mammals*: i.e., insectivorous bats.

(4) *Middle-zone scansorial mammals*: ranging up and down the trunks and lianes, entering both the canopy and ground layers. Smaller members are mixed feeders on fruit and invertebrate prey, the larger are carnivorous predators:

(a) *Diurnal*: e.g., treeshrews *Tupaia* spp., macaques, tree squirrels.

(b) *Nocturnal*: e.g., tree rats, civets, and yellow-throated marten *Martes flavigula*, clouded leopard *Neofelis nebulosa*, sun bear *Helarctos malayanus* and leopard *Panthera pardus*.

(5) *Large ground mammals*: without climbing ability, i.e, herbivores feeding by browsing, scavenging fallen fruit or grubbing in the soil, and carnivorous predators. This group does not lend itself to a division into diurnal and nocturnal sections. Examples are the porcupines, elephant *Elephas maximus*, tapir *Tapirus indicus*, pigs, deer, cattle, wild dog *Cuon alpinus* and tiger *Panthera tigris*.

(6) *Small ground mammals*: which burrow in or search the litter, and perhaps the lower parts of trunks, insectivorous, mixed or vegetable feeders, with attendant carnivores:

(a) *Diurnal*: ground squirrels.

(b) *Nocturnal*: shrews, rats, bamboo rats *Rhizomys* spp., civets, small cats and the Malayan weasel *Mustela nudipes*.

The bats form a large proportion of the forest mammal fauna, comprising over 40% of known species. Exclusively nocturnal, they may be viewed as the chief ecological counterparts of the mainly diurnal birds. The treeshrews, primates and rodents contain both diurnal and nocturnal members. In the first two groups, diurnal activity is normal, although each has one nocturnal member species. The rodents, however, match the 14 species of diurnal (typical) squirrels against a nocturnal assemblage of 11 flying squirrels and 20 rats, with the two bamboo rats and three porcupines also active chiefly by night (although not classified by Harrison). The reciprococity can be very obvious when one watches a fruiting forest tree at dusk: soon after the typical squirrels withdraw, the flying squirrels emerge from their day-time roosts and glide in to feed.

Trapping results in lowland forest show effects of a multiplicity of factors, including local topography, on the distribution of small mammals. In a valley site some 400 m wide, rising from 50 m to 100 m a.s.l., the three *Maxomys* rats present were more or less uniformily distributed, while Mueller's rat *Sundamys muelleri* was confined to the low-lying ground and the long-tailed giant rat trapped mainly on slopes more than 10 m above the stream bed (Harrison, 1957). At Pasoh, in primary and in regenerating forest, trapping success correlated positively with freedom from flooding (except for Mueller's rat, captured only near permanent water) and copious litter, seedlings and rotting logs, and negatively with the presence of stands of bertam palm *Eugeissona tristis* and sedges, and with pig damage. The activity of pigs was thought to reduce the forest litter and food availability for small mammals over wide areas (Kemper and Bell, 1985).

TABLE 12.5. Summary of numbers of species of Peninsular Malaysia mammals by vertical zone and feeding habit; from Harrison (1962). Bats are omitted.

Zone	No. species	Plant	Mixed	'Insect'	Flesh
			DIET		
Canopy	23	15	7	1	0
Middle zone	21	1	17	0	3
Large ground	17	13	1	0	3
Small ground	28	4	9	10	5

12.4.3. Adaptive radiation

The two complementary groups of squirrels are sufficiently similar in general features to be placed in the same family (Sciuridae). The chief distinguishing character of the flying squirrels is a furred gliding membrane ('patagium') joining the fore and hind limbs on each side. Among Peninsular species there is variation in the points of attachment of the patagium, in the detailed anatomy of the alimentary canal and the teeth, and in relative proportions of parts of the skull (notably, the auditory bulla). Despite this evidence of separate evolutionary lines, all flying squirrels are regarded as ultimately sharing a common ancestor and are grouped in one subfamily (Petauristinae), with the typical squirrels including marmots, chipmunks, etc., placed in another (Sciurinae). The several species of flying and typical squirrels of Peninsular Malaysia show comparable variation in size, ranging in adult weight from about 30 g to about 2 kg. Their diets overlap extensively. Seeds and fruit pulp are important foods for all, with the typical squirrels also including specialist bark-consumers and arthropod predators (Payne, in Chivers, 1980) and the flying squirrels including leaf- and shoot-eaters (Muul and Lim, 1978). It seems very likely that their separate activity periods, day *vs.* night, permit the sympatric existence of the two groups of mammals which might otherwise compete more directly for the same resources.

There are other, equally striking examples of adaptive radiation within related groups at the family and lower taxonomic level. For instance, the cats range from the large (tiger) to small (leopard cat *Felis bengalensis*), and include climbing and ground-living species, flesh-eaters and a fish-catcher (the flat-headed cat *Felis planiceps*, see Muul and Lim, 1970). As noted in Harrison's compilation (above), the squirrels include ground-dwelling species, and the forest rats show graded degrees of specialization for arboreal life. Characteristic adaptions found among scansorial rats are as follows: the muzzle is fore-shortened, giving improved binocular vision; the tail is lengthened, often being used in a weakly prehensile manner and provided with a 'bottle-brush' of short, bristly hairs towards the tip, presumably increasing frictional grip; the foot is short and broad. Comparative measurements illustrating these trends are given in Table 12.6.

While squirrels cling by their strong, sharp claws, arboreal rats have prehensile fore and hind feet in which the outer toes are partially opposable. The marmoset rat is among the most accomplished

TABLE 12.6. Body size and relative proportions of some Peninsular Malaysian rats. The species are graded from the most specialized arboreal forms, through those with considerable climbing ability to the last, normally found only at ground level (or in burrows).

	Species	Median head and body length (mm)	Tail length (% H & B)	Hind foot length (% H & B)
↑	Pencil-tailed tree-mouse *Chiropodomys gliroides*	86	135	22
A R B O R E A L	Marmoset rat *Hapalomys longicaudatus*	163	121	18
	Grey tree rat *Lenothrix malaisia*	195	124	17
	Dark-tailed tree rat *Niviventer cremoriventer*	148	119	19
G R O U N D	Long-tailed giant rat *Leopoldamys sabanus*	240	154	20
	Mueller's rat *Sundamys muelleri*	247	117	21
↓	Red spiny rat *Maxomys surifer*	187	99	22

climbers. Its claws are raised on the upper surface of the digits so that contact with the substrate is probably impossible. The tips of the toes are expanded, and their plantar surfaces richly supplied with hypertrophied sweat glands. Fluid from these glands moistens the skin continuously, and presumably assists grip. The marmoset rat can move freely up and down steeply inclined stems of large bamboos, leaving damp foot prints as it goes (Fig. 12.5).

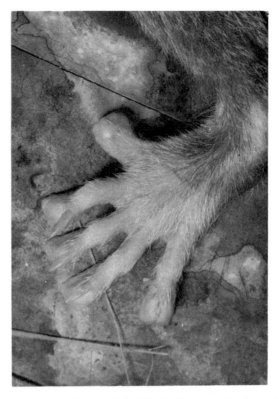

Fig. 12.5. Left hind foot of the marmoset rat, photographed while climbing a large bamboo culm. Note the small claws and the damp footprint. (Photograph by Jane Burton.)

12.4.4. Congeneric sympatrics

Of great theoretical interest are those features that permit similar and very closely related species — members of the same genus — to exist side by side in the rain forest environment. The phenomenom is found among many animal groups and the mammals provide examples, several of which have already been mentioned. Genetic mechanisms presumably play a part in maintaining separate populations (Chapter 11). In some cases field studies have identified ecological factors that appear to delimit distinctive species-specific niches. In others such as the *Crocidura* shrews (Davison, 1984) this is more difficult.

The red and brown spiny rats, *Maxomys surifer* and *M. rajah* are puzzling. Although differing markedly in karyology (Chapter 11), these are difficult to tell apart by external characters. In the past, some mammalogists have combined the two into one species. In a valley trapping area in lowland forest,

Harrison (1957) found that the two overlapped completely in distribution but the brown occurred with significantly greater frequency on the upper slopes while the red was commoner in the valley bottom. He also noted small differences in infestation by trombiculid mites, which he attributed to cross-infestation from other rat species of the slopes and valley bottom, respectively.

Among bats, Cantor's and the common roundleaf horseshoe bats, *Hipposideros galeritus* and *H. cervinus*, have only recently been recognized as distinct (Jenkins and Hill, 1981). Both have been collected at the same sites and because of past confusion there is little reliable information on differences in habit. The former is much rarer in museum collections and is inferred to be solitary or live in small groups, while the latter is known to be among the commonest and most numerous bats of Malaysian caves.Two small *Myotis*, the whiskered bat *M. mystacinus* and Horsfield's bat *M. horsfieldii*, choose very different daytime roosts. The first gathers in small groups in the rolled central leaves of bananas, the second in caves. The two flat-headed bats, *Tylonycteris* spp., show overlapping habits. Both will roost in the internodal spaces of bamboo culms and may, on different occasions, occupy the same sites. In Ulu Gombak Forest Reserve, where both occur, each species selects a different range of roost entrance holes (Figures 12.6 and 12.7). The nectar- and pollen-eating long-tongued bats, *Macroglossus* spp., have similar roosting habits but different diets. *Macroglossus sobrinus* takes pollen chiefly from wild bananas, *Musa malaccensis* and *Musa truncata*, while *M. minimus* depends on a variety of tree species, notably *Sonneratia* spp. of the mangrove. Outside forest habitats, both bats will feed at the flowers of cultivated banana plants but only *M. minimus* takes coconut *Cocos nucifer* pollen (Start and Marshall, 1976).

Among the higher primates sympatric in Peninsular Malaysian forests, gibbons have the least specialized gastro-intestinal tracts, are incapable of digesting fibrous foods and have poor detoxification capacities. Dietic limitations explain their restriction to the most diverse types of forest which provide a suitable variety of food sources. In an area where six primate species of three genera occur (Kerau Game

Fig. 12.6. In a large clump of bamboo *Gigantochloa scortechinii*, the man points to a characteristic entrance to a roost site of flat-headed bats. The hole was originally formed in the wall of the bamboo culm as the pupation chamber of the beetle *Lasiochila goryi*.

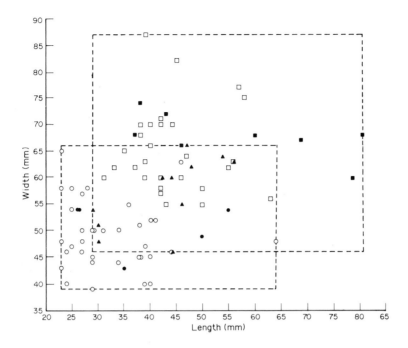

Fig. 12.7. A plot of the dimensions of the entrance apertures of roosts in bamboo culms used by the two species of flat-headed bats at Ulu Gombak, Selangor. The lower rectangle surrounds all roosts occupied at some time by the lesser flat-headed bat O and the larger rectangle those occupied by the larger flat-headed bat □. There are 11 roosts falling in the overlap zone which were used on different occasions by both species △.

Reserve, Pahang), studies by Chivers and his colleagues (Chivers, 1980) found that the diets of lar gibbons and siamang overlapped quantitatively by 53%; lar gibbons ate relatively more fruit and fewer leaves than the siamang. Similar proportional differences were found in the diets of banded leaf monkeys *Presbytis melalophos* (48% fruit, 35% leaves) and dusky leaf monkeys *P.obscura* (32% fruit, 58% leaves). In this case, the different foods chosen reflected the greater use made of emergent trees by the banded leaf monkeys. The long-tailed macaques *Macaca fascicularis* ate on average 52% fruit, 5% flowers, 19% leaves and 23% animal matter; this monkey relied heavily on creepers (lianes) as a source of vegetable foods.

12.5. DAILY ACTIVITY CYCLES

It requires skill and determination, some luck and disregard for personal comforts to follow any single mammal through the forest for long periods. All available evidence suggests that, as in other environments, neither diurnal nor nocturnal mammals are consistently active throughout the light or dark period of the day, respectively. Among bats, the nectarivorous long-tongued bats and the cave fruit bat *Eonycteris spelaea* start feeding at about dusk and are most active within the two subsequent hours, although some individuals continue to forage until dawn (Start and Marshall, 1976). Flying foxes *Pteropus vampyrus* may leave their daytime roosts before darkness; while visiting flowering durian trees *Durio zibethinus*, individuals feed actively in two bouts, first from sunset +1 to +3 h, and again after about 23.30 h (Gould, 1977).

This pattern of peak activity in the early part of the night is probably usual behaviour in all nocturnal small mammals. A brief study of radio-tagged rats in Selangor forests showed that all became restless in their daytime retreats as the light faded and most emerged soon after nightfall. One adult male Mueller's rat followed continuously was active from leaving one den at 19.10 h until retiring to another at 01.15 h; here it remained until re-emerging at 19.00 h the following evening (Sanderson and Sanderson, 1964).

Among diurnal mammals an early start to the day is usual, rising to a mid-morning activity peak. Some show a secondary peak in the late afternoon, after midday quiescence. The calling cycles of gibbon and siamang, surely the most evocative sounds of the Peninsular Malaysian forest, commence with the gibbons (routinely beginning earlier than siamang), males in particular often singing before dawn. The frequency of gibbon calling peaks on average an hour or more before siamang (Fig. 12.2). It is rare to hear either in the afternoon when other activities predominate.

Monkeys are less regular in the pattern of their daily vocalizations. The leaf monkeys also indulge in bouts of nocturnal calling from the roost tree. Among banded leaf monkeys at Kuala Lompat, Kerau Game Reserve, night-time calls appear to be given by single adult males in the roosting troop, and elicit calls in response from other groups within earshot. Two peaks of intense calling, at 19.00–20.00 h and 05.00–06.00 h, coincide with entry and exit from the roost trees and presumably function as territorial advertisement (Curtin, in Chivers, 1980).

Contrasts between the two kinds of primates are also seen in foraging and feeding behaviour. The gibbons and siamang feed most intensively within the first two hours of the day, and thereafter intermittently until middle or late afternoon. Retiring comparatively early to the roost tree, the lar gibbon's active day averages 8.6 hours (range 6.2–10.2 h, $n = 92$ days); that of the siamang is slightly longer, averaging 10.3 hours (range 6.5–11.9 h, $n = 109$ days). The leaf monkeys tend to show two peaks of feeding, one in the morning and another in the late afternoon following a midday rest. Raemaekers and Chivers (in Chivers, 1980) have suggested that this bimodality is imposed by the greater proportion of foliage in the leaf monkey's diet and a slower rate of digestion involving fermentation processes in the specially adapted gut.

12.6. REPRODUCTIVE CYCLES

12.6.1. Frugivores and folivores

The most constant element in the food of frugivorous mammals in Peninsular Malaysian forests is provided by figs *Ficus* spp. Variations in the fruiting cycles of different figs appear to ensure that this resource is more of less continuously available in most lowland forests. Distinctive feeding niches are therefore defined by other items in the diets of different mammal species. Large fluctuations in the abundance of any critical item of food might be reflected in matching changes in the physiological condition of the consuming mammal, including reproductive state. As shown in Chapters 1 and 9, albeit poorly defined, there is an annual cycle of seasonality in the Malaysian rain forest environment. For those mammals that rely chiefly on plants for their food sources, the fluctuations are evidently insufficient to impose a general breeding season. Among the ungulates, for instance, the female Eurasian wild pig, once mature, is capable of producing a continuous series of litters (Diong, 1973). The irregularity of the reproductive patterns of the deer, Cervidae, means that there can be no biological basis for close seasons in hunting law, although their comparatively long gestation periods and sometimes protracted intermissions between successive pregnancies may produce irregular fluctuations in the reproductive activity of local populations. Such apparent cyclicity may be found among local populations of primates, although in a careful study of one, the dusky leaf-monkey, no seasonal variation in sexual activity was detectable in speciments collected in western Perak (Burton, 1983).

Among smaller plant-consuming mammals, the colugo *Cynocephalus variegatus* breeds at all times of year, producing a single young; females may become pregnant again before the previous young is weaned. In the colony of cave fruit bats in Batu Caves, Selangor, breeding is continuous and, once again, lactating females may be pregnant (Beck and Lim, 1973). Among forest rodents, four years' research in Selangor found pregnant females and fertile males in all months of the year among the common species of rats and squirrels (Harrison, 1955). Combined data have indicated seasonal variations in the proportions of pregnant females in the populations of plantain squirrel, black-banded squirrel, pencil-tailed tree-mouse *Chiropodomys gliroides*, Mueller's rat and long-tailed giant rat. On the other hand, these species did not exhibit a common season of heightened reproductive activity, and the effect was not apparent in other forest rats.

Muul and Lim (1974) have warned of potentially fallacious results produced by amalgamations of data from several years, and themselves demonstrated intermittent, non-annual but apparently synchronized breeding among a population of grey-cheeked flying squirrels *Hylopetes lepidus* in Johor. A sympatric population of the smoky flying squirrel *Pteromyscus pulverulentus* showed no comparable periodicity. Undoubtedly more research is needed, but at present it can be concluded that there is no universal breeding season common to those Malaysian forest mammals that rely largely or exclusively on plant matter for food.

12.6.2. Insectivores

Many insectivorous bats show pronounced seasonality. Well studied examples include, in Selangor, the cavernicolous roundleaf horseshoe bats *Hipposideros bicolor, H. cervinus, H. cineraceus* and *H. diadema*, and the non-cavernicolous flat-headed bats *Tylonycteris pachypus* and *T. robustula*. These bats all produce young once annually in the period March–May. Colonies of the horseshoe bats *Rhinolophus affinis, R. macrotis, R. sedulus* and *R. stheno* sampled in Pahang forests, at slightly higher latitude, showed synchronized breeding with this same birth season. Colonies of pregnant or lactating females have also been found of the lesser large-footed bat *Myotis hasseltii* on Langkawi island in January, and of the Malaysian noctule *Pipistrellus stenopterus* on Singapore island in October. On the other hand, collections by B.L.Lim from Batu Caves, Selangor, suggest that *Rhinolophus affinis* may breed twice a year, and at least one normally non-cavernicolous horseshoe bat, *Rhinolophus trifoliatus*, and one vespertilionid, the whiskered bat (which roosts in banana leaves, 12.4.4. above), provide instances of females pregnant in most or all months of the year without discernible pattern (data in Medway, 1983). Again, more information is needed, but in summary it is clear that an annual, synchronized breeding cycle is commonplace but not universal among insect-eating bats, that the breeding season may vary from place to place, even within species, and that contrasting behaviour may be found among closely related species.

Where it exists, intraspecific synchronization can be very close. In *Tylonycteris pachypus* in Ulu Gombak Forest Reserve, the maximum spread of births in 1965 was merely 20 days, and median birth dates in the population in 1965 and 1967 were 14th and 22nd April, respectively (Medway, 1972b). In this genus there is a variable period of delayed fertilization between mating (in November to early December in Selangor) and ovulation (in the second week of January) during which sperm is stored in the female tract, in the fallopian tubes and upper uterus. Synchronization of births is thus ensured by the timing of ovulation rather than mating. The external cue may be the high daytime temperatures experienced within the roost during the brief annual dry period which (in Selangor) can be expected at some time in January each year.

Of other insectivorous small mammals, the Southeast Asian white-toothed shrew *Crocidura fuliginosa* has been found pregnant at all time of the year. Common treeshrews *Tupaia glis* (which take a mixed diet) in Selangor breed at all times of the year, but further north in Kedah and on Penang island (in 1972–75) showed a concentration of births in the first seven months of the year (with confirmed instances of repeated breeding during this season), but also two cases of lactating and pregnant females in November (Langham, 1982). In these cases, environmental constraints restricting the breeding season are evidently weaker.

12.6.3. Reproductive rates

Litter sizes in general are low. One young is normal among most bats, all primates, the colugo, the larger flying squirrels, porcupines, deer and cattle, and twins among some vespertilionid bats, the larger tree squirrels and small flying squirrels and sun bear. Other squirrels, the civets and cats produce 2–3 young, the Eurasian wild pig up to 8 and some forest rats up to 9 (but on average about 4 per litter).

12.7. POPULATION DENSITIES

The greatest concentrations of mammals are found at the roost sites of gregarious cave-dwelling bats. Yet it may be wrong to consider such bats to be 'common'. For instance, the cave fruit bats of the large colony at Batu Caves, Selangor, are known to feed *inter alia* on pollen of *Sonneratia* trees of the coastal mangrove, at its nearest 38 km distant. Even if 40 km is their maximum range, bats from this cave must be capable of exploiting food resources over a circle exceeding 5000 km^2 in area; the true figure may be larger. Clearly, if there is no other occupied cave in the vicinity (and none is known), tens of thousands at the roost site represent no great density over the full feeding range. Non-cavernicolous bats with specialized requirements may also be distributed at roost in a pattern unrepresentative of their dispersal when feeding. For instance, whiskered bats can only find their preferred roost site (the furled, emerging central leaf of a banana plant) for the few days that each leaf remains tightly rolled. Their distribution at roost must be determined by the constantly changing availability of this resource. The flat-headed bats of Ulu Gombak, as shown above, make use of less ephemeral roost sites provided by the internodal spaces of bamboos. The composition of roosting groups changes continually, implying intense competition for the resource, and a low rate of recapture of marked bats indicates that the total population ranges over an area far larger than the few hectares of the study site (Medway and Marshall, 1972). For such mammal species no reliable population densities can be computed. Among bats with no breeding peak, deaths and births may be roughly balanced so that numbers remain more or less constant, but among those with synchronized breeding seasons there must be large annual fluctuations in numbers as young are born into the population.

The non-flying mammals mostly occupy fixed home ranges, individually or as members of social groups. The traditional ranges of herds of large gregarious ungulates, such as elephants or gaur *Bos gaurus*, can be used to derive population estimates based on encounters with recognized herds (Mohamed Khan, 1977). Groups of Sumatran rhinoceros *Dicerorhinus sumatrensis* and tapir also appear to reside within traditional bounds. Densities of the rhinoceros in protected forest habitat in the Endau-Rompin reserve were equivalent to one animal per 40 km^2 (Flynn and Mohd. Tajuddin Abdullah, 1983). One radio-tagged male tapir, relocated on 16 occasions during May–October, ranged over 4.9 sq.miles

(=12.7 km²) (Williams, 1979). Only bearded pigs are known to make long-distance gregarious movements, sometimes termed migrations. Details are scanty, but there are past records of large herds heading eastwards through Johor and across southern Pahang. No observations exist to show that the circle was completed. Game Department statistics, unfortunately, do not distinguish this pig from the more widespread Eurasian wild pig, and it remains the least studied member of its genus.

Among the higher primates, measured home ranges of family groups of siamang vary from 23–48 ha, of lar gibbons 20–58 ha and agile gibbons 25–29 ha. Troops of leaf monkeys, although comprising many more individuals, occupy smaller areas; for example, a group of 17 dusky leaf monkeys, 33 ha, and a group of 18 banded leaf monkeys, 21 ha. Troops of long-tailed macaques numbering 17–27 individuals ranged over 35–46 ha (Chivers, 1980). In all these instances, the major part of the range is defended against conspecifics by territorial behaviour. Corresponding population densities are given in Table 12.7, with estimates of squirrel populations in the same area. The combined biomass of 947 kg/km² represents effectively the full assemblage of diurnal arboreal mammalian consumers of plant tissue inhabiting this tract of lowland forest.

Trapping results have implied that rats of the forest floor are more or less sedentary. Computed lifetime home ranges, derived from recaptures of marked individuals, vary in diameter from 263 m for the brown spiny rat to 476 m for Whitehead's spiny rat (Harrison, 1958). Harrison calculated the total number of rats in his 'SBF area' (disturbed lowland primary forest) as 350 per 100 ha, representing a biomass of 40 kg/km². Because his traps failed to provide an adequate sample of the squirrel community, his estimate of biomass (44 kg/km²) cannot be compared with Payne's (Table 12.7), obtained mainly by

TABLE 12.7. Estimates of population densities in lowland forest sites by various workers, converted to individuals per 100 ha, biomass in kg/km².

Locality	Species	No. individuals	Biomass (kg)	Source
Penang	*Tupaia glis*	478	—	Langham (1982)
Kuala Lompat (averages)	*Hylobates syndactylus*	12.5	97	Raemakers & Chivers, in Chivers (1980)
	H. lar	7.3	29	,,
	Presbytis obscura	46.7	240	,,
	P. melalophos	57.1	286	,,
	Macaca fascicularis	57.5	182	
Kuala Lompat	*Ratufa bicolor*	11	16	Payne, in Chivers (1980)
	R. affinis	26	28	
	Callosciurus prevostii	38	17	
	C. notatus	244	55	
	C. nigrovittatus	5	1.2	
	Sundasciurus tenuis	160	11	
	S. hippurus	2	0.8	
S B F, Selangor	*C. notatus*	57	—	Harrison (1969: Table 3)
	C. nigrovittatus	6	—	
	S. tenuis	23	—	
	Rattus argentiventer	6	—	
	R. exulans	6	—	
	Sundamys muelleri	69	—	
	Berylmys bowersii	6	—	
	Maxomys whiteheadi	51	—	
	M. rajah	120	—	
	M. surifer	131	—	
	Leopoldamys sabanus	11	—	

· visual counts. In another primary forest area, designated 'A', five rat species were trapped, amounting to a combined biomass of 52.5 kg/km². Converted densities, expressed as individuals per 100 ha, were as follows: *Sundamys muelleri* 8, *Maxomys whiteheadi* 70, *M. rajah* 23, *M. surifer* 23, *Leopoldamys sabanus* 195 (from Harrison, 1969). The nature of the habitat at 'A' was not stated, but the composition of the rat fauna suggests that the site included much well-drained slope.

Of other small mammals, a rather high density of common treeshrews at Pantai Aceh, Penang, calculated from mark and recapture data (Table 12.7) may reflect the peculiarities of an island situation. In forest on Tioman island, Pahang, the small mammal fauna is impoverished by comparison with the mainland, but average densities computed from trapping rates are higher (Medway, 1966). Trapping results in forest on the slopes of the Main Range in Selangor indicated a local density of lesser gymnures equivalent to 296 per 100 ha, with individual activity ranges 21 m in diameter (Rudd, 1980). In the same hillside study area, the distribution of pencil-tailed tree-mice was closely associated with bamboo clumps, and it appeared that no meaningful population density could be calculated (Rudd, 1979). In hill dipterocarp forest in Pahang, a survey by mixed methods suggested an average of 20 slow lorises per 100 ha (Barrett, 1981).

For the remainder, general experience suggests that even in remote and undisturbed forest, free from hunting, other insectivores, colugo, porcupines, the assorted Carnivora and ungulates, etc., would together add perhaps half as much again to the biomass of treeshrews, primates, squirrels and other rodents summed above. A reasonable guess for the total biomass of mammals to be expected in lowland primary forest would therefore lie between 1 and 2 tonnes per km².

This figure is much lower than the mammalian biomass of savannah grassland habitat in equatorial Africa, where large herbivorous ungulates may exist at very high densities: for example, 23.5 tonnes per km² in the Rwindi-Rutshuru plain, Congo — a partial computation counting only five species, including hippopotamus *Hippopotamus amphibius*, 11.2 t, and African buffalo *Syncerus caffer*, 6.1 t. The Malaysian figure, however, is of much the same order as biomasses calculated for African forests (see Owen, 1966).

In Peninsular Malaysia, in the simplified habitat of a cultivated tree-crop, densities of small mammal pest species may be enormously higher. For instance, in oil palm plantations, in the absence of control treatment, Malaysian wood rats *Rattus tiomanicus* have been found in numbers corresponding to densities of 28 000–57 350 rats per 100 ha (Wood, 1969). At an average weight of 100 g per trappable (i.e., weaned and independent) rat, this figure represents a biomass of 2.8–5.7 t/km².

12.8. MANAGEMENT AND CONSERVATION

The Malaysian forest mammal community undoubtedly includes some species adapted to exploit the temporarily enriched environment of early successional stages of plant growth and regeneration. Such habitat occurs naturally in small areas following a tree death or windblow, and on a larger scale along river banks or on other sites of soil instability. The environment of plantation tree-crops presents many similarities, and some of these mammals readily become pests. Examples include the Malaysian wood rat (above), the plantain and grey-bellied squirrels and Vordermann's flying squirrel *Petinomys vordermanni* (for the last, see Muul and Lim, 1978).

The Sumatran rhincoceros and tapir feed on plants characteristic of regrowth and early successional forest. The pigs, deer, wild cattle and elephant take a varied vegetable diet and thrive in areas of disturbed or regenerating forest. They can also cause damage to crops on a scale that is commercially intolerable. Given the conservative ranging habits of these large ungulates, the reduction and fragmentation of forest cover and its substitution by plantation inevitably lead to conflict with agricultural interests (Chapter 17). For the succesful conservation of these mammals, some of which are now seriously endangered in Peninsular Malaysia, it may prove necessary to apply sophisticated wildlife management techniques, and to control forested areas so as to ensure the constant availablity of suitable types of subclimax cover.

Yet, although they can assume undue importance, the mammals that are potential pests form only a small proportion of the forest fauna. Most species depend on mature seral or climax vegetation and cannot exist outside this habitat. In order to ensure the perpetuation of this major component of Peninsular Malaysian fauna in full diversity, it is now essential to protect adequate reserves of natural forests, most especially in the lowlands below the hill-foot boundary. Future conservation needs to be based firmly on scientific experience. Relevant biological knowledge is limited to the handful of mammal species that have been studied in the wild. The scope for further investigation is wide, and the results of such research will be of fundamental value for the planned management of this key wildlife resource.

REFERENCES

Barrett, E. (1981). The present distribution and status of the slow loris in Peninsular Malaysia. *Malay. Appl. Biol.* 104, 205–12.

Beck, A. F. and Lim, B. L. (1973). Reproductive biology of *Eonycteris spelaea* Dobson (Megachiroptera) in West Malaysia. *Acta Tropica* 30, 251–60.

Burton, G.J. (1983). Aspects of the reproductive anatomy of the dusky leaf monkey *(Presbytis obscura). Malay. Nat.J.* 36, 285–8.

Caldecott, J. O. (1978). *The effect of altitude on the distribution of the white-handed gibbon and siamang, and on the forest types with which they are associated, in West Malaysia.* B.Sc. (Hons.) Biology project, University of Southampton.

Caldecott, J. O. (1980). Habitat quality and populations of two sympatric gibbons (Hylobatidae) on a mountain in Malaya. *Folia Primatol.* 33, 291–309.

Chasen, F. N. (1940). Handlist of Malaysian mammals. *Bull. Raffles Mus.* 15, i-xx, 1–209.

Chivers, D. J. (ed.) (1980). *Malayan Forest Primates.* Plenum Press, New York & London.

Darlington, P. J. Jr. (1957). *Zoogeography: the geographical distribution of animals.* John Wiley, New York.

Davison, G. W. J. (1984). New records of Peninsular Malaysian and Thai shrews. *Malay. Nat. J.* 36, 211–5.

Diong, C. H. (1973). Studies of the Malayan wild pig in Perak and Johore. *Malay. Nat. J.* 26, 120–51.

Flynn, R. W. and Mohd. Tajuddin Abdullah (1983). Distribution and numbers of Sumatran rhinoceros in the Endau-Rompin region of Peninsular Malaysia. *Malay. Nat. J.* 36, 219–47.

Gould, E.H. (1977). Foraging behaviour of *Pteropus vampyrus* on the flowers of *Durio zibethanus. Malay. Nat. J.* 22, 174–8.

Harrison, J. L. (1955). Data on the reproduction of some Malayan mammals. *Proc. Zool. Soc. Lond.* 125, 445–60.

Harrison, J. L. (1957). Habitat studies of some Malayan rats. *Proc. Zool. Soc. Lond.* 128, 1–21.

Harrison, J. L. (1962). The distribution of feeding habits among animals in a tropical rain rorest. *J. Anim. Ecol.* 31, 53–64.

Harrison, J. L. (1966). *Introduction to Mammals of Singapore and Malaya.* Malayan Nature Society, Singapore.

Harrison, J. L. (1969). The abundance and population density of mammals in Malayan lowland forests. *Malay. Nat. J.* 22, 174–8.

Jenkins, P. D. and Hill, J. E. (1981). The status of *Hipposideros galeritus* Cantor, 1846 and *Hipposideros cervinus* (Gould, 1854) (Chiroptera, Hipposideridae). *Bull. Br. Mus. Nat. Hist. (Zool)* 41, 279–94.

Kemper, C. and Bell, D. T. (1985). Small mammals and habitat structure in lowland rain forest of Peninsular Malaysia. *J. Tropical Ecol.* 1, 5–22.

Langham, N. P. E. (1982). The ecology of the common tree shrew, *Tupaia glis* in peninsular Malaysia. J. Zool 197, 323–344.

Marsh, C. W. and Wilson, W. L. (1981). *A Survey of Primates in Peninsular Malaysian Forests.* Universiti Kebangsaan, Malaysia.

Medway, Lord (1964). The marmoset rat. *Malay. Nat. J.* 18, 104–10.

Medway, Lord (1966). Fauna of Pulau Tioman : the mammals. *Bull. Natl. Mus., Singapore* 34, 9–32.

Medway, Lord (1972a). The distribution and altitudinal zonation of birds and mammals on Gunong Benom. *Bull. Br. Mus. Nat. Hist. (Zool).* 23, 105–54.

Medway, Lord (1976b). Reproductive cycles of the flat-headed bats *Tylonycteris pachypus* and *T. robustula* (Chiroptera, Vespertilionidae) in a humid equatorial environment. *Zool. J. Linn. Soc., Lond.* 51, 33–61.

Medway, Lord (1983). *The Wild Mammals of Malaya (Peninsular Malaysia) and Singapore*. 3rd ed. Oxford University Press, Kuala Lumpur.

Medway, Lord and Marshall, A. G. (1970). Roost-site selection among flat-headed bats (*Tylonycteris* spp.). *J. Zool., Lond.* 161, 237–45.

Medway, Lord and Marshall, A. G. (1972). Roosting associations of flat-headed bats, *Tylonycteris* species (Chiroptera, Vespertilionidae) in Malaysia. *J. Zool., Lond.* 168, 463–82.

Medway, Lord and Yong, H. S. (1976). Problems in the systematics of the rats (Muridae) of Peninsular Malaysia. *Malay. J. Sci.* 4(A), 43–53.

Muul, I. and Lim, B. L. (1970). Ecologial and morphologial observations of *Felis planiceps. J. Mammal.* 51, 806–8.

Muul, U. and Lim. B. L. (1974). Reproductive frequency in Malaysian flying squirrels *Hylopetes* and *Pteromyscus. J. Mammal.* 55, 393–400.

Muul, I. and Lim, B. L. (1978). Comparative morphology, food habits and ecology of some Malaysian arboreal rodents. In *The Ecology of Arboreal Folivores*, ed. G. G. Montgomery, pp. 361–8. Smithsonian Institution, Washington, D. C.

Owen, D. F. (1966). *Animal Ecology in Tropical Africa*. Oliver & Boyd.

Rudd, R. L. (1979). Niche dimension in the bamboo mouse, *Chiropodomys gliroides* (Rodentia : Muridae). *Malay. Nat. J.* 32, 347–9.

Rudd, R. L. (1980). Population density and movements of the lesser gymnure, *Hylomys suillus. Malay. Nat. J.* 34, 111–2.

Sanderson, G. C. and Sanderson, B. C. (1964). Radio-tracking rats in Malaya. *J. Wildl. Mgt.* 28, 752–68.

Start, A. N. and Marshall, A. G. (1976). Nectarivorous bats as pollinators of trees in West Malaysia. *Linn. Soc. Symp. Ser.* 2, 141–50.

Steven, W. E. (1968). *The Conservation of Wild Life in West Malaysia*. Cyclostyled report, Office of the Chief Game Warden, Seremban, Malaysia.

Tweedie, M. W. F. (1978). *Mammals of Malaysia*. Longman, Malaysia.

Williams, K. D. (1979). Radio-tracking tapirs in the primary rain forest of West Malaysia. *Malay. Nat. J.* 32, 247–52.

Wood, B. J. (1969). Population studies on the Malaysian wood rat (*Rattus tiomanicus*) in oil palms. *The Planter* 45, 510–26.

Completed October 1985

CHAPTER 13

Birds

David Wells

Department of Zoology, University of Malaya, 59100 Kuala Lumpur, Malaysia

CONTENTS

13.1. INTRODUCTION

Wild birds have long been part of Southeast Asian trade and skins from what is now Peninsular Malaysia began to reach museums in Europe and India in number in the early 1800s. Mid- to late nineteenth century collectors of note included Drs. Cantor and Maingay, F. Stoliczka, A. R. Wallace, H. R. Kelham and W. R. Davison. Between them, these men covered much of the lowlands of the west coast and southern Johor. The first montane collections were made by L. Wray, in the Larut hills in 1886-7, for the newly founded Perak State Museum at Taiping; much the same ground was worked in 1899 by A. L. Butler from the Selangor Museum. In 1903, H. C. Robinson was appointed to the Selangor Museum (soon thereafter the Federated Malay States Museum). He was joined in 1908, and later succeeded by, C. B. Kloss. Together, these two were responsible for a wide ranging ornithological exploration of the peninsular region.

In 1927 and 1928, Robinson published the first two of a projected five-volume treatise on the birds of the faunistic subregion recognized as the 'Malay Peninsula'. After his death, volumes 3 and 4 were produced by F.N. Chasen (d. 1942). The final volume was ultimately written by Medway and Wells (1976), who took the opportunity to review current ornithology in the Peninsular, with special chapters on the resident birds, migrants, and the eastern Palaearctic migration system.

Meanwhile, Chasen (1935) had compiled a checklist of the avifauna of the Sunda region (= western Malesia). Gibson-Hill (1949) drew on this to produce an annotated list of Peninsular Malaysian birds, used in turn by Glenister (1951) as the taxonomic basis of the first comprehensive bird book, still in print. Introductory texts by Madoc (1956) and Tweedie (1960) are also still available, as is the field identification guide by King, Woodcock and Dickinson (1975).

Mist-nets began to be used in the country in significant numbers in 1958-9, heralding the start of formal population and migration studies. In 1963 a nation-wide bird-ringing project was founded at the University of Malaya, where it continues to operate (now alongside an independent scheme of the Department of Wildlife and National Parks). Since 1962, annual or biennial bird reports have been published in the *Malayan Nature Journal*.

13.2. HABITAT COMMUNITIES

Approximately 370 bird species make heavy or exclusive use, for all or part of the year, of forests or the forest fringe in Peninsular Malaysia (including mangrove, and terrestrial forest on islands). Forest birds form over 60% of the total recorded avifauna and include all Sunda subregional endemics occurring in the area.

Three centres of evolutionary diversity coincide with the principal forest formations. By far the largest such habitat community is that of lowland (mixed dipterocarp) rain forest. Including birds of the forest fringe or 'edge' and the aerial realm above the canopy, it amounts to 282 species. Montane forest follows with 96 species. About 50 species occur habitually in mangrove (Nisbet, 1968), but only 25 are heavily dependent on this habitat and — discounting mudflat foragers, such as storks and herons — only six are locally exclusive to it (two kingfishers, a pitta, a tit, a flycatcher and a sunbird). For reasons which may have to do with adaptation to its restricted flora and simple structure (Ward, 1968), mangrove has contributed disproportionately to the recent non-forest bird community. Such familiar garden birds as magpie robin *Copsychus saularis* are probably of mangrove origin in the Peninsula and there are signs that mangrove will contribute yet more species to the cultivated country which it now adjoins (Wells, 1985).

Among the families represented in forest by residents and/or migrants, Peninsular Malaysia has a notable share of treeswifts (Hemiprocnidae) with 2 species, hornbills (Bucerotidae) with 9, fairy bluebirds and leafbirds (Irenidae/Chloropseidae) 6, broadbills (Eurylaimidae) 7 (Figure 13.1), pittas (Pittidae) 7, cuckoo-shrikes (Campephagidae) 9 and flowerpeckers (Dicaeidae) 9. Widespread families contributing a dozen species or more are the hawks (Accipitridae) 14, phasianids 13, pigeons

(Columbidae) 14, cuckoos (Cuculidae) 18, owls (Tytonidae/ Strigidae) 13, woodpeckers (Picidae) 20, bulbuls (Pycnonotidae) 19, thrushes (Turdidae) 19, warblers (Sylviidae) 16, flycatchers (Muscicapidae) in the sense of Sibley and Ahlquist (1985) 25, sunbirds (Nectariniidae) 15 and, above all, babblers (Timaliidae) with 47 species. Babblers contribute up to 25 % of community richness by number of species, but this dominance carries no connotation of ecological bias. The 'jobs' that babblers do in Malaysian forests are diverse.

Fig. 13.1. The green broadbill *Calyptomena viridis*. (Photograph by Frank Lambert.)

13.3. ORIGINS AND AFFINITIES

Although they meet on forested mountain slopes, 75 % of resident lowland genera and 50 % of montane genera are unshared. At species level, 57 % of the resident lowland inland forest community is endemic to the Sunda subregion, but only 16 % of the montane community and 10 % of that of mangrove. Exclusively Peninsular Malaysian endemics comprise, in lowland rain forest, only the Malayan peacock pheasant *Polyplectron malacense* (unless it can also be found in Sumatra) and, in montane forest, the mountain peacock pheasant *P. inopinatum* (accepting that the whistling thrush *Myiophoneus robinsoni* is not specifically distinct from Greater Sunda island relatives). In mangrove, there are none. That peacock pheasants are the only endemics, in both inland communities, may reflect their polygynous social systems (Davison, 1983) under which sexual selection of one or more kinds could have driven the evolution of special plumage comparatively rapidly.

Evidently, much of the richness of the lowland mixed dipterocarp forest bird community owes its origin to events and circumstances peculiar to Sundaland. The habitat itself has a similarly restricted range (Chapter 2). This is untrue of montane forests and mangrove which, together with representative avifaunas, occur well beyond the limits of the subregion. All forest types are none the less likely to have attained their characteristic bird communities, through species multiplication or immigration, over the later Quaternary period. In the Sunda lowlands, opportunities for speciation have probably been provided both by high sea levels striking across the continental shelf (as now) and, during periods of eustatic recession of the sea, by fragmention of the forest cover into 'habitat islands' by savanna intruding under the influence of a more continental climate (Chapter 1; Batchelor, 1983). The assumption is that an alien vegetation, with its own competing avifauna, would have been as effective in restricting the dispersal of rain forest birds, hence gene flow, between such habitat islands as the sea is between real islands.

One still restricted species range, that of Gurney's pitta *Pitta gurneyi*, suggests the past existence of a rain forest refuge in the Burmese/Thai part of the peninsula. Several Indochinese birds occurring no further south have developed local subspecies that also hint at long residence in the same area (Wells, 1985). Implied in this reconstruction is a reduction in lowland rain forest cover, leading to faunal losses, in the Malaysian part of the Peninsula because of its more central position in the exposed landmass of the glaciations and hence more severely seasonal climate. It follows that the lowland forest, now the richest habitat, may have acquired or re-aquired some of its birds by colonization only rather recently. Support for this interpretation is provided by many examples of northern *vs.* southern subspecies among lowland rain forest birds, a few of which (e.g., pied hornbill *Anthracoceros albirostris*) show introgressive plumage patterns symptomatic of recent contact (Wells and Medway, 1976). Supportive, too, is the anomalous distribution of the fulvous-chested flycatcher *Rhinomyias olivacea*. Though widespread in the Sunda Islands, it occurs in the Peninsula only in semi-evergreen rain forest over the same range as Gurney's pitta, as though cut off by an extinction in the south which Thai birds have been unable to replace and which, by chance, has never been made good from the more humid forests of Sumatra or elsewhere.

Certain other species absent from the Peninsula, but present in Borneo and Sumatra, have evidently also failed to gain or regain a foothold.

Despite supposedly long persistence, montane forest habitat in Peninsular Malaysia has not favoured local speciation. Of the true montane avifauna, 81% is in common with the uplands of tropical Indochina/China, and loss of species southward towards the Peninsula points to the principal historical direction of dispersal. Conversely, no montane Sunda endemic in the Peninsula occurs north of the Thai border. The endemic mountain peacock pheasant is most closely related to a species in the highlands of Sumatra. All twelve more widespread Sundaic species are also shared with that island, fire-tufted barbet *Psilopogon pyrolophus* and Sumatran niltava *N. sumatrana* entirely so. These two, and three others, are not differentiated from their Sumatran relatives, indicating a rather recent invasion, probably from the larger montane block of Sumatra to the smaller one of the Peninsula. With the exception of the localized population of hill prinia *Prinia atrogularis* on the summit plateau of Mt Tahan (its nearest related populations in Sumatra), no outlying peak in Peninsular Malaysia supports independent divergence even to subspecies level.

13.4. FOREST BIRDS ON SEA ISLANDS

Forest birds are known from 28 Peninsular Malaysian islands, amounting to 117 species. Of these, pied imperial pigeons *Ducula bicolor* nest exclusively on islands, but only Nicobar pigeon *Caloenas nicobarica* and an unnamed scops owl are true island 'tramp' species. There is a single montane representative, streaked wren babbler *Napothera brevicaudata* on Tioman. Island forest faunas otherwise comprise impoverished subsets of mangrove and mainland lowland forest birds. Morphological divergence has occurred only on islands that are both comparatively large and remote. On Tioman, forms of red-eyed bulbul *Pycnonotus brunneus*, grey-throated babbler *Stachyris nigriceps* and black-throated tailorbird *Orthotomus atrogularis* have been distinguished and, in the Langkawi group, collared scops owl *Otus bakkamoena*, laced woodpecker *Picus vittatus* and white-rumped shama *Copsychus malabaricus*. The owl, shama and tailorbird are larger overall, or have larger foraging organs than mainland counterparts, conforming with the common trend of character release on islands world-wide. Expansion of niche is implied, as it is in white-chested babblers *Trichastoma rostratum* on Pangkor, Perak, which have spread from their normal valley-bottom habitat to colonize the island's hillsides.

Purple-throated sunbird *Nectarinia sperata* on Redang, Terengganu, provides a possible example of recent invasion. The species was not recorded in 1950 (Gibson-Hill, 1952), but in 1977 was the commonest and most conspicuous passerine in all wooded habitats on the island. A congener, olive-backed sunbird *N. jugularis*, has meanwhile disappeared, except from a nearby islet not at present inhabited by the purple-throated.

13.5. DIVERSITY

13.5.1. Variations in species richness

Countrywide observations adequately demonstrate that dipterocarp forest of the lowland plains is the richest of all bird habitats in Peninsular Malaysia. Within-habitat or alpha-diversity is high, with large numbers of species found in homogeneous forest vegetation: for instance, at two well studied localities, 196 species in roughly 2 km^2 of gently undulating land at Pasoh, Negeri Sembilan, and 202 in 2 km^2 of the Kerau Game Reserve at Kuala Lompat, Pahang. Peat swamp forest has been less closely investigated. A six-day exploratory census of about 1 km^2 near Tanjung Karang, Selangor, found only 71 species. While this total is unlikely to have been exhaustive, several groups of easily detected birds were notably under-represented by comparison with dryland dipterocarp forest. Thus, of two usually sympatric forest shamas, only rufous-tailed *Copsychus pyrropyga* was found, and of aerial sallying insectivores only one flycatcher and one monarch, compared with 15 in the same month at Pasoh. Comparable impoverishment has been noted in lowland peat forests of southeast Sumatra (Silvius *et al.*, 1984). It is also believed that the semi-evergreen rain forest of the northwest (Chapter 2) is relatively poor in bird species, although an inland population of greater goldenback *Chrysocolaptes lucidus* and resident hooded pitta *Pitta sordida* are unique there. By comparison, in a very well studied area of montane habitat at Fraser's Hill, the total of 77 species excluded only six strictly montane species present elsewhere in the Main Range.

Alpha-diversity declines with increasing altitude (Fig. 13.2). Of the lowland community, the ranges of 41 species terminate more or less at the hill-foot boundary. This important break correlates with a marked change of slope in the terrain although not with an obvious vegetational ecotone. Losses again increase sharply with the approach to mossy forest which almost everywhere begins close to the lower limit of lower montane forest.

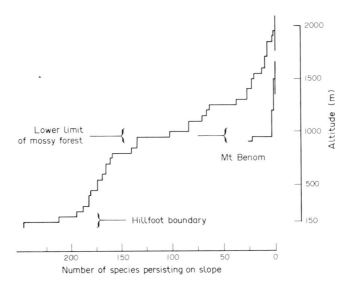

Fig. 13.2. Limits to the range of lowland forest birds on mountain slopes in Peninsular Malaysia; pooled records versus those from a single sample locality (Mt Benom, Pahang). (Brackets denote the approximate position of altitudinal habitat boundaries).

Lowland species are not entirely excluded from montane forest but, apart from mobile nectarivores, frugivores and predators, occurrences are rare at least in pristine habitat, such as Mt Benom (Fig. 13.2, inset). Indeed, forested summits too small to support true montane avifaunas, such as Kedah peak (=Gunung Jerai) or Ledang, remain comparatively birdless. It seems that few lowland species other than these possess the behavioural plasticity to fill the vacant ecological space.

In the central and southern parts of Peninsular Malaysia, 14 bird species are confined to middle slopes (Wells, 1985). Five of these replace lowland congeners or close relatives. A dependable example is black-browed barbet *Megalaima oorti* which supplants the lowland species, yellow-crowned barbet *M. henricii*, at about 800 m and is in turn replaced by golden-throated barbet *M. franklinii* at about 1350 m. Relic populations of crested argus *Rheinardia ocellata*, isolated at the upper limit of hill dipterocarp forest on a few sandstone mountains along the Pahang-Kelantan-Terengganu border, exist in similar relationship to the lowland great argus *Argusianus argus* (Davison, 1979). Grey-cheeked and ochraceous bulbuls (*Criniger* spp.) occupy mutually exclusive ranges but do not always directly replace each other. On the northeast slope of Mt Benom, they are separated by some 200 m elevation (Medway, 1972). Much wider altitudinal gaps between other potentially parapatric pairs have been found on the submontane slope of Mt Mulu, Sarawak, suggesting other limitations.

In the Main Range and on large outliers, the lower edge of mossy forest rather rigorously restricts the distribution of montane birds. Recognition of the boundary at the level of the individual is obvious in the behaviour of multi-species foraging flocks. It is habitually crossed only by a few medium to large frugivores and predators. Again, among the montane species delimited, few meet congeneric lowland forest counterparts. Upwards, most montane birds reach peak abundance within characteristic bands of habitat, often spaced between congeners such as in the warblers *Seicercus* and laughing-thrushes *Garrulax*. Alpha-diversity reaches minimum in elfin summit vegetation. On Mt Benom, the largest pristine mountain yet examined in detail, only 11 species have been identified on the summit ridge. Countrywide, just two species appear to be exclusive to this zone.

13.5.2. Abundance and biomass

The number of individuals living in a defined area has been measured only at Pasoh, in lowland dipterocarp forest below the hill-foot boundary, and within this habitat only in the understorey. Excluding as far as possible non-territorial itinerants and young of the year, a 15 ha plot supported 381 individuals of 69 species, of aggregate biomass (calculated from average body weights) close to 17 kg. Within this sample, the distribution of abundance was log-normal (Fig. 13.3). 52 species (76 percent of the assemblage) were represented by seven individuals or fewer (2% or less of the sample) and 25 by a maximum of two individuals. Scaly-crowned babbler *Malacopteron cinereum* was commonest, at 24 individuals or 6.3% of the sample, followed by short-tailed babbler *Trichastoma malaccense* and Siberian blue robin *Erithacus cyane* (both among the few litter-foraging passerines of lowland forest), then yellow-bellied bulbul *Criniger phaeocephalus*.

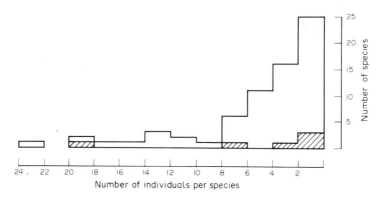

Fig. 13.3. The abundance of sedentary adult birds in 15 ha of mature lowland forest understorey at Pasoh, Negeri Sembilan. (Hatched parts of the columns denote non-breeding migrants).

There are no data from other Peninsular Malaysian forest habitats that can be compared directly but at Pasoh again Wong (1985) found a lower density of territorial birds in the understorey of old regenerating forest than in a plot in the adjacent primary forest. On Mt Benom, mist-net captures fell from a mean 1.8 per standard net/day at the base of the mountain to 1.0 close to the mossy forest boundary (Medway, 1972), implying a decline in density. On Mt Mulu, Sarawak, Wells *et al.* (1978) found that capture rate overall declined more steeply than loss of species, concordant with the proposition that most lowland species encounter progressively less suitable living conditions with increasing elevation. On the flank of Mulu, except among mid-elevation specialists, proportionally fewer young were fledged than within the community in adjacent forest below the hill-foot boundary.

13.6. PATTERNS IN THE USE OF FOREST RESOURCES

Although there are some data on birds operating near the ground in lowland rain forest, little has been recorded of foraging activities in the avifauna as a whole. The following analysis, carried out separately for each habitat community, is restricted to main items of diet and to one space component, namely, the position in the vegetation column at which a species typically finds food. Not in all cases has this been easy to fix; for instance, some large hawks wait on high branches but make some captures on the ground. Never the less, for most bird species there is a regular relationship between the structural layers of the forest and expectation of occurrence.

The vertical divisions recognized in Tables 13.1 and 13.2 are: the *understorey*, being the litter on the ground through herbs, seedlings, saplings, stemless palms and boles to a height of, very roughly, 3 m; the *canopy*, here defined as the uneven leafy 'skin' of the forest, elevated into the often rather open crowns of giant emergents; and a more diffuse *interior* zone of trunks, big branches, sub-canopy crowns and a good deal of generally shaded space above the understorey and below the canopy. Interior and understorey together are conveniently termed the *shade layer*. The shade layer makes up the greater habitat volume, but the canopy varies much in height, around gaps and at natural forest boundaries descending almost to ground level. The typical 'edge' bird subcommunity includes canopy species that are able to make this descent with their habitat. Because mangrove mostly lacks a distinctive leafy understorey, Table 13.3 recognizes only two layers: *crown* and *sub-crown*.

The diet groupings are approximations that include a maximum of two components deemed to contribute importantly. They are tabulated in trophic order, from primary consumption (nectar, fruit, seeds) through mixed diets to secondary and tertiary consumers. The tabulations show numbers of species and niches occupied in the habitat communities, without regard to relative density or biomass of the populations involved. Other than strictly edge and small-island tramp species, they cover all birds of each habitat community hence perhaps 10 percent more than are likely to be found at any one site.

13.6.1. Animalivores

Invertebrates (essentially arthropods, and chiefly insects) offer the greatest number of compatible ways of existence for birds. In montane and lowland forests and mangrove, insectivores outnumber other single component specialists by factors of 7, 5 and 2, respectively. Species incorporating arthropods into mixed diets outnumber all other kinds of generalist, and only among insectivores do specialists outnumber generalists.

Vertebrate prey appear to offer a smaller range of niches and, except in mangrove where fish are important, there are fewer specialist predators than generalists. Values are lowest in montane habitat where the only truly indigenous predators of vertebrates are two small owls and the green magpie *Cissa chinensis*, which are also all partially insectivorous. The four hawks forage also in the lowlands — indeed, none has yet been found breeding in montane forest. The two hornbills are also visitors from the lowlands. This poor carrying capacity is plausibly attributed to the comparative scarcity of reptiles, widely taken by these hawks and hornbills.

13.6.2. Nectarivores

At the lowest of the trophic levels, eaters of flower parts, fruit or seeds form one third or less of community species totals (Tables 13.1-3) and in each habitat 19-23 % of diets (Table 13.4).

Sunbirds (Nectariniidae) are the principal nectarivores. Most seem to take nectar whenever it is available, but no Malaysian member of the family is a total nectarivore. Thus, the long bills of spiderhunters *Arachnothera* spp. are tools for extracting arthropods (including spiders from webs) as

TABLE 13.1. Exploitation of food and space by the mainland lowland forest bird community. (Numbers are species per taxonomic group; forest-edge and island tramp species are excluded).

Space: Diet:	Sheltered stream	Understorey	Interior	Canopy	Above canopy
Land vertebrates (9)	—	?	hawks 1 owls 1	hawks 5	hawks 1 falcons 1
Fish (5)	hawks 2 owls 1 kingfishers 2	—	—	—	—
Small verts/ large inverts (17)	herons 1 storks 1 finfoots 1 rails 1 broadbills 1	cuckoos 1 owls 2 kingfishers 1	owls 2 kingfishers 1 broadbills 2	hawks 2 falcons 1	—
Invertebrates (139)	thrushes 1 flycatchers 1	trogons 2 kingfishers 1 woodpeckers 3 pittas 4 crows 1 babblers 23 thrushes 6 flycatchers/ monarchs 6 wagtails 1 sunbirds 1	cuckoos 9 frogmouths 3 trogons 3 honeyguides 1 woodpeckers 7 broadbills 2 drongos 2 nuthatches 1 babblers 3 thrushes 1 warblers 2 flycatchers/ monarchs 12	cuckoos 4 bee-eaters 2 woodpeckers 3 cuckooshrikes 7 leafbirds 2 drongos 2 crows 2 tits 1 babblers 2 warblers 4 flycatchers 3	nightjars 1 swifts 9 treeswifts 2 rollers 1 swallows 2
Small verts/fruit (9)	—	—	hornbills 2	hornbills 7	—
Invertebrates/ fruit (50)	—	pheasants 10 pigeons 1 bulbuls 2 flowerpeckers 1	barbets 1 woodpeckers 1 bulbuls 6 flowerpeckers 2	barbets 5 cuckooshrikes 1 leafbirds 3 bulbuls 9 orioles 1 starlings 1 flowerpeckers 5 white-eyes 1	—
Invertebrates/nectar (19)	—	sunbirds 1	sunbirds 3	sunbirds 6	—
Fruit (13)	—	—	pigeons 1 broadbills 1 parrotfinches 1	pigeons 6 parrots 3 fairy bluebirds 1	—
Nectar (0)	—	—	—	—	—

much as the housing for highly protrusible tongues. Bill-lengths of the larger lowland forest canopy species are in rough proportion to body mass (Table 13.5), although it is likely that the different lengths and curvatures have evolved at least partly in relation to flower geometry. It must nevertheless be added that, so far, there is no confirmed case of effective pollination by birds of any Malaysian forest plant.

All spiderhunters and *Aethopyga* and *Nectarinia* sunbirds of lowland forest also range deeply into montane forest. If the large individual movements of ringed little spiderhunters *Arachnothera longirostra*

TABLE 13.2. Exploitation of food and space by the montane forest bird community.

Space: Diet:	Sheltered stream	Understorey	Interior	Canopy	Above canopy
Land vertebrates (4)	—	—	—	hawks 4	—
Fish (0)	—	—	—	—	—
Small verts/large inverts (3)	—	owls 1	owls 1 crows 1	—	—
Invertebrates (58)	thrushes 2	pittas 1 babblers 5 thrushes 3 flycatchers 2	cuckoos 1 trogons 1 woodpeckers 5 broadbills 1 drongos 1 nuthatches 1 babblers 6 warblers 5 flycatchers/ monarchs 5	cuckooshrikes 1 babblers 1 flycatchers 1 warblers 2	swifts 10 swallows 3
Small verts/fruit (2)	—	—	—	hornbills 2	—
Invertebrates/ fruit (22)	—	pheasants 3 babblers 1 thrushes 1	barbets 1 bulbuls 1 babblers 3	barbets 2 cuckooshrikes 1 leafbirds 1 bulbuls 1 orioles 1 babblers 3 white-eyes 1 flowerpeckers 1 finches 1	—
Invertebrates/ nectar (2)	—	—	—	sunbirds 2	—
Fruit (6)	—	pigeons 2	parrotfinches 1	pigeons 3	—
Nectar (0)	—	—	—	—	—

(as much as 5 km in distance and 490 m in elevation) are typical of the group, then it appears that their economy is geared primarily to finding nectar sources. The more fully insectivorous sunbirds of the lowlands (*Anthreptes* and *Hypogramma* species) rarely transgress the montane forest ecotone. At the same time, the montane forest also supports species of its own (streaked spiderhunter *Arachnothera magna* and black-throated sunbird *Aethopyga saturata*) as does mangrove.

Elsewhere in the Oriental region, hanging parrots (*Loriculus*), fairy bluebirds (*Irena*), leafbirds (*Chloropsis*) and white-eyes (*Zosterops*), are known to take nectar in forest but observations in Peninsular Malaysia are not confirmed.

13.6.3. Frugivores

Many families contribute to the generalist frugivore totals (Tables 13.1-3), but specialist or obligate frugivores (taking no or few other kinds of food) are represented only in five. In both inland forest

TABLE 13.3. Exploitation of food and space by the mangrove bird community on the mainland.

Space: Diet:	Below crown (incl. water)	Crown	Above crown
Land vertebrates (2)	—	hawks 2	—
Fish (5)	herons 1 owls 1 kingfishers 3	—	—
Small verts/large inverts (4)	kingfishers 3 crows 1	—	—
Invertebrates (32)	pittas 1 babblers 1 warblers 1	cuckoos 5 woodpeckers 5 cuckooshrikes 2 leafbirds 1 drongos 1 tits 1 babblers 1 thrushes 1 warblers 5 flycatchers/monarchs 4 whistlers 1	swallows 2
Small verts/fruit (0)	—	—	—
Invertebrates/fruit (3)	—	white-eyes 1 flowerpeckers 2	—
Invertebrates/nectar (4)	—	sunbirds 4	—
Fruit (8)	—	pigeons 4 parrots 3 starlings 1	—
Nectar (0)	—	—	—

TABLE 13.4. Actual and proportionate (percentage) representation of diets among bird species using mature forests. Note that the half-components of a non-specialist diet contribute 0.5 points each and a whole diet one point (after Karr, 1975).

Food: Community:	Nectar	Fruit/seeds	Invertebrates	Vertebrates
Mangrove	2 (3)	9.5 (16)	37.5 (65)	9 (16)
Inland lowland forest	5 (2)	42.5 (18)	166.5 (71)	25 (11)
Montane forest	1 (1)	18 (22)	57.5 (69)	6.5 (7)

communities, generalist frugivores (among which have been included the barbets, Capitonidae; Figure 13.4) outnumber specialists by a factor of four. Of the 13 identified specialists in lowland forests and six in montane forest, probably only 10 pigeons of the genera *Treron, Ducula* and *Ptilinopus* and the three parrots feed solely on fruits, seeds or flowers. Peninsular Malaysian parrots are mainly flower and/or

TABLE 13.5. Relationship between average bill-length and body-weight in spiderhunters (*Arachnothera* spp, Nectariniidae) of the inland lowland forest canopy.

Species	Bill from gape (mm)	Body weight (g)	Weight/bill ratio
Thick-billed spiderhunter	31	11	0.35
Grey-breasted spiderhunter	37	19	0.51
Yellow-eared spiderhunter	46	22	0.47
Long-billed spiderhunter	58	28	0.48

seed predators. Most other frugivores are likely to disperse viable seeds of some of the plants they attend, either by carrying fruit to eat at a distance or after passage through the gut. This implies co-evolution between plant and bird, and the widely recognized dichotomy of fruiting strategies by bird-dispersed plants in the tropics (Snow, 1981) seems to hold in Peninsular Malaysian forests.

The great majority of small generalist frugivore-insectivores concentrates on the semi-continuous or frequent fruitings of pioneer to building-phase trees at disturbed sites. It is presumed (though no-one seems to have checked, locally) that individually these typically small fruits present a low nutrient reward, but can be gathered in large amounts with little energy expended — once the source has been located. In the lowland rain forest, their chief consumers are bulbuls (notably *Pycnonotus* and *Hypsipetes* species), leafbirds and flowerpeckers and, in montane forest, bulbuls, a leafbird, a flowerpecker, a white-eye and babblers (especially in the genera *Garrulax, Leiothrix* and *Heterophasia*).

Fig. 13.4. Barbets are generalist frugivores. (Photograph by Frank Lambert.)

Many figs, *Ficus* spp., exhibit a similar strategy. Those that fruit at canopy level and produce small, soft synconia attract a wide range of frugivores. Over 30 bird species have been identified at individual strangling figs (McClure, 1966; Wells, 1975). The few Peninsular Malaysian fig syconia tested (Vellayan, 1981; F.R. Lambert, pers. comm.) offer low protein and fat rewards. Nevertheless, within the fruit part of their diet some birds may preferentially eat figs, including lowland forest *Megalaima* barbets and helmeted hornbill *Rhinoplax vigil*.

Among obligate frugivores, the green pigeons *Treron* spp., green broadbill *Calyptomena viridis* and fairy bluebird are notable fig visitors. For these birds, the relatively frequent, ill-synchronized fruitings of figs are liable to provide an important stop-gap when other food is in short supply. Equally, for a number of fig species, birds are undoubtedly the principal diurnal dispersers of seed. As among all fruit, size, hardness, firmness of attachment and the weight-bearing capacity of surrounding stems (i.e., accessibility) must be important ways in which the dispersers and, conversely, the resource are partitioned. Thus, among the commonest fig-visiting green pigeons, *Treron olax, T. curvirostra* (Figure 13.5) and *T. capellei,* consecutive weight differences approach or exceed a factor of two. These pigeons also differ in the size, shape and strength of the bill.

Obligate frugivores that may not habitually depend on figs are the imperial pigeons, *Ducula* spp., and jambu fruit pigeon *Ptilinopus jambu*. Like the green pigeons, imperials tend to move about in groups. The jambu fruit pigeon makes communal nocturnal dispersal movements, but seems in the main to forage singly. In lowland rain forest it has been seen taking sparse, inconspicuous, dark-coloured fruits from an interior crown that was attended by no other birds. Possibly, it represents the opposite extreme of frugivory by favouring scarce but high reward items, evolved to attract or be accessible to few other dispersers.

The volume of edible fruit available to birds fluctuates in space and time. Even among figs there are short periods when, locally, no tree is producing. The phenology of few other kinds of bird trees has yet been elucidated. If most of those at canopy and interior levels participate in the general pattern of weak

Fig. 13.5. The thick-billed green pigeon *Treron curvirostra*. (Photograph by Frank Lambert.)

annual periodicity, with greater than annual intervals between episodes of heavy, gregarious flowering (Chapter 9.4), there will be long periods – at least, in the lowland forests – when fruit is scarce over large areas. As among nectarivores, the evolutionary response of most frugivores has been to turn to animal prey, mainly invertebrates but in the case of hornbills also small vertebrates. All Malaysian hornbills that have been tested respond enthusiastically to live prey, and their bills are clearly adapted not only for manipulating fruit but also for seeking and capturing arboreal reptiles.

A complementary behavioural strategy in response to the clumped dispersal of fruit sources is the patrolling of a large activity space. Complete individual ranges have so far been mapped only for radio-tagged adult male great argus (Davison, 1981). Unusual reproductive behaviour, tying the male to the vicinity of a display court, makes it unlikely that the maximum 6.2 ha of this generalist pheasant is typical. Indeed, tagged great argus spent much of the time roosting quietly – one way to minimize energy needs in an environment where ambient temperatures remain close to the thermoneutral range. Among canopy frugivores, wreathed and wrinkled hornbills, *Rhyticeros* spp., make long flights high over the forest, in the case of the wreathed sometimes between distant isolated forest patches or to unusual elevations. For some bulbuls, such as the black-headed *Pycnonotus atriceps*, ranges that are large by comparison with those of wholly insectivorous passerines have been inferred from very low rates of recapture of ringed individuals at regular netting sites (Medway & Wells, 1976). An extreme example is the black-and-white bulbul *P. melanoleucos*, occuring in all inland forest formations but appearing at any one site in small flocks at erratic intervals: four times in 54 months of regular watching at Pasoh forest reserve.

In this case, as for emerald dove *Chalcophaps indica* and most if not all specialist frugivores, true nomadism is implied. Ringed pigeons have been recovered as far away as 800 km (emerald dove) and have crossed the sea to Sumatra (emerald dove, jambu fruit pigeon). Disperal by these latter species, green pigeons and occasionally green broadbills and blue-crowned hanging parrots *Loriculus galgulus* occurs at night. Other parrots may be seen moving by day, and all were quick to discover the food bonanza offered by the spread of oil-palm estates (Ward & Wood, 1967). The pin-tailed parrot finch *Erythrura prasina*, a specialist predator of seeds of the large bamboos of forest gaps and the forest edge (Chapter 5), is another such nomad.

The four pigeons of montane forest, all obligate frugivores, do not make nocturnal dispersal movements. Two of them, little cuckoo dove *Macropygia ruficeps* and mountain imperial pigeon *Ducula badia*, on the other hand, habitually make diurnal sorties into lowland habitat. Here they feed, but may also obtain other items such as minerals. Thus imperial pigeons commonly visit mangroves, which otherwise appear to offer frugivores only the benefit of safe communal roost sites (Nisbet, 1968), and little cuckoo doves resort to mineral springs (G. C. Madoc, in litt.). As far as is known, all species nest exclusively in montane forest.

13.6.4. Vertical stratification

Table 13.6 summarizes the distribution of diet components through the forest column. Both in lowland and montane forests, nectar and fruit eating are strongly skewed towards the canopy. As in other rain forests, understorey values are boosted by the presence of birds taking fallen fruit that is not the product of the layer in which they forage. These birds are all generalists. Among them, the emerald dove is nomadic, but pheasants and partridges are not. Their social organization and patterns of dispersal (Davison, 1981) seem more strongly influenced by the animal component of their diets.

By contrast, insectivory predominates in the shade layers, peaking in the understorey of lowland forest and interior of montane forest, although some difficulty was experienced in separating these

TABLE 13.6. Vertical distribution of feeding niches through the vegetation column of inland forests. (L = lowland forest and M = montane forest. The point system is as in Table 13.4 and values are real numbers. Above-canopy foraging is discounted).

Space: Diet:		Understorey	Interior	Canopy
Vertebrates	L	9.5	5.5	10
	M	0.5	1	5
Invertebrates	L	62	55	49.5
	M	16	29.5	12
Fruit/seeds	L	7	9	26.5
	M	4.5	3.5	10
Nectar	L	0.5	1.5	3
	M	0	0	1

layers within the latter habitat. In the understorey especially, environmental conditions may favour narrow niche specialization among the arthropods themselves and provide more kinds of arthropod microhabitats on which different bird species may concentrate (Figure 13.6).

Fig. 13.6. Large frogmouths *(Batrachostomus auritus)* are insectivores of the shade layer, specialized for nocturnalism. (Photograph by Frank Lambert.)

For vertebrate predation, the high understorey figure in lowland forest partly reflects fishing, and fish resources are non-existent in montane forest streams (Chapter 16). Figures could be deceptive because, in both forest types, *Spilornis* and *Spizaetus* eagles spend most time at canopy level but catch some prey on the forest floor.

13.6.5. Interregional comparisons

Directly comparable data from other parts of the tropics are available only for lowland forests, in Gabon and Papua New Guinea (Table 13.7). The similarity between Gabon and Peninsular Malaysia is startling. With habitat communities not greatly different in size, it suggests that resources in these two regions have provided parallel opportunities for evolution. In Papua New Guinea, representing East Malesia, the similarity stops at vertebrate predation. Within the somewhat smaller sized community of the chosen example, arthropods are proportionally only half as important as in Peninsular Malaysia, and primary products 2–3 times as important. This bird fauna, taxonomically more remote than the African example, has clearly evolved along a substantially different pathway. Thus, in a sample of 6 ha of rich submontane forest in eastern Papua New Guinea, 50 % of the tree species produced seeds dispersed by birds (Pratt and Stiles, 1985). Raemakers *et al.* (1980) estimated that 67 percent of Peninsular Malaysian forest trees are dispersed by "pulp eaters". In a sample of 33 species on the submontane slope at Ulu Gombak, Selangor, however, McClure (1966) identified only four (two of them figs) with fruit taken by birds. In Sarawak, Fogden (1972) put the proportion of lowland tree species bearing fruit edible by birds at below 10 percent.

TABLE 13.7. Percentage representation of food niches occupied by birds in inland lowland forest in three regions of the inner tropics. (Gabon data are from Anon. (1979) for the Makokou area, Ivindo valley, and the African literature; Papua New Guinea data are from Bell (1977) for the Brown river area, Port Moresby).

Food: Locality:	Nectar	Fruit/seeds	Invertebrates	Vertebrates
Peninsular Malaysia	2	18	70	10
Gabon, W.–Central Africa	2	17	70	10
Papua New Guinea	9	41	39	11

13.7. SEASONALITY

13.7.1. Food and foraging

The nocturnal dispersal movements of obligate frugivores of lowland forest can be interpreted as part of a food-searching response to the patterns of forest phenology (Chapter 9). There is evidence of seasonal enhancement of activity among jambu fruit doves and *Treron* pigeons; at floodlights at Fraser's Hill, more birds were caught in relation to trapping effort during September–December than in other months (Medway and Wells, 1976), but rarely was no movement detected.

While frugivores seek localized abundances, at which they may gather in large numbers, insectivorous birds behave as though food in forest is never abundant or easily procured. With the exception of ground-layer foragers (e.g., pittas, some babblers, Siberian blue robin), small to medium-sized species commonly forage in multi-species flocks. These form at least as frequently in montane as in lowland forest, but have not been confirmed to occur in mangrove.

One benefit of flocking is to enhance the foraging efficiency of vegetation searchers and surface gleaners especially by allowing participants to exploit the search movements of neighbours. It is assumed that more useful prey items are captured than could be located by the lone forager in equivalent time, while the flock also provides access to an enlarged area of habitat. The extent to which the flock member obtains the additional benefit of a reduced need for vigilance is uncertain in Peninsular Malaysian forests, where diurnal predators of birds are scarce.

Routes taken through the forest show some consistency. From a vantage point in the canopy, McClure (1967) noted that total space covered by a flock is predictable from day to day. The implied territoriality suggests interactions between equivalent flocks (cf. Munn and Terborgh, 1979), but this has not been studied. Vertically, canopy and shade-layer flocks span the full height of the forest column and at times they move in unison, the adaptations of individual participants ensuring that the different assemblages maintain their typical levels.

In these flocks, some species occur at a higher incidence than others, constituting the nuclear membership. Birds such as drongos typically form the core around which the flock reassembles daily (from solitary roost sites), suggesting continuity of individual as well as species membership. From studies in Sarawak, Croxall (1976) has suggested that there may be a limit to the overlap of modes of foraging by flock members. This has not been investigated locally.

A further uncertainty is the extent to which flock size in Peninsular Malaysian forests is influenced by species *versus* number of individuals. Many species often occur at a rate of more than two individuals per flock, including even such aggressive birds as greater or lesser racket-tailed drongos *Dicrurus paradiseus* and *D. remifer* (respectively in lowland and montane forests). Drongos take prey raised by the predominant vegetation searchers in swift, far-reaching, aerial sallies; in exchange, perhaps, they offer a protective or sentinel role. Intuitively, it would not seem likely that many individuals could coexist with profit. Other species associate only temporarily with flocks. Thus, a female scarlet-rumped trogon *Harpactes duvaucelii* noted quietly flycatching in step with but 2–3 m above a group of understorey babblers, without her mate (seen in the area), abandoned the group as soon as the latter vacated her territory.

Although multi-species flocks form throughout the year, participation by local birds intensifies in the later months, typically from August. In most years, this is the period of peak recruitment of recently independent young and of incoming migrant conspecifics and others (13.9), hence plausibly the time of greatest pressure on food resources. It is also a season of reduced abundance of most arthropods of interest to insectivorous birds, reaching a low point between September and November (Ward, 1969; Davison, 1981). The decline is not more than a few fold, but a rising incidence of prolonged rain in late months of the year may exacerbate its effects by curtailing the foraging day.

13.7.2. Breeding and moulting

Where data are adequate, reproduction and renewal of the plumage have mostly been found to be discreet, cyclic events, and more or less exclusive. Only in swiftlets, *Aerodramus* and *Collocalia* spp. (Langham, 1980; Hails and Turner, 1985), which have aerodynamic requirements that dictate a slow rate of moult, in male crested argus on their display courts, and in female hornbills, confined to the nest chamber during incubation and brooding, are breeding and moult of the primaries known to overlap in the individual. Hornbills, of course, are fed by one or several other adults during this period and expend no energy in foraging.

Typically, in forest birds, breeding and moulting are seasonal, mostly less synchronized and with greater year-to-year variation locally than, for example, the birth dates of microchiropteran bats (Chapter 12.6.2). Yet the first egg date of the mangrove and open country Pacific swallow *Hirundo tahitica*, is as predictable as bat seasons (C.J. Hails, pers.comm.) and, within a broader population spread, individual yellow-bellied bulbuls at Pasoh have moulted the equivalent primary feather within one week of a twelve-month interval.

In general, schedules vary with the principal component of diet. They are least defined among frugivorous pigeons, feeding their young on crop secretions. Mountain imperial pigeons show year-round breeding, with eggs or chicks recorded in all months except May, August and November, as do little cuckoo doves with records in all except April, November and December. Nothing is known of the moult periodicities of these two species, but records of primary moult among adult emerald doves in May and from August through February suggest a spread approaching the non-seasonal.

The few nest records available for diurnal raptors of forest are combined in Figure 13.7(a). On barely adequate evidence, hawk eagles (*Spizaetus* spp.) lay principally in November or December and fledge most young in February or March, though a downy chick of Blyth's hawk eagle *S. alboniger* at the end of March is unlikely to have flown until May and nestling changeable hawk eagles *S. cirrhatus* have been found in May. It may be some advantage to the nestling to be raised during the relatively dry months of the early part of the year, but food benefits, either in or after leaving the nest, have still to be identified. Though it forages entirely outside forest, the white-bellied sea eagle *Haliaeetus leucogaster* starts at least as early, with no young recorded after early June (G.C. Madoc, *in litt.*). On the other hand, crested serpent eagles, more specialized snake hunters than hawk eagles, have been found at the nest in September. Nothing is known of the moult schedules of any of these species, but crested goshawks *Accipiter trivirgatus* moulting in the primary tract have been mist-netted in July and October. Nests of this goshawk, partly a bird-predator, have been found in January and April, and young may be on the wing when the fledglings of insectivorous birds make comparatively easy prey. The seasonality of nocturnal owls taking vertebrates is too poorly known to be considered, although nests with eggs of buffy fish owls *Ketupa ketupu* have been found in all months except February, March, May, June and August.

Far more information has been gathered on the largest dietary grouping detailed in Tables 13.1–13.3, birds making important use of arthropods and primary consumers whose chicks are likely to be dependent on animal protein for their early growth. Records of breeding among 127 species and primary tract moult (as an index of total moult) among 103, excluding migrants, even though drawn from many localities and pooled over many years, both show a marked and unimodal seasonality (Figure 13.7b). This figure makes no distinction between months of usual versus exceptional activity. Moreover, pooled data have broadened, hence overlapped, the bases of the curves well beyond the limits expected in a single locality in any one year. It is, nevertheless, easy to see that most breeding occurs within the season of generally greater abundance of arthropods in forest and that, even though moult is staggered by a minimum four months, its incidence declines sharply after October. Breeding and subsequent moult thus fall outside the lean season and wettest weather, during which significant feather loss could also add to the demands of thermoregulation.

The curve of breeding incidence within montane forest shows a slightly sharper peak and no activity in November-December. The seasonality of montane arthropods has never been separately studied, but an enhanced weather factor may operate. While wet everywhere, November and December are also the months of most persistent mountain mist cover, producing conditions of coolness and reduced visibility that are avoided by birds on mountains elsewhere in the tropics (Serle, 1981; Round, 1982). The slightly earlier peak, in comparison with the lowlands, is possibly an artifact since March has been a favoured month for visits to the hill stations.

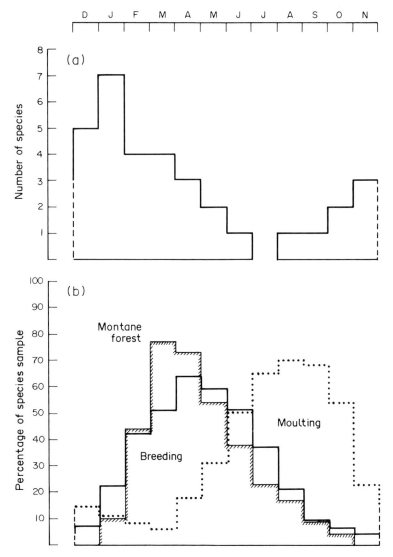

Fig. 13.7. Seasonality (a) of breeding by forest raptors (species per month, n = 8); (b) of breeding and moulting by forest birds taking arthropods (percentage species per month, n = 127 and 103 respectively).

Exceptions to this seasonal pattern among insectivores are few but interesting. Roulrouls *Rollulus roulol*, which nest on the ground, have been found breeding in all months except January, March, June, August and November. Eggs of the stream-haunting chestnut-naped forktail *Enicurus ruficapillus* have been found in all months except November, with a peak during July–September when most insectivorous passerines are well into their annual moult (Medway and Wells, 1976). In this and the white-crowned forktail *E. leschenaultii* moult occurs principally in the early months of the year, when the only other birds likely to be moulting are individuals that had been forced to suspend in the previous season. This eccentric schedule has been explained in terms of foraging advantage. Forktails take much food from the rocks and margins of forest streams, and it has been speculated that debris brought down by spates may increase their supply. Recently independent young may, therefore, obtain advantage from the more frequent spates towards the end of the year (Chapter 16.2).

That not all individuals attempt to breed every year has been observed among great argus. Suboptimal, flat land at Pasoh held more than twice the number of males occupying display courts, and the calling season lasted significantly longer in a year of exceptional fruiting than in preceding and following years (Davison, 1981). Here, proximate and ultimate timers of breeding appear to merge (cf. Immelmann, 1973) and the suggestion of Ward (1969), that reproductive development is initiated by seasonal attainment of a critical plane of nutrition in the individual, remains an attractive idea. There is no neater explanation of year-to-year variation in the onset of the breeding seasons of forest birds, or of individual variation in any one year. In yellow-bellied bulbuls at Pasoh, females attained a greater proportionate bulk of pectoral musculature before breeding than males. This makes sense in terms of egg-formation and allows the not too surprising suggestion that females provide the more critical timing response. It has been assumed that the nutritional cue is a reserve of protein but, in their studies of white-bellied swiftlets, Hails and Turner (1985) proposed a role also for stored fat. Fat is important during swiftlet egg formation and the laying season may continue only as long as daily feeding conditions permit females to replenish stores.

13.8. LIFE HISTORY PARAMETERS

13.8.1. Longevity

Resident birds that have reached the stage of independence in tropical environments often have a high life expectancy (Fry, 1980). No full life table has yet been constructed for a forest bird in Peninsular Malaysia but 46 passerines and ten non-passerines are represented in the files of the University of Malaya Bird Ringing Project by at least one individual that has survived five years from initial ringing (frequently when already adult). Because most are understorey species, accessible to mist-netting, these birds tend to belong to the dominating insectivorous diet group but they include specialists and generalists, range in mass from 9g to 115g and represent a variety of foraging behaviour guilds in all three habitat communities. Recoveries beyond five years are summarized in Figure 13.8, leading to record intervals of 173 and 156 months, respectively, in grey-cheeked bulbul *Criniger bres* and cinnamon-rumped trogon *Harpactes orrhophaeus*.

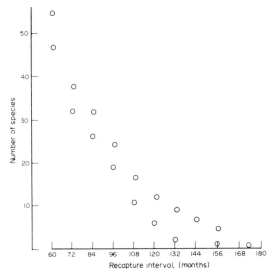

Fig. 13.8. Forest species represented by at least one individual surviving the indicated interval after ringing. (Data from the records of the University of Malaya Bird Ringing Project).

At local community level, mist-netting has shown that a minimum three percent of the marked sedentary residents of 15 ha of understorey at Pasoh remained after an interval of eight years. However, the marked individuals recovered in successive surveys were mostly different, suggesting that percentage survival is likely to have been under-estimated. From recaptures and sightings of colour-marked birds in a community in Sarawak containing many of the same species, Fogden (1972) estimated a mean year-to-year survival rate among adults of 90 percent. Both in Sarawak and at Pasoh, over twelve months within the initial census period, some species showed 100 percent survival.

13.8.2. Productivity

High survival implies commensurately little recruitment to fill gaps, and forest birds typically produce few young in any one year. Clutch-size is adequately recorded for 170 species, i.e., for between one third and one half of those characteristic of each of the three habitat communities. The full range of normal clutches among non-passerines is 1–6 eggs and among passerines 2–4. Ninety percent of all species lay three or less and the modal value is two, unbiassed by habitat (Figure 13.9). This value cuts across all diet groups as defined in Tables 13.1–13.3, except for predators taking mainly land vertebrates, among which as many species lay one as two eggs. Most are large birds. The higher modal value of three in the small vertebrate/invertebrate animalivore group reflects the large clutches of eurylaimine broadbills that build pendant, globular nests (Figure 13.10) and certain hole-nesting species such as black-thighed falconet *Microhierax fringillarius*, the broods of which may have greater than average security from predators. Construction of 'safe' nests has been noted as correlating with a proportionately bigger investment per clutch (Snow, 1975).

Within taxonomic groups, among the specialist insectivores, frogmouths and treeswifts invariably lay only one egg, and build nests designed to contain no more. Pheasants, in the invertebrate/fruit group, show the greatest variation of all, in relation to the degree of parental care accorded to chicks: from one or two in peacock pheasants (*Polyplectron* spp.) and argues that feed young bill to bill, up to six in firebacks (*Lophura* spp.) whose chicks forage for themselves from sources merely indicated by the hen, via an intermediate situation in roulroul where the brood in divided between the parents (G.W.H. Davison, pers.comm.). Among obligate frugivores, a difference exists between specialized, seed-dispersing imperial pigeons and jambu fruit doves, with clutches of one, and the sometimes seed-predating green pigeons (*Treron* spp.) with clutches of two. Flower and seed consuming parrots all lay still larger clutches, but they also use tree-holes.

Medway and Wells (1976) cited an estimated mean fledging success of 0.4 young per pair over a single breeding season in the lowland understorey community at Pasoh. A more detailed census of the same 15 ha plot in a subsequent season found 80 young among 125 pairs of 29 species, or 0.64 per pair. With other mist-nettable species, comprising an estimated 41 additional pairs, among which no young were found this figure fell to 0.48 per pair, close to that for a mix of lowland forest and forest edge species in Sarawak (Fogden, 1972). No good direct estimate of ultimate new recruitment into the breeding community was obtained at Pasoh. In his Sarawak study, Fogden put this in the region of ten percent per annum. Whilst in evolutionary terms normal clutch-size must be that which, on average, maximizes the life-time output of fledged young, there has been controversy over the link between productivity and mortality. In general, environmental factors affecting the one are likely to have a parallel effect on the other. Equable conditions that promote survival of adults are thought also to permit narrowing of the foraging niche as a way of avoiding competition, making food rarer. That this may limit the resources parents are able to invest in young, and presumably also in eggs, in Peninsular Malaysia is supported by the results of experimental brood enlargement (Bryant and Hails, 1983).

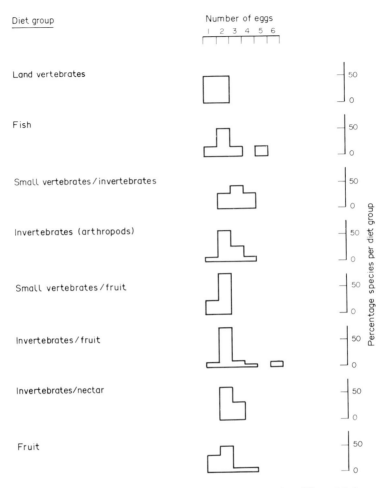

Fig. 13.9. Normal clutch-size per diet-group in 170 species of forest birds.

13.9. MIGRANTS

In none of the foregoing sections have special reasons been found for separate treatment of long-distance migrants. They claim one here only because of their unusual diversity in peninsular forests.

Among bird species, other than shorebirds, that winter in Peninsular Malaysia, near one third frequent woodlands of some kind and approximately one sixth (26 species) depend more or less exclusively on forests. Among 7000 migrants captured at floodlights over several boreal autumns at Fraser's Hill, 90 percent fell within the former category and as much as 50 percent in the latter. Flycatchers contribute disproportionately to the species total in natural forests but in the years concerned were represented in the migration stream by proportionately high numbers of only one species, brown-chested flycatcher *Rhinomyias brunneata*. This and the Chinese subspecies of the blue-throated flycatcher *Cyornis rubeculoides* may winter exclusively in Peninsular Malaysia, which is a lacuna in the regional distributions of their closest resident relatives. The rare narcissus flycatcher *Ficedula narcissina elizae* and even rarer rufous-headed robin *Erithacus ruficeps* have also been found wintering only

Fig. 13.10. The pendant nest of the green broadbill. (Photograph by Frank Lambert.)

in the Peninsula (Nisbet, 1976).Except for black baza *Aviceda leuophotes* and crested honey-buzzard that are assumed to take some vertebrates (both mostly at the forest edge) and the larger thrushes that seek local abundances of small fruit, migrants resemble the majority of residents in taking mostly arthropods. They are also faithful to the same gross divisions of forest habitat (Table 13.8). Most, by far, winter with the richest community, in lowland forest. Montane forest is used by frugivorous thrushes and appears critical also for two leaf warblers (*Phylloscopus* spp.) and one flycatcher; mangrove for a big kingfisher and a third leaf warbler. Within the favoured habitat a minority of (mainly understorey) species is delimited by the hillfoot boundary and, as among residents, more are found below than above it. Vertical column space is likewise partitioned into the familiar canopy, interior and understorey layers.

These restrictions imply spatial isolation. Within the genus *Phylloscopus*, pale-legged leaf warbler *P. tenellipes*, where it occurs in the north-west, winters lower in the vegetation column of lowland forest than arctic and eastern crowned warblers *P. borealis* and *P. coronatus*. Arctic warblers also frequent the crown layer of mangrove while the exclusively mangrove-haunting dusky warbler *P. fuscatus* forages in the sub-crown layer. In the canopy of lowland forest, arctic and eastern crowned warblers co-occur in

TABLE 13.8. Habitats selected by the 40 non-breeding migrants considered to winter regularly in mature forests. (In cases of overlap between habitats, species have been entered more than once. Habitats of a few species have been inferred from those of resident conspecifics).

GROUP SITE	Wetland non-passerines	Raptors	Cuckoos	Other non-passerines	Flycatchers/monarchs	Other passerines	TOTAL SPECIES	Percentage distribution
Mangrove								
Sub-crown	0	0	0	1	0	1	2	21
Crown	0	1	2	1	2	2	8	
Inland forest								
Understorey								
(a) below hillfoot	2	0	0	0	2	1	5	
(b) above hillfoot	0	0	0	0	1	0	1	
(c) not limited	1	0	0	1	2	3	7	
Interior								
(a) below hillfoot	0	0	0	0	0	0	0	
(b) above hillfoot	0	0	0	0	0	0	0	67
(c) not limited	0	0	1	1	3	1	6	
Canopy								
(a) below hillfoot	0	1	0	0	0	1	2	
(b) above hillfoot	0	0	0	0	1	0	1	
(c) not limited	0	0	3	2	1	3	9	
Montane forest								
Shade-layer	0	0	0	0	0	1	1	13
Canopy	0	0	0	0	1	4	5	

multi-species foraging flocks (McClure, 1967). They hunt in subtly different ways — eastern crowned pursuing dislodged arthropods further into the airspace around crown foliage — but both are displaced by the resident flyeater *Gerygone sulphurea*. Inornate and two-barred greenish warblers *P. inornatus* and *P. plumbeitarsus* share montane forest with the resident mountain leaf warbler *P. trivirgatus*, but their ecological relationships are unknown. Pasoh forest reserve has provided some demographic information at community level. 2 km² of mature lowland forest contained 14 species definitely identified during the mid-winter period, with a probability of seven others not readily separated in the field from resident conspecifics. Over two consecutive mid-winter seasons, the six species in 15 ha of understorey averaged 8.1 percent of total sedentary individuals and approximately 3 percent of aggregate biomass. In a single mist-net sample taken elsewhere in the same reserve, Karr (1976) obtained 8.3 percent migrants (a figure of 17.8 percent, from a second sample, in his paper is erroneous). At Ulu Gombak, inclusive monthly catches in regenerating submontane forest during the winter months averaged 10–15 percent migrants but included a number of typical edge species (Medway and Wells, 1976). Wintering migrant population densities at Pasoh spanned the characteristic range of densities of residents in the understorey, from a typical 1–4 individuals per 15 ha to (in separate winters) 15 and 20 Siberian blue robins, making this the second commonest of all understorey species. As the smallest specialist litter forager in the community, the Siberian blue robin may hold an unusual resource monopoly. Rate and location of recapture of marked individuals within the one season in this and other species, and between-season return to the vicinity of sites of earlier capture, implied a degree of side fidelity not different from understorey residents.

Migrants to the lowland forest understorey may differ as a class from residents in their survival and nutritional status. No individual of any species has been taken in more than three separate winters. Among well over 100 Siberian blue robins handled at Pasoh, only four were indisputably in their third or later winter of life. Year-to-year survival among territory-holders of this species averaged 46 percent, a value much lower than among comparable-sized passerines performing similar migratory journeys but wintering in open country in Peninsular Malaysia (Nisbet, 1967; Nisbet and Medway, 1972).

The few Palaearctic migrants (including Siberian blue robin) that have been tested assume a typically tropical basal metabolic rate during their wintering sojourn, some 30 percent below the expected breeding range level (Hails, 1983), and are unlikely to be at a disadvantage in terms of maintenance needs. Their performance is, nevertheless, different from that of residents over the same timespan. Those intercepted before reaching their wintering habitat are characteristically below mean winter weight, carry little or no visible fat reserves and have reduced pectoral muscles. On arrival they rapidly make good, attaining a mid-winter maximum weight between October and December even though feeding conditions are supposedly poor. Weight then declines to a mean minimum a month or two before spring departure. Some individuals of all species left the census plot at Pasoh without further fattening. Among others, fat gain represented not more than 20 percent of the best estimate of departure weight of Siberian blue robins and brown-chested flycatchers. Fattening is clearly not related to the total length of the spring journey (very different in these two) and the supposition is that constraints on feeding in forest force northbound migrants to collect most of their reserves at intermediate staging points, the first of which must lie well within the tropics.

13.10. RESPONSE TO HABITAT DISTURBANCE

13.10.1. Natural gaps

Tree-falls provide the most frequent natural gaps. Fresh gaps that are large enough to expose the ground to full sun are avoided by all shade-layer birds, but canopy species are less impeded. Gaps offer immediate benefits, perhaps, only to certain aerial salliers (black-thighed falconet, whiskered tree-swift *Hemiprocne comata*, blue-throated bee-eater *Merops viridis*, brown flycatcher *Muscicapa latirostris*) that gain from increased visibility. Within the light-demanding pioneer vegetation that responds to gaps, bananas and gingers attract spiderhunters to their copious, semi-continuous production of nectar. In lowland habitat, these are accompanied by small, foliage searching insectivores (striped tit babbler *Macronous gularis*, black-throated tailorbird) that follow the canopy interface down to a low level, plus a few gap seekers (fluffy-backed babbler *Macronous ptilosus*, white-bellied munia *Lonchura leucogastra*) that disperse directly through the understorey. In montane forest, landslips add to gap habitat and colonizing species again include sunbirds and skulking insectivores.

At a later stage of regrowth, in both formations, concentrations of mostly passerine generalists, conspicuously denser than occur in any stratum of mature forest, gather at the fruitings of pioneer tree associations. Such trees are often small and the return of the shade-layer community awaits a still later, as yet ill-defined stage of regrowth. Two gaps of 0.5–1 ha extent in the lowlands at Pasoh took about ten years to acquire their first true shade-adapted birds. Factors contributing to the avoidance of pioneer growth by such birds may include sensitivity to high temperatures and a well known behavioural avoidance of bright light. More important still must be the re-establishment of appropriate microclimate and fine vegetation structure, limiting the rate at which their invertebrate prey can colonize specific foraging sites.

13.10.2. Logging

Natural tree-falls open out an isolated fragment of the floor of forest for a limited period of time and draw upon a surrounding hinterland of colonists. 'Selective' logging as now practised in Peninsular Malaysia (Chapter 10), by removing about 50 percent of the basal area of the stand and baring at least this much of the soil surface, inverts this relationship. Gap, with its detrimental features, ceases to be the habitat patch and becomes instead the matrix — with predictable results. Near Jelebu, Negeri Sembilan, Manning (in Medway and Nisbet, 1967) noted the rapid disappearance of dusky broadbill *Corydon sumatranus* and all trogons (*Harpactes* spp.). Trogons were also among many shade-layer species that vanished after selective logging from censused compartments of Tekam forest reserve, Pahang (Johns, 1985). The fate of individuals was not discovered, but their expectation of finding alternative living space must have been low (cf. Lovejoy *et al.*, 1984). Of over 1000 ringed birds displaced by clear felling at Pasoh in 1970, none is known to have reached a second study site 2.5 km distant in adjacent untouched forest. At Subang, Selangor, after half of their site had been cleared, McClure and Hussein (1965) found only about five percent of the displaced individuals of normally territorial species in immediately surrounding forest.

Plots of known logging history at Tekam showed resurgence of species richness and biomass per unit area. This was chiefly due to generalists that naturally exploit pioneer and building-phase vegetation. Diversity and abundance of the specialist, shade-adapted insectivore species, notably babblers and flycatchers, remained depressed (Johns, 1985). Importantly, up to the time of Johns' study, sources of potential colonists had remained intact close by.

At least in the short term, canopy birds seem better able to cope with the effects of logging, perhaps because they and their foods are less susceptible to resulting changes in microclimate. For some leaf-searching insectivores, lost foraging space may be temporarily offset by a flush of new foliage in the residual stand after removal of neighbours. On the other hand, not all canopy foragers normally nest in the canopy and breeding may be disrupted by loss of suitable shaded sites. Non-excavating cavity nesters may find tree-holes particularly difficult to replace in the short term. At Tekam, a pair of rhinoceros hornbills *Buceros rhinoceros* (Figure 13.11) nested in a large, holed *Shorea pauciflora* even after its use as a spar to remove many surrounding trees in a high lead logging operation (Johns, 1982; pers.comm.).

The time elapsing to complete recovery of the original bird community after logging is not known. At Pasoh, 20–25 year old regenerating forest was used by all bird species of the immediately adjacent untouched forest, but with demographic differences. In an apparently restored understorey, population densities varied over short distances from greater to less than in mature forest. More importantly, Wong (1985) noted more itinerancy, less territory-holding and reduced breeding success compared with mature forest. Finally, their sparseness even after this interval suggested that two species, rufous-tailed shama and brown-chested flycatcher, had barely begun to recolonize; indeed, the flycatcher occurred only on passage (wintering a short distance away in mature forest).

Fig. 13.11. A rhinoceros hornbill *Buceros rhinoceros*. (Photograph Frank Lambert.)

13.10.3. Conservation implications

While the mangrove community bears a special relationship to man-made environments, fewer than 20 birds of inland forest have effectively established themselves beyond the limits of original habitat. Plantations of some exotic timbers, e.g., *Albizzia falcataria*, appear to provide foraging opportunities for some forest birds at canopy level (Sheldon, 1986) but no clean monoculture has yet been shown to support a breeding population. The implications of converting forest lands to other uses in Peninsular Malaysia are, therefore, self-evident. Pressure on the lowland plains is particularly heavy and the assumption that submontane habitat adequately caters to the needs of lowland forest species is compromized by the possibility that populations, other than of slope specialists, need to be supplemented from below the hillfoot boundary (above, 13.5.2).

Logged-over forest now greatly exceeds the area of primary habitat remaining in the lowlands (Chapter 10), and its value to generalist frugivores has already been made clear. Strangling figs, the great source of small fruit in mature forest, have no commercial timber value but, on the other hand, are at risk when young through the felling of their hosts. Though many are mobile, fig specialists are, therefore, less well insulated from the effects of logging.

Logging coupes are defined by licencing but the rate of cutting over the past 25 years has been such that most logged forest is now at best only in remote contact with sources of the original bird community. Given that some of this forest will be allowed to regenerate its carrying capacity for birds, there remains the question of how fast colonists can spread over the landscape. This varies predictably. Of 33 forest species recorded at least once in 15 years crossing now marooned patches of overgrown plantation-cum-secondary forest around Kuala Lumpur (Yorke, 1984; author's unpublished data) only four have been shade-layer birds (and two of these regular at the forest edge everywhere). In combination with findings at Pasoh, and the likely economics of logging to come, it is suggested that the time required for full faunal recovery in typical production forest may prove to exceed an acceptable interval between fellings.

No facts, therefore, seriously challenge the commonsense view that conservation of the forest avifauna in Peninsular Malaysia will require the setting aside and planned management of adequate nature reserves. In summary, the most critically assailed community is that of lowland forest, particularly in its zone of peak species richness. At least one endemic is at risk. With continued conversion of its habitat to other land uses, via bunding and drainage, the mangrove community is next. Even though it has furnished man-made habitats with more species than any other it is the only community already to have suffered losses from extinction, mostly in the form of large waterbirds breeding colonially at traditional sites. By contrast, montane forest in the Peninsula must rank as one of the least disturbed bird habitats remaining in the humid tropics — virtually unlogged, settled only on a highly local scale and over large areas scarcely explored. Perhaps for these reasons, no part of its area of maximum bird species diversity, on the Main Range, has yet been proposed for conservation purposes. The message has a familiar ring.

REFERENCES

Anon. (1979). *Liste des vertébres du bassin de l'Ivindo (République Gabonaise)*. Institut de Biologie Tropicale, C.E.N.A.R.E.S.T.

Batchelor, B. C. (1983). *Sundaland tin placer genesis and late Cainozoic coastal stratigraphy in western Malaysia and Indonesia*. Ph.D. thesis, University of Malaya.

Bell, H. L. (1977). *The vertical distribution of a lowland rain forest bird community in New Guinea*. M.Sc. thesis, University of Papua New Guinea.

Bryant, D.M. and Hails, C.J. (1983). Energetics and growth patterns of three tropical bird species. *Auk* 100, 425–39.

Chasen, F. N. (1935). A handlist of Malaysian birds. *Bull. Raffles Mus.* 11, i–xx, 1–389.

Croxall, J. P. (1976). The composition and behaviour of some mixed-species bird flocks in Sarawak. *Ibis* 118, 333–46.

Davison, G. W. H. (1979). The evolution of Crested argus. *World Pheasant Assoc. J.* 5, 91–7.

Davison, G. W. H. (1981). Diet and dispersion of the Great argus *Argusianus argus. Ibis* 123, 485–94.

Davison, G. W. H. (1983). Behaviour of the Malay Peacock Pheasant, *Polyplectron malacense* (Aves:Phasianidae). *J. Zool., Lond.* 201, 57–66.

Fogden, M. P. L. (1972). The seasonality and population dynamics of equatorial forest birds in Sarawak. *Ibis* 114, 307–43.

Fry, C. H. (1980). Survival and longevity among tropical land birds. *Proc. 4th Pan-African Ornithol. Congr.*, 333–43.

Gibson-Hill, C. A. (1949). An annotated checklist of the birds of Malaya. *Bull. Raffles Mus.* 20, 1–299.

Gibson-Hill, C. A. (1952). Ornithological notes from the Raffles Museum 15, Notes on the avifauna of Great Redang island (Trengganu). *Bull. Raffles Mus.* 24, 220–40.

Glenister, A. G. (1951). *The Birds of the Malay Peninsula, Singapore and Penang*. Oxford University Press, London.

Hails, C. J. (1983). The metabolic rate of tropical birds. *Condor* 85, 61–5.

Hails, C. J. and Turner, A. K. (1985). The role of fat and protein during breeding in the White-bellied swiftlet (*Collocalia esculenta*). *J. Zool., Lond.* 206, 469–84.

Immelmann, K. (1973). Role of the environment in reproduction as a source of "predictive" information. In *Breeding Biology of Birds*, ed. D. S. Farner, pp.121–47. National Academy of Sciences, Washington.

Johns, A. D. (1982). Observations on nesting behaviour in the Rhinoceros hornbill *Buceros rhinoceros. Malay. Nat. J.* 35, 173–7.

Johns, A. D. (1985). Selective logging and wildlife conservation in tropical forest: problems and recommendations. *Biol. Conservation* 31, 355–75.

Karr, J. R. (1975). Production, energy pathways, and community diversity in forest birds. In *Tropical Ecological Systems*, eds. F. B. Golley and F. Medina, pp.161–76. Springer Verlag, New York.

Karr, J. R. (1976). On the relative abundance of migrants from the north temperate zone in tropical habitats. *Wilson Bull.* 88, 433–58.

King, B., Woodcock, M. and Dickinson, E. C. (1975). *A Field Guide to the Birds of South East Asia.* Collins, London.

Langham, N. (1980). Breeding biology of the Edible-nest swiftlet *Aerodramus fuciphagus. Ibis* 122, 447–61.

Lovejoy, T. E., Rankin, J. M., Bierregard, R. O., Brown, K. S., Emmons, L. H. and Van der Voort, M. E. (1984). Ecosystem decay of Amazon forest remnants. In *Extinctions*, ed. M. H. Nitecki, pp.295–325. University of Chicago Press, Chicago.

Madoc, G. C. (1956). *An Introduction to Malayan Birds.* Malayan Nature Society, Kuala Lumpur.

McClure, H. E. (1966). Flowering, fruiting and animals in the canopy of a tropical rainforest. *Malay. Forester* 29, 182–203.

McClure, H. E. (1967). The composition of mixed species flocks in lowland and submontane forests of Malaya. *Wilson Bull.* 79, 131–54.

McClure, H. E. and Hussein bin Othman (1965). Avian bionomics of Malaya 2. The effects of forest destruction upon a local population. *Bird-Banding* 36, 242–69.

Medway, Lord (1972). The Gunong Benom Expedition 1967, 6. The distribution and altitudinal zonation of birds and mammals on Gunong Benom. *Bull. Brit. Mus. (Nat. Hist.) Zool.* 23, 105–54.

Medway, Lord and Nisbet, I. C. T. (1967). Bird Report: 1965. *Malay. Nat. J.* 20, 59–80.

Medway, Lord and Wells, D. R. (1971). Diversity and density of birds and mammals at Kuala Lompat, Pahang. *Malay. Nat. J.* 24, 238–47.

Medway, Lord and Wells, D. R. (1976). *The Birds of the Malay Peninsula.* Vol.5. H. F. and G. Witherby, London.

Munn, C. A. and Terborgh, J. W. (1979). Multi-species territoriality in neotropical foraging flocks. *Condor* 81, 338–47.

Nisbet, I. C. T. (1967). Migration and moult in Pallas's grasshopper warbler. *Bird Study* 14, 96–103.

Nisbet, I. C. T. (1968). Utilization of mangroves by Malayan birds. *Ibis* 110, 348–52.

Nisbet, I. C. T. (1976). The eastern Palaearctic migration system in operation. In *The Birds of the Malay Peninsula*, Vol.5. eds. Lord, Medway and D. R., Wells, pp.57–69. H. F. and G. Witherby, London.

Nisbet, I. C. T. and Medway, Lord (1972). Dispersion, population ecology and migration of Eastern great reed warblers *Acrocephalus orientalis* wintering in Malaysia. *Ibis* 114, 451–94.

Pratt, T. K. and Stiles, E. W. (1985). The influence of fruit size and structure on composition of frugivore assemblages in New Guinea. *Biotropica* 17, 314–21.

J. J., Raemakers, F. P. G., Aldrich-Blake and J. B., Payne, (1980). The forest. In *Malayan Forest Primates*, ed. Chivers, D. J., pp.29–61. Plenum, New York.

Round, P. D. (1982). Notes on breeding birds in north-west Thailand. *Nat. Hist. Bull. Siam Soc.* 30, 1–14.

Serle, W. (1981). The breeding season of birds in the lowland rainforest and the mountain forest of West Cameroon. *Ibis* 123, 62–74.

Sheldon, F. H. (1986). Habitat changes potentially affecting birdlife in Sabah, east Malaysia. *Ibis* 128, 174.

Sibley, C. G. and Ahlquist, J. E. (1985). The phylogeny and classification of the Australo-Papuan passerines. *Emu* 85, 1–14.

Silvius, M. J., Simons, H. W. and Verheugt, W. J. M. (1984). *Soils, Vegetation, Fauna and Nature Conservation of the Berbak Game Reserve, Sumatra, Indonesia.* Research Institute for Nature Management, Arnhem.

Snow, D. W. (1975). *The Web of Adaptation.* Collins, London.

Snow, D. W. (1981). Tropical frugivorous birds and their food plants: a world survey. *Biotropica* 13, 1–14.

Tweedie, M. W. F. (1960). *Common Malayan Birds.* Longmans, London.

Vellayan, S. (1981). The nutritive value of *Ficus* in the diet of Lar gibbon *(Hylobates lar). Malay. Appl. Biol.* 10, 177–81.

Ward, P. (1968). Origin of the avifauna of urban and suburban Singapore. *Ibis* 110, 239–55.

Ward, P. (1969). The annual cycle of the Yellow-vented bulbul *Pycnonotus goiavier* in a humid equatorial environment. *J. Zool., Lond.* 157, 25–45.

Ward, P. and Wood, B. (1967). Parrot damage to oil-palm fruit in Johore. *The Planter* 43, 1–3.

Wells, D. R. (1975). Bird Report: 1972 and 1973. *Malay. Nat. J.* 28, 186–213.

Wells, D. R. (1985). The forest avifauna of western Malesia and its conservation. In *Conservation of Tropical Forest Birds*, eds. A. W. Diamond and T. E. Lovejoy, pp.213–32. International Council for Bird Preservation, Cambridge.

Wells, D. R. and Medway, Lord (1976). Taxonomic and faunistic notes on birds of the Malay Peninsula. *Bull. Br. Ornithol. Club* 96, 20–34.

Wells, D. R., Hails, C. J. and Hails, A. J. (1978). *A study of birds of the Gunung Mulu National Park, Sarawak.* Cyclostyled report deposited at the Royal Geographical Society, London.

Wong, M. (1985). Understorey birds as indicators of regeneration in a patch of selectively logged West Malaysian rainforest. In *Conservation of Tropical Forest Birds*, eds. A. W., Diamond and T. E., Lovejoy, pp.249–63. International Council for Bird Preservation, Cambridge.

Yorke, C. D. (1984). Avian community structure in two modified Malaysian habitats. *Biol. Conservation.* 29, 345–62.

Completed August 1986

CHAPTER 14

Termites

N.M. Collins

IUCN Conservation Monitoring Centre, 219c Huntingdon Road, Cambridge CB3 0DL, U.K.

CONTENTS

14.1. INTRODUCTION

There are over 2300 described species of termites (Insecta: Isoptera), the majority occurring within tropical latitudes. Their stronghold is tropical rain forest where up to 60 species may occur in a community (Fig. 14.1). Although termites are secretive insects, they are one of the most abundant forms of animal life in Malaysian forests, rivalled only by the ants, many of which prey upon them. Although some of the termites of Malaysia are well known, much research remains to be done, even into basic taxonomy and systematics.

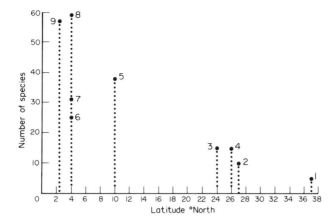

Fig. 14.1. Species richness of termites in Malaysia and other Asian countries. Note that the sampling areas differ very widely. Collections at latitudes above 8°N are wide-ranging, those below 8°N are from restricted localities within one forest type. Counts from large areas would be considerably higher. The total for all forest types in the Mulu National Park, Sarawak, was 72 species (Collins, 1984), while Borneo as a whole would probably contain twice as many. The count for Pasoh Forest was 57 (Abe, 1978), while the total for Peninsular Malaysia exceeds 150. 1. Japanese Islands. 2. Ryukyu Islands. 3. Formosa (Taiwan) (in Abe, 1979). 4. south-eastern China (Light, 1929). 5. Philippines (Light, 1930). 6. Heath (kerangas) forest, Mulu, Sarawak. 7. Swampy alluvial forest, Mulu, Sarawak. 8. Lowland dipterocarp forest, Mulu, Sarawak (Collins, 1984). 9. Lowland dipterocarp forest, Pasoh, Negeri Sembilan (Abe, 1978).

14.1.1. Social structure

Termites are social insects that live in colonies of several hundred to seven million individuals. They not only feed, groom and protect each other, but the offspring of one generation assists the parents in raising the next, the mark of truly social animals. Each society is divided between several castes, the winged and sighted reproductives and the wingless and usually blind workers and soldiers, specially adapted to feed, maintain and protect the colony. One reproductive king and queen, or sometimes several pairs, are tended by the sterile workers and soldiers. Queens of more advanced termites can grow to 14 cm long, weigh over 60 g and lay up to 30 000 eggs per day. Once a year flying reproductives (alates) are released in a swarm for the short nuptial flight which precedes the founding of new colonies. At this stage the termites are at their most vulnerable and are preyed upon by ants, spiders, lizards, small mammals and birds, not to mention man. Mortalities are high, with perhaps less than 0.1 per cent surviving to raise a mature colony. The male alate is attracted to the female by a pheromone secreted from a gland on her underside. Once the male has found a mate, they both shed their wings and run off in search of a suitable location to build a nest.

The Isoptera evolved from a common stem with cockroaches during the early Mesozoic or late Palaeozoic about 220 million years ago. The colonial, wood-feeding North American cockroach *Cryptocercus punctulatus* Scudder shares a number of characteristics with primitive termites, including colonies of mutualistic gut protozoa. The protozoa are passed from old to young in semi-digested food, a key factor in the evolution of social behaviour. Building upon this heritage, evolutionary advances in the termites have occurred in two main areas: feeding and defence.

14.1.2. Feeding

The basic foodstuff of termites is plant material, generally dead and decaying, even down to the organic remnants in soil. Development of mutualistic associations not only with protozoa, but also with bacteria and fungi, has improved the capacity of termites to digest and assimilate cellulose, lignin and other plant components. The Isoptera is the only order of insects with a general ability to digest cellulose, although certain Diptera and Coleoptera have a very limited capacity to do so. The evidence for digestion of lignin is poor, but there is no doubt that some species can digest at least a proportion of their dietary lignin. Because of these digestive capabilities termites are an important component of most tropical ecosystems.

In rain forest the wide range of foodstuffs and feeding sites permits very varied feeding habits. Rotten wood is the most important food resource, but rotting leaves are consumed by a number of specialized species. Fresh litter is utilized in seasonal forests but in humid forests decay sets in very quickly. Soil-feeders are common in Malaysian forests, but for zoogeographical reasons there are fewer species than in the forests of Africa (Collins, 1988).

Foraging patterns are also very variable. A few species may be seen foraging in the open during the day, but the majority feed underground, inside dead wood or under covered foraging galleries (Abe, 1979). Primitive species, like *Neotermes* and *Glyptotermes*, never leave the branch in which they are nesting. Others, such as the fungus-growing Macrotermitinae, may travel over 50 m from the nest through a semi-permanent underground network of tunnels in order to reach new sources of food.

With the exception of their annual swarms, most termites are rarely seen, but the damage that results from their feeding is all too familiar to people in rain forest regions. Although only about one in ten species are of economic importance (Harris, 1961; 1969), these are often the species which survive in man-made environments. They cause serious damage to plantations of exotic trees such as rubber, pine and teak, to crops such as tea and sugar cane, and to building timbers, furniture and books. It has been suggested that management to maintain the high diversity of termites could, through competitive exclusion, help to reduce the impact of damaging species (Tho, 1974).

14.1.3. Defence

Novel methods of defence have been vital in the arms race against the hordes of termite predators, notably ants (Deligne *et al.*, 1981; Prestwich, 1984). Termites defend themselves in a number of ways: by building fortified nests, by covering their food in a protective sheet of mud while they feed underneath, and with defensive workers and soldiers (Fig. 14.2). The majority of termites have soldier castes with large armoured jaws, but in the most advanced form of soldier these have become redundant. The Nasutitermitinae (snouted termites) have a specially developed frontal gland that produces sticky and irritating chemicals from the tip of an extended snout on the front of the soldier's head. This strategy is so successful at warding off the attentions of ants that most snouted termite soldiers have very reduced jaws, and a number of nasute genera forage in open columns, defended at the sides by the snouted soldiers. In Malaysian forests the only genus of termites that lacks the soldier caste is *Protohamitermes* (Ahmad, 1976).

14.2. THE TERMITES OF PENINSULAR MALAYSIA

About 150 species of termites have been recognized in Malaysia, divided between 44 genera. The number of species per genus shown in Table 14.1 is only approximate since it includes over 35 species that have been collected but not yet formally described (Tho, 1982a). It is unfortunate that even this economically and ecologically important group, which has been studied in Malaysia for well over a century, is still inadequately known.

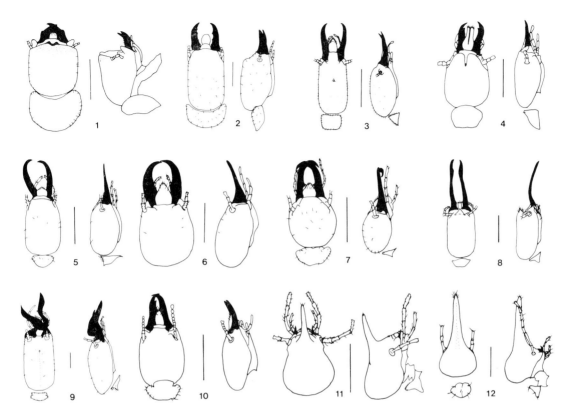

Fig. 14.2. A selection of Malaysian termites demonstrating the range of adaptations in the soldier head. 1. *Cryptotermes* with phragmotic head for blocking tunnels. 2. *Glyptotermes*, toothed mandibles. 3. *Heterotermes*, shearing mandibles. 4. *Schedorhinotermes* minor soldier, toothed mandibles and extended labrum to channel defensive fluids from the frontal gland. 5. *Microcerotermes* with finely serrated mandibles. 6. *Prohamitermes*, heavy shearing mandibles. 7. *Globitermes*, strongly curved. 8. *Termes*, snapping, slightly asymmetrical. 9. *Pericapritermes*, snapping, strongly asymmetrical. 10. *Odontotermes*, toothed, cutting. 11. *Hospitalitermes*, reduced mandibles, extended nasus with frontal gland opening at tip. 12. *Hirtitermes*, reduced mandibles, robust nasus. All scale bars represent 1 mm in length.

There are seven living families of termites, six of which, the Mastotermitidae, Kalotermitidae, Termopsidae, Hodotermitidae, Rhinotermitidae and Serritermitidae, are known collectively as the 'lower termites' while the seventh, the Termitidae, comprises the 'higher termites'. The lower termites have mutualistic flagellate protozoa living in their hindguts. These are vital to the termites since without them they are unable to digest the cellulose that is a major component of their diet (Breznak, 1982). The Termitidae do not have special gut protozoa, but use either anaerobic bacterial cultures in their hindguts, or mutualistic fungi grown on their faeces (Macrotermitinae only), to assist their digestive processes. Four of the seven families of termites, the Mastotermitidae, Termopsidae, Hodotermitidae and Serritermitidae, do not occur in Malaysia. In the paragraphs that follow, diagnostic features of the three Malaysian termite families will be described, giving examples of a small selection of common species.

14.2.1. Kalotermitidae

The Kalotermitidae (dry wood termites) are often difficult to find, but probably occur in small numbers in all rain forests. In Malaysia *Neotermes* and *Glyptotermes* are virtually confined to dead limbs

TABLE 14.1. The genera of termites in Peninsular Malaysia, with estimates of the number of species per genus (in brackets). These estimates include many new species awaiting formal description. Data after Ahmad (1968, 1971, 1976); Ahmad and Akhtar (1981); Harris (1957) and Tho (1982a).

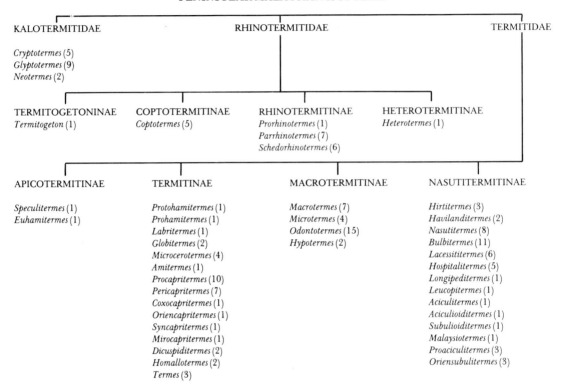

PENINSULAR MALAYSIAN ISOPTERA

KALOTERMITIDAE

Cryptotermes (5)
Glyptotermes (9)
Neotermes (2)

RHINOTERMITIDAE

TERMITIDAE

TERMITOGETONINAE
Termitogeton (1)

COPTOTERMITINAE
Coptotermes (5)

RHINOTERMITINAE
Prorhinotermes (1)
Parrhinotermes (7)
Schedorhinotermes (6)

HETEROTERMITINAE
Heterotermes (1)

APICOTERMITINAE

Speculitermes (1)
Euhamitermes (1)

TERMITINAE

Protohamitermes (1)
Prohamitermes (1)
Labritermes (1)
Globitermes (2)
Microcerotermes (4)
Amitermes (1)
Procapritermes (10)
Pericapritermes (7)
Coxocapritermes (1)
Oriencapritermes (1)
Syncapritermes (1)
Mirocapritermes (1)
Dicuspiditermes (2)
Homallotermes (2)
Termes (3)

MACROTERMITINAE

Macrotermes (7)
Microtermes (4)
Odontotermes (15)
Hypotermes (2)

NASUTITERMITINAE

Hirtitermes (3)
Havilanditermes (2)
Nasutitermes (8)
Bulbitermes (11)
Lacessititermes (6)
Hospitalitermes (5)
Longipeditermes (1)
Leucopitermes (1)
Aciculitermes (1)
Aciculioiditermes (1)
Subulioiditermes (1)
Malaysiotermes (1)
Proaciculitermes (3)
Oriensubulitermes (3)

and trunks in the forest canopy, perhaps as a result of competitive exclusion from other niches by the more advanced Termitidae. They are occasionally found in fallen logs on the ground but probably do not survive long in contact with the forest floor.

Both *Cryptotermes cynocephalus* Light and *C. domesticus* (Haviland and Sharp) are serious pests of furniture and buildings throughout Malaysia. The soldiers have a characteristic foreshortened and heavily armoured head, believed to be of particular value in blocking and defending nest galleries (phragmosis) (Fig. 14.2). The origin of these species, as of *C. dudleyi* Banks, is uncertain. That they were originally introduced by man is suggested by their distribution, which is in mangrove swamps along coastlines and in building timbers in towns.

14.2.2. Rhinotermitidae

The most important family of lower termites in the Malaysian forests is the Rhinotermitidae (damp wood termites). They feed on rotten wood, mainly in standing or fallen trunks and limbs. The Malaysian genera are *Coptotermes* and *Heterotermes*, which are found throughout the tropics, and *Termitogeton*, *Prorhinotermes*, *Parrhinotermes* and *Schedorhinotermes*, which have an Indomalayan distribution.

Coptotermes curvignathus Holmgren is a common termite of Malaysian jungle, where it feeds on dead wood, often in standing trees. Nests are in large chambers hollowed out in boles or tap roots below ground level, and are divided into stories by thin horizontal layers of carton, i.e., chewed and partly digested wood mixed with faecal material. The termites may hollow out the middle of old and over-mature trees to a height of several metres and a single colony may spread to a number of trees (Harris, 1957). In the rain forest this destruction of old trees is a natural part of the cycle of forest growth, but in plantations it becomes a serious problem. *C. curvignathus* is a notorious pest of rubber trees (*Hevea brasiliensis*) and pine trees, particularly *Pinus caribaea* (Tho, 1974). *Coptotermes* soldiers exude a characteristic white defensive fluid from the protruding orifice of the frontal gland on the top of the head *Termitogeton planus* (Haviland) is confined to Peninsular Malaysia and Borneo. Although not uncommon in some forests, it is a little-known species. The dorso-ventrally flattened head of both the soldier and worker is quite distinctive, and possibly allows greater manoeuvrability in the lamellar galleries which the species excavates in the centre of rotting logs. *Schedorhinotermes* species such as *S. malaccensis* (Holmgren) and *S. medioobscurus* (Holmgren) are common in the forest and have similar habits to *Parrhinotermes*, generally nesting and feeding in rotten logs and the boles of standing dead trees. *Schedorhinotermes* has large, conspicuously dimorphic soldiers with orange heads, whereas *Parrhinotermes* soldiers are monomorphic and much smaller. *Prorhinotermes flavus* Bugnion and Popoff is also small and occurs only in mangrove forests, mainly on the west coast (Tho, 1982a).

14.2.3. Termitidae

Over 73 per cent of termites are in the single family Termitidae, the higher termites. In terms of species richness and population density they predominate in all tropical forests. The family is divided into four subfamilies (Table 14.1), each with specialist feeding and defence strategies.

The soil-feeding subfamily Apicotermitinae has its centre of origin in Africa and is represented on the edge of its range in Malaysia only by two species. *Euhamitermes hamatus* Holmgren, a very rare and poorly known species, has not been recorded since 1913 (Roonwal, 1970; Tho, 1982a). *Speculitermes* sp. has been recorded only in a single collection from a teak plantation in northern Peninsular Malaysia (Tho, 1982a).

The subfamily Termitinae includes both wood-feeding and soil-feeding species. Amongst the wood-feeders, two genera (*Microcerotermes* and *Amitermes*) are tropicopolitan, and four genera (*Protohamitermes, Prohamitermes, Labritermes, Globitermes*) are endemic to the Orient (Krishna, 1970; Ahmad, 1976). The other genera of Termitinae listed in Table 14.1 are all soil-feeders with snapping mandibles (Fig. 14.2). This group has diverged widely in Peninsular Malaysia, where nine of the twelve Oriental genera are to be found (Ahmad and Akhtar, 1981).

Protohamitermes globiceps Holmgren was only recently recorded from Peninsular Malaysia (Tho, 1982a), but is a common species in Borneo (Ahmad, 1976). It lacks the soldier caste and builds a diffuse nest in dead wood or soil, feeding on highly decomposed wood.

Prohamitermes mirabilis (Haviland) is a common species that feeds in the organic surface layer of the soil or in very rotten stumps. The subterranean nest is dark brown and brittle, consisting of rounded cells with small inter-connecting holes. It has been demonstrated that the termites keep a store of spherical pellets in the cells, ready to block the entrance holes against enemies (Tho, 1981).

Globitermes sulphureus (Haviland) is common, highly destructive of dead wood and sometimes attacks living trees (Roonwal, 1970). It can become very common in cultivated areas and rubber plantations, but is not present in all forests (Abe, 1978). The cylindrical mound can reach 1–1.5 m in height and is covered with rough earth; the inside and below-ground parts are made of hard, black carton (Harris, 1957). The soldiers are easily identified by their strongly curved, slender mandibles, and the yellow coloration of the body, caused by the defensive labial gland secretions. The secretions are forced explosively through a weakness in the body wall and sprayed over attackers (Deligne *et al.*, 1981; Prestwich, 1984).

Dicuspiditermes nemorosus (Haviland), from the soil-feeding branch of the Termitinae, has a wide distribution in the primary forests of Sri Lanka, Sumatra and Borneo, as well as Peninsular Malaysia (Tho, 1972). The above-ground nests are made of the soil-like faeces of the termites, but are highly variable in shape (Haviland, 1898; John, 1925; Tho, 1972; Matsumoto, 1976; Abe, 1978). The most common design is a pillar-like mound up to 35 cm high, with a bulbous head section. The density of nests may reach over 200 per hectare, each with a population of 7000–67 000 termites (Matsumoto, 1976).

The closely related *Homallotermes foraminifer* (Haviland) is also common in rain forest, with discus-shaped nests protruding only a few centimetres above the soil surface (Abe, 1978). Nest populations of this soil-feeder are in the range of 7000–56 000 individuals, and nest densities of 85–165 ha^{-1} have been recorded (Matsumoto, 1976).

Members of the subfamily Macrotermitinae grow basidiomycete fungi of the genus *Termitomyces* upon combs built from faeces. The workers build the combs up on one side and consume them from the other. During the turnover time of one or two months the fungus breaks down the faecal material, degrading complex polysaccharides and concentrating nitrogenous and other nutrients through respiratory loss of organic carbon (Collins, 1983a). These termites tend to have high weight-specific consumption rates and a correspondingly greater impact on decomposition processes than other termites (Collins, 1981; 1983b). The centre of origin of the Macrotermitinae is Africa, and of the 13 genera, only *Macrotermes*, *Odontotermes*, *Microtermes* and *Hypotermes* have reached Malaysia. They feed on dead wood and leaves, often in a relatively fresh condition, being enabled to do so by the action of their mutualistic fungi. Most other termites require the action of free-living fungi to make their food palatable. *Odontotermes*, *Hypotermes* and *Microtermes* all nest underground, their fungus combs built in scattered soil cavities linked by passages. *Odontotermes* and *Microtermes* can become serious pests in Malaysia, causing expensive damage to crops and buildings (Harris, 1957; 1961; 1969). *Microtermes obesi* Holmgren and *M. pakistanicus* Ahmad are pests of sugar cane and other plantation species (Tho, 1982a). The genus *Macrotermes* includes the largest termites in Malaysia and the soldiers can inflict a painful bite. The nest of *M. carbonarius* (Hagen) is at ground level, but covered by a domed mound of compact earth that can be up to 2 m high and 4 m across (Abe, 1978). Nest densities have been recorded at 15–41 per hectare in primary lowland forest, each with a population of 72 000–106 000 termites (Matsumoto, 1976). The inside of the nest consists of numerous chambers linked by narrow passages and containing stores of chewed food and fungus combs. Foraging parties travel through underground galleries, emerge from exit holes and feed on twigs and leaf litter, cutting pieces for transport back to the nest (Abe, 1979). Unlike the other species of *Macrotermes*, workers and soldiers of *M. carbonarius* are almost black in colour, an adaptation to their unusual habit of foraging during the daytime as well as at night. At Pasoh Forest Reserve it has been estimated that this species alone can process over 20 per cent of annual leaf litter production (Abe, 1982; Matsumoto and Abe, 1979). *M. malaccensis* (Haviland) is a significant consumer of dead wood in the forest, but does not build conspicuous mounds (Abe, 1980). *M. gilvus* (Hagen) builds small mounds up to about 1 m high and can survive equally well inside the forest or on farms and roadsides (Tho, 1978). While feeding mainly on twigs and plant debris, *M. gilvus* can attack tea, coffee and other crops, as well as young trees (Harris, 1957).

The subfamily Nasutitermitinae is the largest, and in many respects the most specialized, of the higher termite subfamilies. In all Malaysian genera the mandibles are partially or wholly redundant because of the development of a defensive nasus, or snout, from which irritating, sticky and topically poisonous fluids may be ejected (Fig. 14.2) (Deligne *et al.*, 1981; Prestwich, 1984). Nasutes are an important element of the rain forest. Along with the Rhinotermitidae and Macrotermitinae, they destroy large quantities of dead wood. However, their feeding habits are extremely catholic, ranging from the organic remnants and humus utilized by *Subulioiditermes* and its relatives, through rotten wood (*Bulbitermes*, *Nasutitermes*) and rotten leaves (*Havilanditermes, Longipeditermes*) to lichens and mosses (*Hospitalitermes*).

Bulbitermes is a diverse genus whose taxonomy is sadly neglected. Both *Bulbitermes* and its close relative *Nasutitermes* need thorough revision. The *Bulbitermes* species most commonly encountered are those which build a spherical nest, up to 35 cm across, round the branch of a tree or sapling (Abe, 1978; Collins, 1984). Quite often the nest is only 1·2 m from the ground. *Bulbitermes* feeds mainly on dead and rotten wood, using faecal carton to build the nest, with its fine network of internal galleries. The outer shell of the nest is brittle, easily damaged and constantly replaced as the nest expands, but the inner portion is very hard and resistant.

Longipeditermes longipes (Haviland) is a characteristic and easily recognised member of the rain forest termite community. The soldiers and workers are black in colour, a common feature of species that forage in the day, as well as at night (Abe, 1978). The fragile carton nest is subterranean, often under rotten tree stumps, and riddled with plant roots (Abe, 1978; Collins, 1984). Exit holes for foraging parties open on the surface nearby; columns generally file no further than 5 m or so, feeding on rotten leaves and carrying them back to the nest in balls carried by the workers (Abe, 1979). *Longipeditermes* is distinguished from the other day-foraging black or dark brown termites, *Hospitalitermes* and *Lacessititermes*, not only by its different foodstuff and shorter columns, but also by the presence of two distinct sizes of soldier, clearly visible guarding the flanks of the foraging column. *Hospitalitermes* soldiers are all of one size and although *Lacessititermes* may have one, two, or even three sizes of soldier, the size difference between them is only slight (Tho, 1982a).

Hospitalitermes hospitalis (Haviland) and *H. umbrinus* (Haviland) are very characteristic termites of the Malaysian forests. Like *Macrotermes carbonarius, Longipeditermes longipes* and *Lacessititermes* spp., they are black species and can often still be seen foraging in the morning. However, unlike *Macrotermes, Hospitalitermes* has no underground foraging galleries. Instead it leaves its nest, a low mound often built alongside or under the base of a tree, in long columns that can reach 40 m or more in length (Abe, 1979; Collins, 1979). Reports of columns up to 100 m long require verification (e.g., Harris, 1957). The moving ribbon of workers, guarded on the flanks by stationary nasute soldiers, tramples narrow pathways up trees and across logs, clearly visible even when the insects are absent. *Hospitalitermes* has the unusual habit of feeding on mosses and lichens, transporting them back to the nest in balls carried by individual workers. The lichens are believed to be a good source of dietary nitrogen (Collins, 1979). At a density of three nests per hectare, up to 50 kg of mosses and lichens may be consumed per hectare per year (Collins, 1979). *Hospitalitermes* (the hospitable termite) is so named because it invariably shares its nest with *Termes*, several species of which are found nowhere else (Tho, 1982a). It is believed that the two genera benefit from each other's specialized mode of defence. *Lacessititermes* species are generally very dark brown, sometimes with an orange tip to the soldier's nasus. The nest is made of fragile carton, often situated in the fork of a tree, around a small branch or even on a palm frond (Abe, 1978). Foraging for leaf litter, twigs and bark with lichens attached occurs at night and in the early morning (Tho, 1982a).

14.2.4. Community structure

Termite communities vary quite considerably in their complexity, both in terms of species richness and guilds. Guilds may be defined as groups of species that exploit the same class of environmental resources in a similar way. Feeding guilds include branch, bole, leaf and humus feeders; nesting guilds include subterranean, mound, arboreal and wood nesters. The composition of termite communities is correlated with forest and soil types (Salick and Tho, 1984). The richest diversities of guilds and species are present in well-drained lowland primary forests, but some of them disappear in forests that are limited by the suboptimal conditions found at higher altitudes, or by adverse hydro-edaphic conditions.

Thus, in the montane forests of Gunung Mulu, in Sarawak, mound-builders and leaf-feeders become rare or absent (Collins, 1980a). In swamp forests subterranean nests and soil feeders disappear, while bole feeders and arboreal nesters predominate (Collins, 1984; Salick and Tho, 1984).

With a complex three-dimensional structure and equable climatic conditions, lowland rain forest offers a greater variety of nesting and feeding sites for termites than any other biome. Fig. 14.3 is a diagrammatic illustration of the nesting locations and feeding habits of termite genera in lowland rain forest. With the exception of unusual genera such as *Hospitalitermes*, which can feed on living lichens, few termites feed in the forest canopy. Dead boughs may be occupied by primitive Kalotermitidae such as *Neotermes* and *Glyptotermes*, but this is the only group of termites needing no contact with the forest floor. Similarly, the zone between the canopy and the soil contains few nesting sites, although some genera, such as *Microcerotermes*, *Bulbitermes*, *Lacessititermes* and *Nasutitermes*, build nests on branches and trunks. The bases and boles of trees are used for support by a number of genera, e.g. *Hospitalitermes* and *Dicuspiditermes*. Large mounds are built by *Macrotermes* and *Globitermes*. Builders of small mounds are mainly soil-feeding Termitinae such as *Homallotermes* and *Dicuspiditermes*. The Rhinotermitidae feed and nest entirely within dead wood, and many genera of Termitidae feed and nest entirely underground.

High rainfall and steady temperature regimes have led to significant behavioural adaptations in rain forest termites (Collins, 1988). Free-standing nests and those attached to tree trunks are vulnerable to heavy storms and stem flow respectively, and some species have developed devices for minimizing the impact. The globose head on the mound of *Dicuspiditermes nemorosus* may serve as a simple roof. Arguably the globular, arboreal nest of *Bulbitermes* assists non-erosive flow of rain water. *Amitermes* and *Microcerotermes* build tree-side carton nests covered irregularly with small hanging protuberances that shed rain, thus conserving nest material. Steady air temperatures within the forest permit the building of relatively poorly insulated nests, and allow foraging for longer periods, even in the open during the day for a number of species.

Predictably, deforestation has very serious effects on termite communities (Collins, 1980b). Nesting sites are lost, plant litter production is reduced, the microclimate becomes more severe, and the levels of soil organic matter are reduced by rapid oxidation. These factors inevitably cause a drastic reduction in species richness. The groups most seriously affected are those that lose their foodstuffs (e.g., humus-feeders and rotten wood-feeders), those that forage in open columns and lose the shelter of the forest canopy, and those that build simple nests that are poorly protected from extremes of microclimate. These categories include most rain forest species. Concomitant with a general downward trend in species richness is the development of large populations of certain pre-adapted or generalist species that may become pests of agriculture or silviculture. *Macrotermes gilvus*, a common roadside species, survives well outside forests (Tho, 1978) and *Coptotermes* survives to become a pest in plantations (Harris, 1969). Such impoverished communities of termites resemble those found in hydro-edaphically limited locations such as in swamps and mangroves.

14.3. ECOLOGICAL IMPACT

Termites have an impact on ecosystems in three main ways, with a fourth possibility poorly understood. They are: 1) feeding on plant material, often of a very poor nutritional quality, 2) through reproduction and providing material for predatory food chains, 3) physically translocating and chemically altering soils through their building activities, 4) fixing atmospheric nitrogen. All of these factors are related to the abundance and biomass of termites, measurement of which is fundamental to a greater understanding of termite ecology.

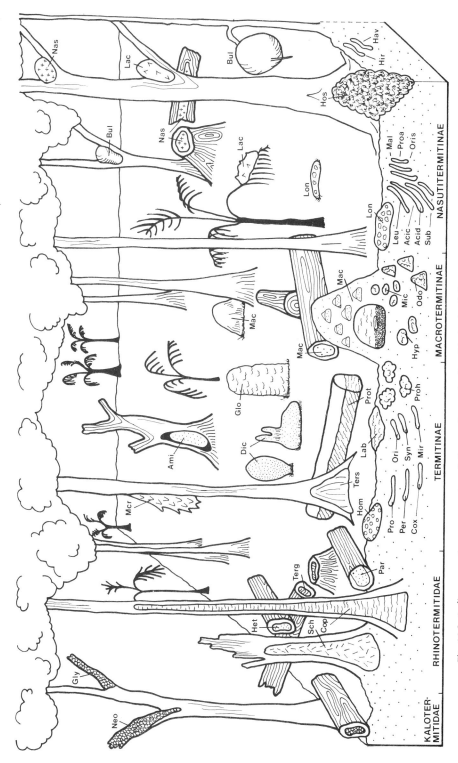

Fig. 14.3. A diagrammatic representation of the termites to be found in lowland dipterocarp forest in Peninsular Malaysia. Abbreviations using three or four letters from the name refer to the following genera:Kalotermitidae: *Glyptotermes*[1], *Neotermes*[1]. Rhinotermitidae: *Termitogeton*[3], *Coptotermes*[3], *Parrhinotermes*[3], *Schedorhinotermes*[3], *Heterotermes*[3]. Termitinae: *Protohamitermes*[3], *Prohamitermes*[3], *Labritermes*[3], *Globitermes*[3], *Microcerotermes*[3], *Amitermes*[3], *Procapritermes*[4], *Pericapritermes*[4], *Coxocapritermes*[4], *Oriencapritermes*[4], *Syncapritermes*[4], *Dicuspiditermes*[4], *Homallotermes*[4], *Termes*[3,4] (in Hospitalitermes nest). Macrotermitinae: *Macrotermes*[1], *Microtermes*[1], *Odontotermes*[1], *Hypotermes*[1]. Nasutitermitinae: *Hirtitermes*[3], *Havilanditermes*[3], Nasutitermes*[3], *Bulbitermes*[3], *Lacessititermes*[2,3], *Hospitalitermes*[2], *Longipeditermes*[3], *Leucopitermes*[4], *Aciculioditermes*[4], *Subulioditermes*[4], *Malaystotermes*[4], *Proaciculitermes*[4], *Oriensubulitermes*[4]. Principal foodstuffs are indicated by [1]fresh litter, [2]lichens and mosses, [3]rotten wood and leaf litter, [4]soil.

14.3.1. Populations and biomass

In any natural ecosystem it is essential that dead plant material should be broken down, incorporated into the soil and once again made available for the growth of living plants. Bacteria and fungi play the largest part in this process of decay, but in rain forest termites are the most important animals involved. Data on termite populations and biomass are few, but in Table 14.2 data from Sarawak and Negeri Sembilan provide some indications. As may be seen, termite populations in Malaysia can reach 2–4000 m^{-2}, dominating other soil animals. Their biomass is usually in the range 1–10 g m^{-1}. The data are rather variable in the extent of sampling. Termites are distributed in soil, timber and nests, each requiring a separate sampling procedure. All the forests listed in Table 14.2 were sampled for subterranean populations and most for above-ground nests, but populations in tree trunks, limbs and logs are rarely estimated. Nevertheless, comparison of the data permits a few observations. The Sarawak data show that termite populations decrease with rising altitude (montane sites) or poor drainage (swampy site), but rise in highly organic soils (heath site) despite lower species diversity (Collins, 1984). Comparison between the Pasoh and Mulu dipterocarp sites show how variable populations may be. The fungus-growing Macrotermitinae are co-dominant in Pasoh but virtually absent in Mulu. Termitinae and Nasutitermitinae soil populations were broadly similar in the two sites (Abe, 1979; Collins, 1983b), and the higher total populations are due mainly to more numerous epigeal and arboreal nests at Pasoh. The reason for this is obscure, particularly since the species compositions of the two sites were broadly similar in these two subfamilies (Collins, 1984).

Some perspective on the significance of termites may be obtained by comparing their biomass levels with other groups of animals. All studies in tropical rain forests indicate that the biomass of the decomposer fauna greatly exceeds that of herbivores and predators. Analyses from Malaysia show that termites comprise a larger component of the soil fauna than any other group (with ants invariably in second place), and by far the largest component of the decomposer soil fauna (Collins, 1980a; 1983b; Kondoh et al., 1980). Combined biomass values for birds and mammals in Malaysian forests are in the range 1–3 g m^{-2}, about one third of the biomass of termites.

TABLE 14.2. The species richness, populations and biomass of termites in various Malaysian forests. Separate sampling procedures are needed for nest, soil and dead wood populations; [1] denotes dead wood populations not estimated; [2] denotes nest and dead wood populations not estimated.

| Localities and forest types | Number of species | Populations m^{-2} | | | | | Biomass g m^{-2} (w.w.) | Authors |
		Rhino-termit-inae	Term-itinae-inae	Macro-termit-inae	Nasuti-termit-inae	Total		
Negeri Sembilan, Pasoh Forest Reserve:								Abe; 1978; 1979
lowland dipterocarp[1]	57	n.d.	1280–1730	934–984	943–1093	3160–3810	8.69–10.13	
Sarawak, Gunung Mulu N. P. forests:								Collins, 1980a; 1983b; 1984
lowland dipterocarp	59	293	612	5	617	1527	2.4	
swampy alluvial	31	96	103	21	170	390	0.72	
heath (kerangas)	25	600	1127	0	544	2271	4.1	
lower montane (1130 m)[2]	10	0	33	0	5	38	0.09	
upper montane (1130–1860 m)[2]	<10	0	10–286	0	9–182	99–295	0.01–0.78	
upper montane (1970–2376 m)[2]	0	0	0	0	0	0	0	

14.3.2. Consumption

No measurements of total consumption of organic matter and humus by termites have been made in any rain forest and quantitative estimates have been achieved for only three rain forest species, *Macrotermes carbonarius, Longipeditermes longipes* (Matsumoto and Abe, 1979) and *Hospitalitermes umbrinus* (Collins, 1979). Studies in tropical savannas are more advanced and these, coupled with laboratory experiments, have provided estimates of feeding rates for all the major taxonomic groups except the soil-feeding Termitinae, for which new data are needed. Using estimates of consumption rates broadly supported by these data Collins (1983b) calculated consumption by termite communities in Peninsular and Eastern Malaysia. The differences between the two localities are quite striking (Table 14.3). The fauna at Gunung Mulu in Sarawak consumes 7–35 g m^{-2} yr^{-1} depending upon the soil conditions, but at Pasoh Forest, in Negeri Sembilan, termites take 155–173 g m^{-2} yr^{-1}. The main difference is due to the voracious Macrotermitinae, abundant at Pasoh but very rare at Mulu (Table 14.2). The Macrotermitinae have consumption rates 2–6 times higher than other groups. The Pasoh consumption data resemble very closely the 188 g m^{-2} yr^{-1} consumed by termites at Mokwa in the southern Guinea savanna of Nigeria (Collins, 1981), where the fauna was also dominated by the Macrotermitinae. The role of termites in consuming litter depends to a large extent on the proportions of the different groups in the fauna. Data from Old World sites indicate that high rainfall (or groundwater availability) may encourage free-living fungal activity and soil-feeding termite groups, and discourage the fungus-growing Macrotermitinae. In this event, the termite faunas of the wetter forests (such as Mulu) may take a smaller proportion of litter input, even though populations may be high.

TABLE 14.3. Estimated consumption of organic matter by termites in Malaysian forests (after Collins, 1983b).

Localities and forest types	Consumption of organic matter g m^{-2} yr^{-1}					Proportion of annual litter production (%)
	Rhino-termit-inae	Term-itinae	Macro-termit-inae	Nasuti-termit-inae	All termites	
Negeri Sembilan, Pasoh Forest Reserve:						
lowland dipterocarp	n.d.	23.8–31.3	129.4–138.4	2.2–3.5	155.4–173.2	14.7–16.3
Sarawak, Gunung Mulu N.P. forests:						
lowland dipterocarp	6.4	11.2	0.7	1.7	20.0	2.1
swampy alluvial	2.1	1.9	2.9	0.5	7.4	0.6
heath (kerangas)	13.1	20.7	0	1.5	35.3	3.4

The weight of organic matter consumed by termites is only one aspect of their impact on decomposition. Processing or translocating organic matter without consuming it may also be significant. Termites do not consume whole trees, but they may assist tree-fall, thereby creating access for other decomposer organisms. It is known that termites nest and feed inside tree trunks, sometimes to a height of 10 m or more (Becker, 1972; Abe and Matsumoto, 1979). Recently, one species at least, *Microcerotermes dubius* (Haviland), has been implicated in conjunction with fungal pathogens as a cause of tree falls and gap formation in virgin forest (Tho, 1982b). Comminution as well as consumption of leaf litter may be an important impact of foraging by leaf-feeding termites at Pasoh. Abe (1980; 1982) has demonstrated that Macrotermitinae at Pasoh encourage microbial decay by increasing the surface area of woody litter available to microbes, and by importing microorganisms with the soil they use for building.

14.3.3. Production

No quantitative estimates of termite production have been made in any rain forest ecosystem. Production:biomass ratios for a wide range of invertebrates, including termites, is about 3:1, although production by the Macrotermitinae may be 2–4 times higher. Erring on the conservative side and using this lower figure for all groups, the biomass figures in Table 14.2 imply the potential for production of up to about 30 g m^{-2} yr^{-1} (300 kg ha^{-1} yr^{-1}). In terms of weight this is equivalent to the production of one cow per hectare and in a steady state population the biomass would be lost annually to predatory and decomposer food chains. In view of this, it is perhaps not surprising that termites support an army of specialized and opportunistic predators, from ants to pangolins, that attack termite nests, alate swarms and foraging columns anywhere between their underground burrows and the forest canopy.

14.3.4. Building

All termites except the Macrotermitinae use some or all of their faeces in building nests or foraging galleries. As these are abandoned and decompose, the components are recycled. Wood-feeding species produce carton, a woody substance with low nutrient concentrations and high levels of lignin and other undigested components. Nests made from carton support low populations of bacteria and are slow to decay (Lee and Wood, 1971). The soil-feeding termites of the Apicotermitinae, Termitinae and Nasutitermitinae all build their nests from faeces augmented with soil and saliva. It has often been observed that abandoned, or even occupied, nests are taken over by plant roots, e.g., *Longipeditermes* (Abe, 1978). In Nigerian and Venezuelan forests nests have been shown to form fertile patches of significance in determining the availabilty of nutrients to germinating seedlings (Anderson and Wood, 1984; Salick *et al.*, 1983). The nests contained higher concentrations of carbon, nitrogen, phosphorous, potassium and other elements than the soil. Such processes are very likely to be important in Malaysian forests also, although no experimental work has yet been done on this subject.

The role of termites in the mixing of soil profiles is not clear. As might be expected, subterranean termite populations are characteristically found in the organic layers of soil profiles. In Pasoh, Malaysia, termites were mainly found in the 0–15 cm layer, and were rare below 25 cm (Abe and Matsumoto, 1979). There was evidence of stratification, with Termitinae mainly in the 0–10 cm layer, Nasutitermitinae in the 5–10 cm layer and the Macrotermitinae at 15–20 cm. In some forests the combined building activities of the Macrotermitinae probably represents a considerable annual turnover of forest soils (Harris, 1957; Abe, 1980).

14.3.5. Nitrogen fixation

Certain bacteria have enabled termites (except Macrotermitinae) to fix atmospheric nitrogen, enhancing the low nitrogen content of their food. The only other insect known to have this ability is the cockroach *Cryptocercus punctulatus*, furnishing more evidence of its close relationship with termites. Estimates of nitrogen fixation rates by termites vary widely and although many rain forest species are now known to have this capacity, there are insufficient data to calculate the impact of a whole termite community on the nitrogen cycle of a forest. Recent reviews of the data on nitrogen fixation have demonstrated the need for much more research in this field (Collins, 1983a; Breznak, 1982).

14.4. CONCLUSIONS

Malaysian rain forests support the largest recorded termite communities, with up to 59 species in small plots ($<$ 5 ha). Of the total of 150 or so species known from Peninsular Malaysia, three-quarters are in the family Termitidae (Table 14.1). The Termitidae play a significant role in all the main feeding guilds, consuming wood, leaves and humus. The Rhinotermitidae, although not particularly diverse, play a significant role in the destruction of dead timber. The Kalotermitidae are scarce, but are the only group that can live in the canopy with no connection to the soil. Most termites live and nest in or on the soil, or within a few metres of its surface (Fig. 14.3).

Data on populations and biomass are few, and although the ecological impact of termites on tropical forests is believed to be considerable, the mechanisms are still poorly understood. It seems that the taxonomic composition of the community has a significant impact on the overall levels of litter consumption. The Macrotermitinae consume more than other groups and where this subfamily is common, the impact of termites in removing litter is correspondingly high. Consumption of litter is not the only impact of termites on the forest ecosystem. They also comminute litter and thereby speed up bacterial and fungal decay. By causing large trees to fall they facilitate their breakdown and replacement by young stock. Termite nests, particularly abandoned ones, can be an important source of nutrients for seedlings. Termites themselves are an important source of food for a wide variety of forest animals. These and other aspects of the ecological role of termites in rain forest require further study if their full impact is to be understood.

The distribution of termites reflects the general increase in diversity of insects with proximity to the equator (Fig. 14.1). Regrettably, this diversity is inversely proportional to our taxonomic knowledge, and it is likely that thousands, perhaps millions, of insect species await discovery in tropical rain forests. Termites are not a particularly diverse group, a reflection of the limited opportunities for specialization within the detritivorous habit, yet over 25 per cent of the known Malaysian fauna still remains formally undescribed (Tho, 1982a). Many termite species have widespread distributions within Malaysia and the Indomalayan region, but some appear to be very restricted and rare. It is notable, for example, that *Glyptotermes buttelreepeni* (Holmgren), *Schedorhinotermes translucens* (Haviland) and *Euhamitermes hamatus* (Holmgren) have not been recorded in Peninsular Malaysia since 1912–14 (Tho, 1982a), despite extensive faunistic studies (e.g., Salick and Tho, 1984). Whether they have been overlooked, misidentified or extirpated remains unknown.

The impact of deforestation on termites has been little-studied and the poor state of the systematics of the group makes it difficult to undertake a thorough assessment. However, it is known that the species diversity of these cryptic and sensitive insects can be decimated by destruction of the forest canopy and exposure to direct sunlight. Only a few pre-adapted species are able to survive, some of which subsequently become regarded as pests. Similar losses are likely for other invertebrates in the decomposer community, whose members are generally ill-adapted to withstand desiccation and wide temperature fluctuations. There are no data to show whether such sensitive animals all manage to survive in pockets of forest and protected areas, or whether, as may be suspected, clearance of the Malaysian forests is causing a proportion of them to become extinct.

ACKNOWLEDGEMENTS

I should like to thank Melanie M. Collins for the art-work in Fig. 14.3.

REFERENCES

Abe, T. (1978). Studies on the distribution and ecological role of termites in a lowland rain forest of West Malaysia 1) Faunal composition, size, colouration and nest of termites in Pasoh Forest Reserve. *Kontyu* 46, 273–290.

Abe, T. (1979). Studies on the distribution and ecological role of termites in a lowland rain forest of West Malaysia 2) Food and feeding habits of termites in Pasoh Forest Reserve. *Jpn J. Ecol.* 29, 121–135.

Abe, T. (1980). Studies on the distribution and ecological role of termites in a lowland rain forest of West Malaysia 4) The role of termites in the process of wood decomposition in Pasoh Forest Reserve. *Rev. Ecol. Biol. Sol* 17, 23–40.

Abe, T. (1982). Ecological role of termites in a tropical rain forest. In *The Biology of Social Insects* edited by M.D. Breed, C.D. Michener & H.E. Evans, pp. 71–75. Westview Press, Boulder.

Abe, T. and Matsumoto, T. (1979). Studies on the distribution and ecological role of termites in a lowland rain forest of West Malaysia 3) Distribution and abundance of termites in Pasoh Forest Reserve. *Jpn J. Ecol.* 29, 337–351.

Ahmad, M. (1968). Termites of Malaysia I. Nasute genera related to *Subulitermes* (Isoptera, Termitidae, Nasutitermitinae). *Bull. Dept. Zool., Univ. Punjab (N.S.)* Art. 3, 34 pp.

Ahmad, M. (1971). Termites of Malaysia. II. Genus *Hirtitermes* Holmgren (Isoptera, Termitidae, Nasutitermitinae). *Proc. Pak. Acad. Sci.* 8, 27–36.

Ahmad, M. (1976). The soldierless termite genera of the Oriental region, with a note on their phylogeny (Isoptera: Termitidae). *Pak. J. Zool.* 8, 105–123.

Ahmad, M. and Akhtar, M.S. (1981). New termite genera of the *Capritermes* complex from Malaysia, with a note on the status of *Pseudocapritermes* (Isoptera: Termitidae). *Pak. J. Zool.* 13, 1–21.

Anderson, J.M. and Wood, T.G. (1984). Mound composition and soil modification by two soil-feeding termites (Termitinae, Termitidae) in Nigerian riparian forest. *Pedobiologia* 26, 77–82.

Becker, G. (1972). Termiten im Regenwald des Magdalenenstromtals in Kolumbien. *Zeitschr. Angew. Entomol.* 71, 431–441.

Breznak, J.A. (1982). Intestinal microbiota of termites and other xylophagous insects. *A. Rev. Microbiol.* 36, 323–343.

Collins, N.M. (1979). Observations on the foraging activity of *Hospitalitermes umbrinus* (Haviland), (Isoptera: Termitidae) in the Gunong Mulu National Park, Sarawak. *Ecol. Entomol.* 4, 231–238.

Collins, N.M. (1980a). The distribution of soil macrofauna on the West Ridge of Gunung (Mt.) Mulu, Sarawak. *Oecologia* 44, 263–275.

Collins, N.M (1980b). The effect of logging on termite (Isoptera) diversity and decomposition processes in lowland dipterocarp forests. In *Tropical Ecology and Development* edited by J.I. Furtado, pp. 113–121. International Society of Tropical Ecology, Kuala Lumpur.

Collins, N.M (1981). The role of termites in the decomposition of wood and leaf litter in the southern Guinea savanna of Nigeria. *Oecologia* 51, 389–399.

Collins, N.M. (1983a). The utilization of nitrogen resources by termites (Isoptera). In *Nitrogen as an Ecological Factor*, 22nd Symposium of the British Ecological Society edited by J.A. Lee, S. McNeill & I.H. Rorison. Blackwell Scientific Publications, Oxford.

Collins, N.M. (1983b). Termite populations and their role in litter removal in Malaysian rain forests. In *Tropical Rain Forest: Ecology and Management* edited by S.L. Sutton, T.C. Whitmore and A.C. Chadwick. Blackwell Scientific Publications, Oxford.

Collins, N.M. (1984). The termites (Isoptera) of the Gunung Mulu National Park, with a key to the genera known from Sarawak. *Sarawak Mus. J.* 30(51), *Special Issue No. 2. Gunung Mulu National Park* edited by A.C. Jermy and K.P. Kavanagh, pp. 65–87.

Collins, N.M. (1988). Zoogeography, diversity and ecological patterns of termites in rain forests. In *Tropical Rain Forest Ecosystems* edited by H. Leith and M.J.A. Werger. Ecosystems of the World Vol. 14B. Elsevier, Amsterdam.

Deligne, J., Quennedy, A. and Blum, M.S. (1981). The enemies and defense mechanisms of termites. In *Social Insects* Vol. 2 edited by H.R. Hermann. Academic Press, New York and London.

Harris, W.V. (1957). An introduction to Malayan termites. *Malay. Nat. J.* 12, 20-32.

Harris, W.V. (1961). *Termites, Their Recognition and Control.* Longmans, London, 187 pp.

Harris, W.V. (1969). *Termites as Pests of Crops and Trees.* Commonwealth Institute of Entomology, London.

Haviland, G.D. (1898). Observations on termites, with descriptions of new species. *J. Linn. Soc. Lond., Zool.* 26, 358–442.

John, O. (1925). Termiten von Ceylon, der Malayischen Halbinsel, Sumatra, Java und Aru-Inseln. *Treubia* D, 360–419.

Kondoh, M., Watanabe, S., Chiba, S., Abe, T., Shiba, M. and Saito, S. (1980). Studies on the productivity of soil animals in Pasoh Forest Reserve, West Malaysia. V. Seasonal change in the density and biomass of soil macrofauna: Oligochaeta, Hirudinea and Arthropoda. *Mem. Shiraume Gakuen Coll.* 16, 1–26.

Krishna, K. (1970). Taxonomy, phylogeny and distribution of termites. In *Biology of Termites* Vol. II edited by K. Krishna and F.M. Weesner. Academic Press, London and New York. 643 pp.

Lee, K.E. and Wood, T.G. (1971). *Termites and Soils.* Academic Press, New York & London. 251 pp.

Light, S.F. (1929). Present status of our knowledge of the termites of China. *Lingnan Sci. J.* 7, 581–600.

Light, S.F. (1930). Notes on Philippine termites, IV. *Philipp. J. Sci.* 42, 13–58.

Matsumoto, T. (1976). The role of termites in an equatorial rain forest ecosystem of West Malaysia. I. Population density, biomass, carbon, nitrogen and calorific content and respiration rate. *Oecologia* 22, 153–178.

Matsumoto, T. and Abe, T. (1979). The role of termites in an equatorial rain forest ecosystem of West Malaysia. II. Leaf litter consumption on the forest floor. *Oecologia* 38, 261–274.

Prestwich, G.D. (1984). Defence mechanisms of termites. *A. Rev. Entomol.* 29, 201–232.

Roonwal, M.L. (1970). Termites of the Oriental region. In *Biology of Termites* Vol. II edited by K. Krishna and F.M. Weesner. Academic Press, London and New York. 643 pp.

Salick, J., Herrera, R. and Jordan, C.F. (1983). Termitaria: nutrient patchiness in nutrient deficient forests. *Biotropica* 15, 1–7.

Salick, J. and Tho. Y.P. (1984). An analysis of termite faunae in Malayan rainforests. *J. Appl. Ecol.* 21, 547–561.

Tho, Y.P. (1972). Observations on the biology of the termite *Dicuspiditermes nemorosus* (Haviland). *Malay. Nat. J.* 25, 10–17.

Tho, Y.P. (1974). The termite problem in plantation forestry in Peninsular Malaysia. *Malay. Forester* 37, 278–283.

Tho, Y.P. (1978). The common mound-building termite. *Nature Malaysiana* 3, 16–24.

Tho, Y.P. (1981). A unique defense strategy in the termite *Prohamitermes mirabilis* (Haviland) of Peninsular Malaysia. *Biotropica* 13, 236–238.

Tho, Y.P. (1982a). *Studies on the taxonomy and biology of termites (Isoptera) of Peninsular Malaysia.* Ph.D. thesis, University of Aberdeen.

Tho, Y.P. (1982b). Gap formation by the termite *Microcerotermes dubius* in lowland forests of Peninsular Malaysia. *Malay. Forester* 45, 184–192.

Completed February 1985

CHAPTER 15

Forest Lepidoptera

H.S. Barlow

P.O. Box 10139, 50704 Kuala Lumpur, Malaysia

CONTENTS

15.1. INTRODUCTION

The Lepidoptera (butterflies and moths) of Peninsular Malaysia and its offshore islands are probably better known than any other insect group in the area. Even so, knowledge is extremely patchy on the interactions of these insects with the forest environment. Extensive taxonomic work on the butterflies, including the Hesperiidae (skippers), has been summarized by Eliot in Corbet and Pendlebury (1978). We have records of 1014 species, several of which have more than one race within the Peninsula and islands. While there are undoubtedly others still to be recorded, it seems likely that this figure is within 90–95% of the total number of butterfly species occuring here.

The same unfortunately cannot be said of the moths. With the exception of an introduction to some 500 of the commoner and larger moth species occuring in the Peninsula (Barlow, 1982), little if any work has been undertaken since collections were made pre-war by Ridley, working from Singapore, and Corbet in Kuala Lumpur. Much of the Ridley material bears labels indicating Singapore provenance, although later collecting suggests that some of these specimens were more likely to have been taken in the Peninsula, and erroneously labelled. Details of these collections were never published as such, although many of the specimens have now been incorporated into the main collections at the British Museum (Natural History), London, and are therefore available for taxonomists. The lack of work on the moths of the Peninsula contrasts with extensive collecting by Wallace, Shelford and Mjöberg in Sarawak and Waterstradt on Mt Kinabalu (Barlow, 1969). In the last 15 years substantial additional collecting has concentrated on the Bornean fauna (Holloway, 1970, 1976, 1983, 1985b, d and e, 1987, 1988). This is of importance for studies in the Peninsula. It has resulted in the description of many new species now known to occur here also, and suggests patterns of distribution and behaviour which are almost certainly applicable in the Peninsula. Although most collections in Borneo have been made in primary forest, recent work in the Peninsula has been in mixed environments, including advanced secondary associations.

Fig. 15.1. The lemon emigrant *Catopsila pomana* 'pomona' form (Pieridae), a widespread and abundant butterfly often taking part in massed flights. (Photograph by Ken Rubeli.)

15.2. DISTRIBUTION

15.2.1. Regional distributions

The butterfly fauna of the Peninsula is remarkable for the low level of endemicity exhibited: only 23 species out of the 1014 so far recorded are endemic. These are all, in any case, rare species and until substantial additional work has been done within the South East Asian region, it would be unwise to regard them as conclusively endemic.

Table 15.1 lists the Peninsular Malaysian (including Singapore) species of butterflies according to external distribution. From this it will be seen that the representation of species with Sundaic distributions (41.5%) greatly exceeds the Indo-Chinese element (10.5%). This feature has also been discussed by Holloway (1973, 1974).

TABLE 15.1. Peninsular Malaysian species of butterfies, listed according to extra-Malayan distribution (updated from Corbet and Pendlebury, 1978).

	Total Malaysian species in the family	Indo-Chinese (not found south of Singapore)	Sundaic (not found north of Tavoy)	Oriental (found north of Tavoy and in Sundaland or beyond)	Local 'Endemics' (found only between Tavoy and Singapore)[1]
Papilionidae	46	7	12	27	—
Pieridae	45	7	11	27	—
Danaidae	36	4	11	21	—
Satyridae	83	17	34	32	—
Nymphalidae	154	16	47	90	1
Libytheidae	2	—	—	2	—
Riodinidae	16	3	7	6	—
Lycaenidae	385	23	204	147	11
Hesperiidae	247	29	95	110	11
Total	1014	106	421	462	23
Percentage		10.5	41.5	45.5	2.5

The 'Endemics' are: *Euthalia ipona, Deramas alixae, D.anyx, Nacaduba russelli, Niphanda stubbsi, A.cardoni, Catapaecilma lila, C.evansi, Acupicta flemingi, Horaga aractina, Deudorix elioti, Rapala cowani, Bibasis owstoni, Hasora wilcocksi, Celaenorrhinus pahangensis, Zographetus kutu, Isma hislopi, Pyroneura klanga, Zela elioti, Erionota hislopi, Potanthus chloe, Telicota hilda, Pelopidas flavus.*

While very much less is known of the distribution of the moth species, there is no reason to suspect that patterns are substantially different. In many cases of species recorded only from Peninsular Malaysia (Barlow, 1982), it is probable that further collecting will find them elsewhere within South East Asia.

During the successive periods when the polar ice caps advanced and retreated in the Pleistocene, the Peninsula has seldom if ever been cut off from mainland Asia (Chapter 1), but must rather have acted as a funnel down into and up from the Sundaland subcontinent. The funnel effect must have been further facilitated by the substantial swathe of lowland and montane forest which until the last 30 years stretched unbroken down the centre of the Peninsula.

Holloway (bibliography) has considered the distribution of Lepidoptera by altitude within South East Asia. On the basis of work in primary forest, he has identified and discussed the species occurring at higher altitudes. He points out that the major centre of diversity for the group is the area of the eastern Himalaya (Assam, Szechuan and North Burma), while the Sundaland fauna owes its diversity partially to the effects of repeated changes in sea level in the area during the Pleistocene.

15.2.2. Local distributions

Within the Peninsula three major main faunistic regions are recognized:

(a) Kedawi

This area includes the Langkawi islands (with the adjacent Tarutao group of Thailand) and extends as far as the Kedah river. Of the butterfly species in the Peninsular list, 31 occur no further south than the Kedawi region and Upper Perak. Another 55 are represented in the region by distinct subspecies. Until recently, the Kedawi fauna had been thought not to extend East of the main range. Collecting on the East coast has now suggested that some typical Kedawi taxa may extend into this area. The selective factors are believed to reflect the more markedly monsoonal climate which prevails in the northern part of the Peninsula.

(b) Main Peninsular fauna

This extends south from the Kedah river to south-western Johor, but excluding eastern Johor, southeastern Pahang and the Tioman group of islands. The majority of the Peninsular fauna, as might be expected, belongs to this region, with a fairly clearcut distinction between species associated with the lowlands and foothills and those associated with montane biotopes above 1000 m.

(c) Eastern Johor, southeastern Pahang and Tioman island

With the exception of Tioman, this area has been comparatively little collected. However, such work as has been undertaken suggests that it may ultimately be necessary to treat Tioman as a separate island unit, by virtue of the large number of subspecies endemic there. So far 48 subspecies not recorded elsewhere in the Peninsula have been recognized in the Tioman islands, including at least three subspecies superficially identical to subspecies found in Borneo, and two identical to subspecies found in Formosa. The fauna of the remainder of the area also appears to show certain affinities with that of Borneo.

As mentioned above (15.1), although insects in old collections frequently bear labels indicating Singapore provenance, such records must be treated with caution, since material collected before 1900 from South East Asia was frequently inaccurately labelled. On this basis the butterfly fauna of "Singapore" included seven taxa not found in the Peninsula. Unfortunately, the enormous development which has taken place on the island has nearly annihilated all but the commoner garden species. There remain Bukit Timah and various reservoir catchments where access is controlled. Here, some at least of these taxa may perhaps survive.

Knowledge of the distribution of the moth fauna of the Peninsula is not at present adequate to attempt a similar summary. Preliminary work suggests a total of around 10,000 species, including members of

M—O

families traditionally regarded as Microlepidoptera. The state of taxonomy of many groups is such that it is often extremely difficult to identify even some of the commoner moths of economic importance (Holloway in Barlow, 1982), and any major collection is still bound to produce a substantial number of new species.

15.2.3. Habit preferences

Vegetational/altitude distribution of the butterflies has been discussed in Corbet and Pendlebury (1978), largely on the basis of collecting done pre- and immediately post-war. Fleming (1975), collecting by and large some 20–30 years later, recorded several species from a much broader range of localities. Whether this simply reflects an increase in areas accessible and suitable for collecting, or whether is indicates a degree of adaptation by certain species to disturbed habitats is an intriguing question, but possibly one which cannot be satisfactorily answered.

Holloway (1985a and b) has analysed in some detail the distribution by altitude of representatives of subfamilies of the Geometridae on mountains in Borneo, pointing out that there is a marked increase in the proportion of Larentiinae from 1000 m upwards. Similar results would be expected in Peninsular Malaysia. The greatest overall macrolepidopteran diversity in the Gunung Mulu area of Sarawak was observed in samples from lower montane forest (1000 m). Of lowland forest types, the limestone forest was floristically (i.e., as canopy tree species) the least diverse, yet produced the most diverse Lepidoptera sample. There was also significant variation from family to family in Lepidoptera diversities sampled from various forest types. Certain species, e.g., *Buzura insularis* Warren, and *Boarmia* spp. (both Ennominae), *Alex palparia* Walker (Oenochrominae) and certain Eupterotidae, are distinctive of understorey fauna. These details have led Holloway to identify certain families and subfamilies as potential indicators of forest type, which may have a role to play in monitoring processes of forest degradation and regeneration (Table 15.2).

TABLE 15.2. Families having potential as biological indicators (from Holloway, 1985b).

Lymantriidae	Noctuidae	: Chloephorinae
Notodontidae		Hypeninae
Arctiidae	Geometridae	: Ennominae
Nolidae		Larentiinae (High altitudes)
	Lasiocampidae	

A number of workers have remarked that butterflies appear to be commonest on the edges of primary and mature secondary forest. This is partly no doubt because under undisturbed conditions within the forest many butterflies live entirely in the canopy. It is noticeable when walking through primary forest that few butterflies are to be seen, even in sunny conditions. Those most likely to be encountered are of the families Satyridae (which now include the Amathusiinae) and Nymphalidae of the tribe Euthaliinae and Riodinidae. Larvae of the Satyridae, it should be noted, feed primarily on grasses, bamboo and palms, while those of the Riodinidae are reported from trees and shrubs of the family Myrsinaceae.

Species of primary forest associations, particularly those of medium to high altitudes, have frequently been observed on or around small cleared patches of hill-top. Amongst genera particularly well represented under such conditions are *Delias*, species of *Graphium* restricted to higher altitudes and certain lycaenid genera, particularly *Rapala* and *Celastrina*. Common, in Common and Waterhouse (1981) has discussed this feature in relation to Australian butterflies. He has observed that males generally greatly outnumber females. Those females which appeared in an American study were usually virgin, and stayed only long enough to mate. He has suggested that hill-topping "would also help to provide for the mixture of genetic characters of semi-isolated populations of a species in a given region".

It has also been observed that at higher altitudes darker coloured species, particularly amongst the Pieridae, appear to predominate. Thus in the genus *Delias*, *D. descombesi evanthos*, *D. ninus ninus* and *D. belladonna*, all dark-dusted, are found at hill tops. Dark coloration may be associated with the need at such altitudes to be on the wing at short notice to make maximum use of the reduced periods of sunlight available. Darkened wings would absorb the heat and thus warm up the insect more quickly to be ready for flight.

15.3. BLOOD-SUCKERS

A number of moths recorded from Malaysia, primarily in the Noctuoidea, are attracted to the saline secretions from the eyes of animals, and some even suck blood (Banziger, 1972; Buttiker 1964, 1967). In the blood-sucking species, the proboscis has developed a piercing apparatus to penetrate skin, lacking among species which do not exhibit such behaviour. A member of the genus *Calyptra* (subfamily Ophiderinae) is the major blood-sucker. Species of the genus *Othreis* (also Ophiderinae) are capable of piercing the skins of fruit, to suck the juices: clearly a development in the same direction. Other moths exhibiting this behaviour are members of the Geometroidea and Pyraloidea.

Fig. 15.2. A noctuid moth, *Othreis* spp., of the subfamily Ophiderinae. (Photograph by Ken Rubeli.)

15.4. MIMICRY, POLYMORPHISM, DUPLEX SPECIES, ETC.

The incidence of moth/butterfly mimicry associations appears to be most pronounced amongst hill-top frequenting species. Table 15.3 summarizes the major mimetic associations amongst the butterflies and moths of the Peninsula.

TABLE 15.3. Mimetic associations amongst the Lepidoptera of Peninsular Malaysia (expanded from Corbet and Pendlebury, 1978).

Models	Mimics
Atrophaneura varuna ♀	*Papilio memnon* ♀-f. *butlerianus*
Atrophaneura nox ♀	*Papilio memnon* ♀-f. *esperi*
Atrophaneura coon	*Papilio memnon* ♀-f. *distantianus*
Pachliopta aristolochiae	*Papilio polytes* ♀-f. *polytes* *Histia rhodope* (Zygaenidae)
Delias singhapura *Delias baracasa* }	*Cyclosia pieridoides* ♂ (Zygaenidae)
Delias pasithoe *Delias ninus* }	*Elymnias esaca* ♀ *Cyclosia pieridoides* ♀ (Zygaenidae)
Danaus chrysippus *Danaus genutia* *Danaus melanippus* }	*Elymnias hypermnestra tinctoria* ♀ *Hypolimnas misippus* ♀
Tirumala septentrionis	*Chilasa clytia* f. *dissimilis*
Parantica aspasia	*Pareronia valeria* ♀
Parantica agleoides *Radena vulgaris* *Radena similis* }	*Elymnias nesaea*
Parantica melaneus	*Hestina mimetica*
Parantica sita	*Chilasa agestor*
Parantica spp.	*Dysphania* spp. (Geometridae) *Psaphis* spp. (Chalcosiinae) *Longicella mollis* (Agaristidae)
Ideopsis gaura	*Paranticopsis delessertii* ♀ *Elymnias kuenstleri* ♀ *Cyclosia pieridoides virgo* ♀ (Zygaenidae)
Euploea algea *Euploea eyndhovii* }	*Chilasa slateri* *Elymnias kuenstleri* ♂ *Chilasa paradoxa* f. *aenigma*
Euploea mulciber ♂	*Elymnias casiphone* ♂ *Cyclosia midamia* (Zygaenidae)
Euploea mulciber ♀	*Elymnias casiphone* ♀
Euploea midamus	*Elymnias panthera*

(Table 15.3 continued)

Models	Mimics
Euploea klugii	*Chilasa clytia* f. *clytia*
	Chilasa paradoxa f. *aegialus* ♂
Euploea diocletianus ♂	{ *Elymnias penanga* ♀ -f. *penanga*
	Euripus nyctelius ♀ -f. *isina*
	Chilasa paradoxa f. *aegialus* ♀
Euploea diocletianus ♀	{ *Elymnias harterti* ♀
	Euripus nyctelius ♀ -f. *euploeoides*
Euploea spp.	*Dysphania transducta* (Geometridae)
	Psaphis spp. (Chalcosiinae)
Cethosia biblis	*Coryptilum rutilellum* (Tineidae)
Various wasps	Ctenuchidae spp.
Various wasps	Sesiidae spp.
Euploea mulciber mulciber	*Pompelon marginata* (Chalcosiinae)
Various wasps	*Zeuxippa digitata* (Chalcosiinae)
Euploea algea	*Cyclosia inornata cuprea* (Chalcosiinae)

Of the larger day-flying moths, the geometrid genus *Dysphania* would appear to provide the only common species involved in Mullerian mimicry complexes (non-distasteful mimicking distasteful). For example, *Dysphania transducta* feeds on *Carallia eugenoides* (Rhizophoraceae), not noted for the production of toxic substances. Here it appears to be associated in a group with one or more of the chalcosiine genus *Psaphis*, mimicking one of the *Euploea*-group of butterflies. Other species in the genus *Dysphania* may form a blue group with *Arycanda* species (Ennominae) and the agaristid, *Longicella mollis*, possibly mimicking one or more of the *Danaus*-group of butterflies.

Polymorphism in the wing markings of the butterflies is largely confined in the Peninsula to the Papilionoidea, the best known example being the various female forms of *Papilio memnon*. These have been studied in some detail, ultimately for medical purposes, by Clarke *et al.* (1968, 1971). Other examples of polymorphism, primarily in the genus *Chilasa*, were given by Corbet and Pendlebury (1978).

Polymorphism in the moths is confined to a comparatively small number of species, notably those in the noctuid genus *Stictoptera*, and *Ercheia cyllaria*, also a noctuid. These species are not day-flying and appear not to form part of mimicry groups. It is however possible that the more obvious variations in the adult may be genetically linked with other characteristics, physical or behavioural in the early stages, where an element of polymorphism could enhance survival. Sexual dimorphism, which among butterflies almost always contains a mimetic element, is less widespread but also occurs among moths. A number of the more widely encountered species found in the Peninsula exhibiting such dimorphism are given in Table 15.4 below. With the possible exception of the day-flying Chalcosiinae, *Eterusia distincta* and certain *Cyclosia* spp., there appears to be no mimetic advantage in this dimorphism.

Duplex species are widely known throughout the Lepidoptera. Those occurring among the butterflies were listed in Corbet and Pendlebury (1978) and discussed by Holloway (1973). Many more have been observed among the moths, frequently associated with differences in altitude. These have been mentioned by Holloway (1970 and 1985b). Once again, it seems clear that repeated isolation of small

TABLE 15.4. Peninsular Malaysian moths exhibiting sexual dimorphism.

Eterusia distincta	(Chalcosiinae)
Cyclosia spp.	(Chalcosiinae)
Parasa spp.	(Limacodidae)
Chalcocelis albiguttatus	(Limacodidae)
Eilema nebulosa	(Lithosiinae)
Monosyntaxis holman-hunti	(Lithosiinae)
Lymantria spp.	(Lymantriidae)
Euproctis spp.	(Lymantriidae)
Dasychira spp.	(Lymantriidae)
Calliteara spp.	(Lymantriidae)
Locharna strigipennis limbata	(Lymantriidae)
Erebus orion	(Catocalinae)
Erebus ephesperis	(Catocalinae)
Agonista endoleuca	(Catocalinae)
Agonista hypoleuca	(Catocalinae)
Antheraea spp.	(Saturniidae)
Ercheia multilinea	(Noctuidae)
Hydrillodes spp.	(Noctuidae)
Crithote spp.	(Noctuidae)
Iontha spp.	(Noctuidae)

communities during times of high sea-levels in the Pleistocene has facilitated speciation. Thereafter, when sea-levels have fallen, ranges have expanded to overlap geographically. On occasions, however, expansions involving a high altitude relict community expanding downwards and a second, low altitude community, have left uncolonized gaps at intermediate levels. Holloway has remarked that on Mt Kinabalu in Sabah, at least, the higher altitude species in a duplex of this nature is almost always larger than the low altitude representative.

15.5. DIVERSITY

Some indication of the diversity of the moth fauna of the region can be gauged from Holloway's estimate that approximately 4000 species of moths in the families generally regarded as Macrolepidoptera have already been recorded from Borneo. The overall resemblance of the Bornean fauna to that of the Peninsula suggests that similar figures apply here. No estimates have been attempted on the number of Microlepidoptera, but it is not unreasonable to assume that these account for at least an equal number of species. Holloway (1985b), plotting the number of species against the log number of individuals from Mt. Mulu in Sarawak, has already indicated the very high diversity of the moth fauna of the region.

Barlow and Woiwod (in press) have done work on within-habitat diversity (= alpha-diversity) in forest at 2000 ft (610 m) in Peninsular Malaysia. All the moths in the "Macro" groups, together with the Pyralidae were counted, trapping on average 2–3 times a week over one year, to give a total of over 15,000 speciments covering 1426 species. The alpha value was calculated at 386.5 ± 11.9. A large number of species with one representative only were recorded. Holloway (1984) has produced a similarly high alpha value on a large sample of moths from Mt Mulu, Sarawak. However these results are not directly comparable as different trapping techniques were used. The highest value recorded using similar techniques in the U.K. is 60.0, while preliminary results from traps run in Northern Sulawesi in 1985 indicate an alpha value appreciably lower than in the Peninsula. These results confirm the enormous richness and diversity of the rainforest Lepidoptera in Peninsular Malaysia.

In contrast to other animal groups, for instance the mammals and birds (Chapter 13), data so far available suggest that maximum species diversity amongst all Lepidoptera occurs at around 600–1000 m. Possibly the vegetation of hill forest habitats is more favourable than that of the lowlands. Holloway (1985b), noting that from 600–1000 m upwards the Fagaceae begin to predominate, has suggested that these, together with Myrtaceae, may support a more diverse fauna. He has also suggested, in the most tentative terms, that in the lowland forests, alkaloid defences amongst the predominant Dipterocarpaceae may be more frequent. Unfortunately, at present, we lack adequate information on aspects of plant biochemistry and the coevolutionary relationships between such vegetation and the Lepidoptera, crucial to the understanding of these relative diversities (but see next section).

15.6. FOODPLANT RELATIONSHIPS

An analysis of the relevant appendices in Corbet and Pendlebury (1978) and Barlow (1982) suggests that a majority of the Lepidoptera feed on shrubs or low-growing plants often associated with secondary vegetation. While this would go some way to explain the apparent richness of the lepidopteran fauna in areas of secondary growth adjacent to primary forest, it may also reflect our failure so far to establish more than a fraction of the life histories of the Lepidoptera. Under such circumstances it is perhaps not surprising that most of the work should be concentrated on readily accessible plants of secondary vegetation, and that defoliation of forest canopy should have been so little studied.

Preliminary work is at present in hand, primarily studying the relationship of certain danaids to their foodplants, mainly Asclepiadaceae and Apocynaceae, and the apparent need for adult males to ingest certain mineral salts or substances from withered plants before mating. It appears that unless these are ingested by the adult males, their pheromone systems are unable to function adequately to ensure successful mating. In many of the Danaidae, the pheromones are diffused by yellow or white hair-pencils in the terminal segments of the abdomen. These hair-pencils also appear to have a defensive function, as can be seen from the way in which they are extruded by males on capture. This work is now being extended to consider the function of the large and remarkable coremata in *Creatonotos transiens* (Arctiidae), figured in Barlow (1982).

Boppré (1978, 1979) has identified the key role of di-hydropyrrolizidines in the hair-pencil odour and noted that, chemically,these are similar to the heterocyclic moiety of pyrrolizidine alkaloids (PAs), secondary metabolites of a number of plants on which adult danaids feed. This has led to some understanding of the role of the specialized glandular organs on the wings of many male danaids. It has also produced some interesting speculation in the evolution among certain asclepiads of an ability to synthesize poisonous cardiac glycosides, possibly initially as a defence against insect predation. It would appear that some danaids at least have simultaneously developed an ability to absorb such poisons, and convert these into their own defence mechanism against potential predators -- mainly birds. This in turn throws further light on danaid mimicry associations. There remains an immense amount of work to be done on this subject in the danaids alone. A resolution of even some of these problems could provide clues to the functions of many of the specialized scent producing and receiving organs, which are so remarkably and variably developed in the Lepidoptera of the Peninsula.

Until this work is more advanced, it is unlikely that we shall be able to understand the obviously complex relationships between these insects and their hosts. It seems possible that Lepidoptera larvae, which generally feed on the tender growing tips of vegetation, may play a contributory, but seldom key role in the maintenance of vegetational balance in forest ecosystems. Moreover, it is likely that a fuller understanding of the diffusion mechanisms for such pheromones, including infra-red radiation will throw light on the diversity of structure found in antennae of Lepidoptera, and enable us to appreciate some aspects at least of lepidopteran communication systems. Interesting theories on this have recently been advanced by Callahan (1975, 1981, the former providing scientific references). It is clear that many years of research will be required before we can do much more than scratch the surface in our understanding of the full relationships between the Lepidoptera of the Peninsula and their forest environment.

15.7. CONCLUSION

15.7.1. Conservation

The summary given above indicates that knowledge of the foodplants, territorial needs, habitat and nectar requirements of adult butterflies and moths is woefully inadequate. It is therefore desirable that at least one area in each of the major faunal regions of Peninsular Malaya should be retained as a primary forest reserve. While some species appear year after year to be closely confined to a very small area, perhaps by the availability of foodplants, we still know too little of the problems of isolated communities, their pests, diseases and parasites to be able to recommend small areas for retention in preference to large areas. Moreover the pronounced difference in species composition with altitude makes it desirable that representative areas at all altitudes should be preserved. In most cases such areas are already covered by existing national parks.

It is suggested that the following areas of primary vegetation should be treated as minimum requirements on the basis of our limited knowledge at present, and maintained in as undisturbed a condition as possible.

(a) On the main island of Langkawi, inevitably, (in view of recent developments) on the western side of the island.

(b) Kedah Peak, which is already a reserve.

(c) Two or three areas along the main range of the Peninsula, covering all altitudes. Existing reserves (Taman Negara, Kerau Game Reserve and the proposed Endau-Rompin National Park) could well be used for this purpose. Additionally, in view of the considerable amount of work already undertaken in the area, it is suggested that the Gombak and Ulu Kali catchment areas, including Gunung Bunga Buah should be retained.

(d) Pulau Tioman, and a small area of lowland primary forest, if such can be located, near Rompin, Pahang.

(e) An area of mangrove associations, possibly at Kuala Selangor.

15.7.2. A national reference collection

Apart from the need to preserve representative habitats, further study of the Lepidoptera and most other insect groups in South East Asia is seriously hampered by the lack of comprehensive and correctly identified reference collections. Holloway, in Barlow (1982), has expanded on this need, and shown how important it is, even for the correct determination of economically important pest species. Until such a basic tool for the study of the Malaysian fauna within this country is available, it is hard to see how independent Malaysian research into the increasingly important environmental problems on our own doorstep can proceed satisfactorily.

At the same time, in view of the rapid destruction of habitats, the need for very much more detailed research in the key areas of host-plant preferences and life history data is paramount.

ACKNOWLEDGEMENTS

I am grateful to Dr. J. D. Holloway of the Commonwealth Institute of Entomology for his helpful comments on the draft, and to Lt. Col. J. N. Eliot for his assistance in updating tables taken from Corbet and Pendlebury (1978). I am also greatful to Mrs. A. Macartney for information on altitude distribution of the butterflies.

REFERENCES

Banziger, H. (1972). Biologie der lacriphagen Lepidoptera in Thailand und Malaya. *Rev. Suisse Zool.* 79(4), 1381–1469.

Barlow, H. S. (1969). John Waterstradt 1869–1944. *J. Malay Brch. R. Asiat. Soc.* 42, 115–20.

Barlow, H. S. (1982). *An Introduction to the larger Moths of South East Asia.* Malayan Nature Society, Kuala Lumpur.

Barlow, H. S. and Woiwod, I. P. (in press). Moth diversity of a tropical forest in Peninsular *Malaysia, J. Trop Ecol.*

Boppré, M. (1978). Chemical communication, plant relationships, and mimicry in the evolution of danaid butterflies, *Entomol. Exp. & Appl.* 24, 264–77.

Boppré, M. (1979). Lepidoptera and withered plants. *Antenna* 3(1), 7–9.

Buttiker, W. (1964). New observations on the eye-frequenting Lepidoptera from South East Asia. *Verh. Naturforsch. Ges. Basel* 75, 231–6.

Buttiker, W. (1967). Biological notes on eye-frequenting moths from North Thailand. *Mitt. Schweiz. Entomol. Ges.* 39(3–4), 151–79.

Callahan, P. S. (1975). *Tuning in to Nature.* The Devin Adair Co., Conn., U.S.A.

Callahan, P. S. (1981). *The Soul of the Ghost Moth.* The Devin Adair Co., Conn., U.S.A.

Clarke, C. A., Clarke, F. M. M. & Sheppard, P. M. (1968). Mimicry and *Papilio memnon*: some breeding results from England. *Malay. Nat. J.* 21, 201–19.

Clarke, C. A. and Sheppard, P. M. (1971). Further studies on the genetics of the mimetic butterfly *Papilio memnon. Phil. Trans. R. Soc., Lond.* 263, 35–70.

Clarke, C. A., Sheppard, P. M. and Thornton, I. W. B. (1968). The genetics of the mimetic butterfly *Papilio memnon* L. *Phil. Trans. R. Soc. Lond.*, 254, 37–89.

Common, I. F. B. and Waterhouse, D. F. (1981). *Butterflies of Australia.* Revised edition. Angus & Robertson.

Corbet, A. S. and Pendlebury, H. M. (1978). *The Butterflies of the Malay Peninsula.* 3rd edition, edited by J. N. Eliot. Malayan Nature Society, Kuala Lumpur.

Eliot, J. N. (1980). New information on the butterflies of the Malay Peninsula. *Mal. Nat. J.* 33(3&4), 137–55.

Eliot, J. N. (1982). On three swallowtail butterflies from Peninsular Malaysia. *Malay. Nat. J.* 35(1&2), 179–82.

Fleming, W. A. (1975). *Butterflies of West Malaysia and Singapore.* 2 vols. Longman, Kuala Lumpur.

Holloway, J. D. (1970). The biogeographical analysis of a transect sample of the moth fauna of Mount Kinabalu, Sabah, using numerical methods. *Biol. J. Linn. Soc. Lond.* 2, 259–86.

Holloway, J. D. (1973). The taxonomy of four groups of butterflies (Lepidoptera) in relation to general patterns of butterfly distribution in the Indo-Australian area. *Trans. R. Entomol. Soc. Lond.* 125, 125–76.

Holloway, J. D. (1974). The biogeography of Indian butterflies. In *Ecology and Biogeography in India*, ed. M. S. Mani, 473–99. The Hague: W.Junk.

Holloway, J. D. (1976). *Moths of Borneo with Special Reference to Mount Kinabalu*. Malayan Nature Society, Kuala Lumpur.

Holloway, J. D. (1985a). Insect surveys: an approach to environmental monitoring. *Proc. 12 Nat. Ital. Entomol. Congress, Rome, 1980.*

Holloway, J. D. (1983). The Moths of Borneo, Part 4—Family Notodontidae *Mal. Nat.J.* **37** (1 and 2), 1–107.

Holloway, J. D. (1985b). The larger moths of the park: a preliminary assessment of their distribution, ecology and potential as environmental indicators. In *Gunung Mulu National Park, Sarawak: an account of its environment and biota being the results of the Royal Geographical Society - Sarawak Government Expedition and Survey 1977–1978*, edited by A. C. Jermy & K. P. Kavanagh 1981. *Sarawak Mus. J.* Supplement II.

Holloway, J. D. (1985c). Mobile organisms in a geologically complex area: Lepidoptera in the Indo-Australian tropics. In *Time and the Emergence of the Biosphere*. Systematics Association Special Publication.

Holloway, J. D. (1985d). Lepidoptera faunas of high mountains in the Indo-Australian tropics. In *Adaptations and Evolution in Biota of High Tropical Montane Ecosystems*, edited by M. Monasterio & F. Vuilleumier. Springer-Verlag.

Holloway, J. D. (1985e). The Moths of Borneo, Part 14—Family Noctuidae: Subfamilies Euteliinae, Strictoperinae, Plusiinae, Pantheinae. *Mal. Nat. J.* **38** (3 and 4), 157–317.

Holloway, J. D. (1986). The Moths of Borneo, Part 1—Key to Families; Families Cossidae, Metarbelidae, Ratardidae, Dudgeonidae, Epipyropidae and Limacodidae. *Mal. Nat. J.* **40** (1 and 2), 1–116.

Holloway, J. D. (1987). The Moths of Borneo, Part 3—Superfamily Bombycoidea: Families Lasiocampidae, Eupterotidae, Bombycidae, Brahmaeidae, Saturniidae, Sphingidae. pp. 1–199. Southdene Sdn. Bhd., Kuala Lumpur.

Holloway, J. D. (1988). The Moths of Borneo, Part 6—Arctiidae (Part) Southdene Sdn. Bhd., Kuala Lumpur.

Completed September 1986

CHAPTER 16

Freshwaters

Earl of Cranbrook[1] and J. I. Furtado[2]

[1] Glenham House, Great Glenham, Saxmundham, Suffolk, U.K. and
[2] Commonwealth Science Adviser, Marlborough House, London SW1Y 5HX (formerly Professor of Zoology, University of Malaya)

CONTENTS

16.1. INTRODUCTION

The freshwaters of the forest are intrinsically of interest for the living organisms dependent on them. Also important are the linkages of land and water in the forest and the exchange of resources between the aquatic and terrestrial environments, involving processes of physical and biological transfer.

The main pattern of surface drainage has been described and the principal rivers named (Chapter 1). As expected from the landform of the Peninsula, the upper reaches of rivers flowing from the mountain ranges are torrential in character, while the lower reaches may be slow-flowing and meander in their courses across the coastal plains. The limnology of one small river, the Gombak in Selangor, was studied very thoroughly in 1968–69 by J.E.Bishop, whose report (1973) is an important source for this chapter.

Riverine drainage systems in the lowland plains are associated with alluvial peat swamps and a specialized forest vegetation (Chapter 2). Large areas of such swamp have been converted to agriculture or other uses. An important remaining example is Tasik Bera, thought to represent the former headwaters of the proto-Muar river captured at some past time by the Pahang river. The ecology of Tasik Bera was jointly investigated by a Malaysian/Japanese team under the auspices of the International Biological Programme and the report of this survey is a second major source (Furtado and Mori, 1982).

Large natural bodies of standing water are rare in Peninsular Malaysia. Apart from lagoons separating permatang in the coastal strip (Chapter 1), Tasik Cini, an ox-bow lake of the Pahang river system, is the sole example. Artificial pools and lakes have been created in forest, chiefly as impoundments for the abstraction of water supplies. When first established, these show a paucity of aquatic life forms and low productivity. Through colonization and the accumulation of organic matter, their productivity increases naturally after 5–7 years, ultimately stabilizing over 25–30 years (Fernando and Furtado, 1975). The limnology and fisheries potential of reservoirs in forest, such as Ampang and Subang, Selangor, and Bukit Merah, Perak, have been studied by graduate students, although the work remains largely unpublished.

Much smaller, but nonetheless of ecological importance, are ephemeral rain pools or puddles on the forest floor and the reservoirs of water that collect in enclosed places as diverse as holes and hollows in the trunks or limbs of trees, the nodal spaces of damaged bamboo stems (upright or fallen), leaves or leaf-axils of certain epiphytes or the pitchers of *Nepenthes*.

Although the perennial freshwaters are predominantly riverine, the main thrust of limnological research in Peninsular Malaysia has concentrated on still waters. Inland fisheries interest has focussed on pond aquaculture, which flourishes under the traditional Chinese mixed farming system. Investigation of the limnology of flowing freshwaters has been undertaken largely as academic research (see bibliography). In recent years environmental impact assessments of major schemes for hydro-power and/or irrigation and flood control have incorporated study of the flow characteristics and aquatic biota of the large rivers affected, notably the Tembeling and Terengganu. Unfortunately, for reasons including commercial confidentiality, the subsequent reports are not readily available.

A descriptive account of the freshwater fishes of Peninsular Malaysia has been published by Mohsin and Ambak (1983). Studies of the invertebrates are scattered in scientific periodicals. In the local literature reviews can be found of certain groups, for instance desmids (Prowse, 1957), rotifers (Karunakaran and Johnson, 1978), atyid prawns (Johnson, 1961), water fleas (Johnson, 1962), water skaters (Cheng, 1965) and molluscs (A.J.Berry, 1963, 1974b).

16.2. HYDROLOGY

Apart from small inflows along the northern border with Thailand, Peninsular Malaysian freshwaters derive entirely from rain falling within the geographical area which thus forms a natural hydrological unit. Although much of the rain is convective in origin, and therefore often localized in occurrence, periodic variation in the amount and intensity of rainfall is the most important factor in the seasonal division of the year. Year-to-year variation in total rainfall (reviewed in Chapter 1) is reflected in comparable variation in annual discharge from any given catchment, which can exceed $\pm 50\%$ about the average.

Most of the rain falling on forested land is intercepted by the vegetation. From the forest canopy, some rain-water re-evaporates directly to the atmosphere producing the familiar rising wreaths of steam that follow any daytime storm. Because evaporative capacity is a function of the nature and surface area of the vegetation (Sandhu *et al.*, 1980), the water loss above a threshold is independent of the quantity of rain falling. In effect, light rain may fail to wet the soil; from a moderate fall, a relatively large fraction will re-evaporate, while in a storm the proportional loss will be slight. From measurements in the Gombak valley, Kenworthy (1971) calculated that over a year about 45 cm (=18%) of a total precipitation of about 250 cm was lost by direct re-evaporation.

Some of the intercepted rain will trickle down the trunks to reach the ground but, except in heavy storms, the quantity is of little importance (Kenworthy, 1971). The balance will drip from the foliage and twigs. Splash erosion from large drops may cause localized damage but, by interrupting the direct fall of the rain, the main effect of forest cover is to reduce its impact on the soil surface.

Of the water that does reach the ground, again some will be lost by evaporation. Under forest cover, the prevailing conditions of moderate temperature and high humidity limit the amount to about 1% (Kenworthy, 1971). The remainder will soak into the soil up to the point at which its absorptive capacity is reached. This will vary with pre-existing conditions; quoted figures in forested catchments vary from 0.8–1.8 cm (Bishop, 1973). Any excess will then flow over the surface, following natural drainage lines until it reaches a stream.

Since the topsoil under forest permits rapid infiltration, surface sheet flow accounts for a significant proportion of the total precipitation only in heavy storms. Such storms are, however, frequent events (Chapter 1) and for forested catchments in Selangor, it has been estimated that surface run-off accounts for about 25% of total stream-flow (Douglas, 1971). Although visible signs of surface erosion are unusual in the forest environment, there may be localized superficial dispersal of particulate matter. Measurements of sediments transported by surface flow in forest at Pasoh, Negeri Sembilan, and Bukit Lagong, Selangor, varied in the range 0.2–3 cm^3 cm^{-1} per year at interception traps (Peh, 1980). The variation correlated well with rainfall recorded at open sites, but not with slope or soil type. Thus, interception rates at Pasoh were slightly but significantly higher than at Bukit Lagong, although the slope was less pronounced.

Once in the soil, some water will be taken up by plants and subsequently returned to the atmosphere through transpiration. Again, the proportion varies seasonally. In the upper Gombak valley, during 1968–69, monthly differences in volume between rainfall and stream discharge (i.e., the combined water loss through evaporation + transpiration) varied from 69% in August to 35% in January, averaging 58% for the 12-month period (Bishop, 1973, Table 20). Subtraction of the annual figures for immediate re-evaporation (above) indicates that, on average, about 39% of the yearly rainfall is recirculated to the atmosphere through transpiration.

The balance will enter the groundwater store. This is the reservoir from which, in due course, water percolates through drainage channels to emerge at seepage points. The stored groundwater supports the 'basal' flow of streams. There is physical continuity between the groundwater and surface waters and some heterotrophic organisms, such as copepods (see below, 16.5.2), may move from one to the other. In deeply weathered lateritic soils of the interior hills (Chapter 1), some of the groundwater may be out of reach of the vegetation (Douglas, 1971). Elsewhere, there is close contact. For example, in small tributaries of the Gombak, the effect of the transpiration stream on groundwater was observed as a daily drop in basal flow levels between the hours of 10.00 and 15.00, amounting to total cessation in dry spells (Kenworthy, 1971).

Monthly records of stream-flow in forested catchments may vary four-fold within a year, peaking in the wet season (Table 16.1). The curve, however, is smoother and variations less pronounced than in contemporaneous rainfall. It seems that existence of groundwater reserves has a moderating effect, counteracting irregular fluctuations in rainfall.

TABLE 16.1. Monthly run-off in 1968, expressed as average stream depth in mm, for four forested catchments in the Main Range in Selangor; from Douglas (1971, Table 4).

River Catchment	Jan	Feb	Mar	Apr	May	Jun	Jul	Aug	Sep	Oct	Nov	Dec	area (km²)
Selangor	197	93	66	55	60	149	96	85	76	96	113	92	133
Gombak	41	65	53	41	35	53	65	65	65	94	136	118	118
Langat	76	65	50	34	36	67	113	96	63	78	143	122	121
Kenaboi	173	67	42	42	44	63	65	58	46	49	62	69	111

16.3. CLASSIFICATION OF WATERS

On the basis of gradient, morphometry and current velocity, three zones can be recognized: upper, middle and lower. In the forest soil, percolating groundwater collects in natural drainage lines from which lighter particles of clay and even sand may be washed out. Drainage patterns are highly developed and dendritic. High bifurcation ratios and low mean stream-length ratios characterize well-drained watersheds, such as the Gombak (Bishop, 1973). The drainage density in the middle zone is higher than in the upper zone. The effects of lithology on the drainage pattern are poorly understood.

The smallest streams of the upper zone are those that drain the steep (i.e., $> 30°$ slope) valleys of the ridges and hills. They are exceedingly variable in flow rate, springing into spate for a short period after rain and equally rapidly declining (Fig. 16.1), sometimes to a string of disconnected pools. Those at higher elevation have been termed 'montane' streams. They may be bordered by tree-ferns at the source in seepage from *Nepenthes* swamps.

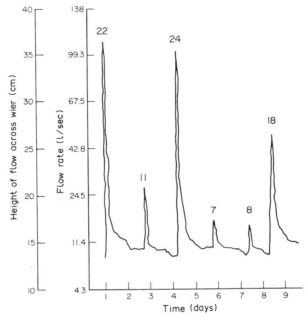

Fig. 16.1. Changes in water level and flow rate in a torrent stream, tributary to the Gombak, during a sample period in October, showing the effects of six successive rain storms. Figures by each peak denote the rainfall in mm at the time of each storm. After Kenworthy (1971, Fig. 4).

Drainage lines and montane streams flow into permanent rocky streams of lesser (but still steep) gradient, cascading from pool to pool through a tunnel of vegetation closing over the water. These have been termed 'saraca streams' (Fig. 16.2), after the dominant tree of the riparian vegetation, the gapis *Saraca thaipingensis* (Corner, 1952). The associated flora includes the following characteristic trees and shrubs: pelawan *Tristania whitiana*, kasai *Pometia pinnata* and the figs *Ficus lepicarpa, F. oligodon* and *F. ischnopoda* (see Chapter 9.3). Tributary streamlets ('upper saraca streams') are characteristically lined by aroids such as *Piptospatha* (Chapter 6).

In Corner's terminology, the larger waters of the middle zone to which saraca streams are tributary are designated 'neram rivers', again from the dominant riparian tree, neram *Dipterocarpus oblongifolius*. In the natural state these are rocky rivers, forming pools, rapids and cascades, with stretches of gravelly shallows intervening. The big riparian trees arch over the river course, their boughs draped with epiphytes. Few pristine stretches remain today outside the Taman Negara, where such as the Tahan river survive in original grandeur and beauty. In the alluvial lowlands, the deep, slow-flowing freshwater stretches of the Lower zone have been termed 'rasau rivers'. Again the name is taken from the most

Fig. 16.2. University of Malaya students sampling the upper Gombak, a saraca stream.

common plant lining the banks, a pandan known locally as rasau, *Pandanus helicopus*. Tasik Bera is important as the only surviving extensive and well-studied example of this habitat, which formerly must have been widespread.

The tributaries of these large rivers, draining the lowland plains, are slow-flowing streams which cut meandering channels through the alluvial soils. In forest such streams are totally shaded by the canopy vegetation. They are flanked by steep, muddy banks sculptured by the frequent rise and fall of water levels, and often partially dammed by fallen trunks or boughs. Where the landform is such that swamps or peaty soils develop, the waters are heavily coloured. These streams are commonly known as 'blackwater streams'. Most alluvial swamps have now been transformed into wet rice fields. The large rivers finally reach the sea through mangrove-lined channels along sheltered coasts, or by way of lagoons and swamps generated by sand-dune formations, as among the permatang of the east coast.

16.4. PHYSICAL AND CHEMICAL FEATURES

16.4.1. Flow rates and water levels

Water levels and flow rates reflect the amount and timing of rainfall. Montane streams and small hill streams of the upper zone are particularly responsive to the vagaries of local precipitation. The run-off from a storm produces a raging torrent, but this rapidly subsides. Basal flow rates may be scarcely more than a trickle, and even in short spells of dry weather the surface stream may be reduced to a string of small pools or disappear altogether. Fluctuations are also pronounced in saraca streams, but with time lags. After storms, surface run-off produces a sharp rise in levels, but this is of short duration and soon followed by a decline to normal (Figure 16.1). In rasau rivers and the slow-flowing streams of alluvial swamps, the depth of water and the rate of flow (even in some cases the direction of flow) may also respond to events in confluent river systems; in Tasik Bera, fluctuations in level exceed 5 m (Furtado and Mori, 1982). After storms or periods of prolonged and heavy rain the banks are overflowed and the surrounding forest flooded. High levels and episodes of flooding correspond with rainfall peaks, and show similar year to year variability. In dry periods, the water may be very low and effectively stagnant.

A rise in level will increase the velocity of flowing water, and hence its erosional and transportational capacity. Within any stream channel, water velocity also varies according to gradient, channel depth and width, and the nature of the bottom. Near the bottom, water velocity may approach zero. The inorganic constituents of the bottom substrate diminish in size locally with flow rates, and overall from the upper to the lower zone. There is thus local and zonal variation in the interstitial pore space in the bottom substrate, affecting this important habitat for many small benthic organisms.

16.4.2. Light penetration

Forest cover filters much of the incident light. In forested reaches of the Gombak, over one year, Bishop (1973) found that only 1.6–15% of visible light reached the water surface, with higher relative values in overcast conditions when diffuse radiation becomes more important than the direct component. In two sites, the mean visible light energy available was 15.6 and 28.8 gcal cm^{-2}, respectively, far below values of 307–419 gcal cm^{-2} measured in open country.

The transparency of water to light is critical for algae and other submerged plants. Attenuation increases with depth and is a function of turbidity. Forest streams are naturally clear. There is inevitably some erosion by storm run-off but, if the soil surface has not been seriously disturbed (as by timber extraction), any temporary rise in turbidity is moderate. For example, in the then largely pristine upper Gombak, a year's sampling found suspended solids at concentrations of 5 ppm upwards, reaching a maximum of 35 ppm in December (Norris and Charlton, 1962). Lower minimum values, but greater variation, have been found in the upper Langat, Selangor: annual range 1–67, mean 18.1 ppm (Goh, 1978).

In peak flooding, silt loads prevent light penetration beyond 5 cm. Under normal conditions, in forested stretches of the Gombak, rapid initial losses with depth are attributable to the diffuse nature of the incident light and the low reflectivity of the banks and bank vegetation (Bishop, 1973). At 40 cm below surface, intensity is reduced to about 60%. In the deeper waters of Tasik Bera measured transparency limits fell in the range 1.5–2.5 m. Penetration is low in the wet season because of the increase in inorganic matter washed from the riparian forest by heavy rain or floods, combined with the agitation of organic particles and debris by water currents. But turbidity also rises at times of extreme low water, as a consequence of the increased density of plankton (Furtado and Mori, 1982).

16.4.3. Water temperatures

The temperature regime of freshwaters reflects the incident radiation. The dense shade of tree cover arching over forest streams lowers average temperatures and moderates diurnal variation. At lowland elevations, Crowther (1982) found that midday temperatures of four small streams draining granite hills in Perak varied only from 24.0 to 24.8°C., and similar values have been recorded in forest streams elsewhere (Johnson, 1967b). At higher elevations, the waters are cooler. Both the moderating effect of shade, and the change due to elevation, are seen in measurements made in the Gombak. Over one year, daily minima in the lower reaches running through open country and settlement were 22.5–25.6°C and maxima 26.8–32.5°C, compared with minima of 20.0–23.5°C and maxima of 23.0–26.5°C in forest (Bishop, 1973).

In Tasik Bera water temperatures average 26.3°C at the surface (range 23.3–31.2°) but only 25.4° at the bottom, reflecting reduced variability (23.2–26.6°). Lower water temperatures are associated with periods of rain and higher temperatures with periods of dry weather. Thermal stratification has been detected in standing waters and the littoral regions of lowland slow-flowing streams or swamps where the water is protected from winds and circulation is impeded, for instance by stands of aquatic plants. Stratification affects metabolic activity and the distribution of aquatic organisms.

16.4.4. pH

The pH of forest streams is affected by the rock or soils over which they flow, by diurnal or seasonal periodicities or by localized topography. Variations are attributable to a combination of factors including rainfall cycles and the production and consumption of carbon dioxide by aquatic organisms which affects the bicarbonate buffering system found in most freshwaters. Lowland forest streams and swamps of southern Peninsular Malaysia are mildly acid: in Tasik Bera, for instance, in the range pH 4.5–6.8. More extreme acid conditions characterize the blackwaters of lowland peat forests and acid sulphate soils: thus, pH 3.6 in gelam *Melaleuca cajuputi* scrub in Melaka State (Johnson, 1967b). In the central and northern parts of the Peninsula, calcium-enriched alkaline waters are associated with limestone outcrops (see Chapter 1). On other bedrock, montane and hill streams have a mean pH around neutrality, as for instance in the upper Gombak (pH 6.7–7.4), although the sedimentary and granitic topsoils are acid (Bishop, 1973).

M–P

16.4.5. Dissolved gases: oxygen and carbon dioxide

Under the prevailing high temperatures at lowland elevations, even fully saturated waters contain only a few milligrammes of oxygen per litre. Oxygen tensions are highest in fast-flowing, unpolluted forest streams. In the upper Gombak, for instance, in spot readings during a year's investigation, dissolved oxygen invariably exceeded 90% saturation, at concentrations in the range 7.0–8.0 mg O_2l^{-1}. By contrast, in the slow-flowing waters of Tasik Bera average dissolved oxygen content was only 1.90 mg O_2l^{-1} (23.6% saturation), although it increased locally in the littoral waters and seasonally during rainy months. In the lowland blackwater streams of Johor and Melaka, Johnson (1967b) found still lower values, in several cases below 12% saturation (< 1 mg O_2l^{-1}). The low oxygen content of such waters may be attributable to the failure of low primary production to compensate for microbial respiration, and the additional uptake of oxygen by humic colouring matter or in bleaching reactions. When absolute quantities are so low, dissolved oxygen can readily become exhausted. Instances of mass fish deaths in lowland slow-flowing streams have been attributed to catastrophic oxygen falls under conditions of warm, rainless weather, with consequent low water levels and stagnation (Johnson, 1967b).

There is great variation in total dissolved carbon dioxide, even in fast-flowing forest streams. Thus, recorded values in the upper Gombak range from 1.4 to 9.3 mg l^{-1}. In all waters, there is a pronounced diurnal cycle, high in the early morning due to overnight respiration and diminishing during the hours of daylight as a consequence of uptake during photosynthesis by aquatic algae and macrophytes.

16.4.6. Salts and nutrients

The concentrations of salts, conventionally measured as the sums of carbonate, halide and other strong acid anions, and sodium, potassium, magnesium and calcium cations, are characteristically low in forest streams (other than blackwaters) and, in many cases, the contribution from rain represents the most important input. Major anions are most often found in the order $HCO_3^- > SO_4^{2-} > Cl^-$, and cations in the order $Na^+ > K^+ > Mg^{2+} > Ca^{2+}$. Very low levels of calcium and magnesium are characteristic of all waters, other than those draining limestone outcrops. Scarcity of dissolved calcium in the water has important consequences for the biota, affecting in particular animals that need large amounts (e.g., molluscs and crustaceans). Average concentrations of cations in the upper Gombak, for example, are (in mg l^{-1}) Na^+ 2.0, K^+ 1.3, Mg^{2+} 0.72, Ca^{2+} 0.45. In Tasik Bera, despite inputs from organic matter, the average calcium content is only 0.41 mg l^{-1}, compared with Na^+ 0.51–4.2 and K^+ 0.19–1.2 mg l^{-1}.

Iron and aluminium concentrations are very low, and sometimes undetectable. Silica, on the other hand, is a major solute in freshwaters. Easily mobilized from granitic bedrock, its concentration is relatively high in forest streams of granitic hills, including the upper Gombak (at 5.6–14.4 mg SiO_2l^{-1}). In lowland waters, especially swamps, it is absorbed by aquatic plants. Concentrations of dissolved silica are low in such waters, including Tasik Bera with a mean value of 3.15 mg l^{-1}. Values fluctuate seasonally with high levels reflecting fresh inflows during rainy months or the mineralization of decomposing organic matter in dry months.

Phosphate is characteristically very scarce (i.e., < 0.1 mg l^{-1}) in unpolluted waters. Its scarcity is attributable to low availablity in the soils and efficient closed recycling in the forest vegetation. Because of its biological origin, levels of nitrogen are typically higher than phosphorus, with organic and ammonium-nitrogen dominating. Absolute values are still low: for example, < 2 mg l^{-1} total N in the upper Gombak.

Chloride concentrations are also low in unpolluted waters, averaging 0.4 mg l^{-1} in the upper Gombak and 1.93 mg l^{-1} in Tasik Bera. Sulphate concentration may be higher: up to 4 mg l^{-1} in the upper Gombak and 3.2 in Tasik Bera. Sulphate levels appear to be related to the release of sulphur during the decomposition of organic matter. If excess is produced above the requirements of the aquatic vegetation, as in the blackwaters on the peaty soils of southern Peninsular Malaysia, concentrations may reach 150 mg l⁻¹ (Johnson, 1967b).

Corresponding with these values, the conductivity of Malaysian freshwaters is low, measuring 14.2 μmhos cm⁻¹ in Tasik Bera and 29.65–41.19 μmhos in the Gombak. Such values accord with other tropical waters (e.g., the Amazon), and emphasize the overall poverty of nutrients in forest freshwater environments.

16.5. BIOTA

16.5.1. Macrophytes

The community structure of aquatic macrophytes (i.e., non-microscopic plants) is less complex than terrestrial rain forest. There are distinctive aquatic communities within forest waters and specialized riparian plant associations. The flora of Tasik Bera has been described by the collaborators of Furtado and Mori (1982). There are seven species of submerged plants, dominated by bladderwort *Utricularia flexuosa* and the swamp aroid *Cryptocoryne griffithii*, two of floating habit and 12 emergents dominated by the reed *Lepironia articulata* and rasau *Pandanus helicopus*. The riparian flora includes an indeterminate number of swamp forest trees, chiefly *Eugenia* spp. (Myrtaceae) and *Palaquium* spp. (Sapotaceae). The swamp forests of southern Peninsular Malaysia have been detailed more specifically by Corner (1978).

The inland rivers and streams of hill country are also bordered by characteristic assemblages of riparian trees (Chapter 9, and 16.3, above). As already noted (16.4.2), shading causes a massive reduction in the visible light reaching the water surface and rapid attenuation at shallow depth. The low levels of light energy severely restrict the community of green aquatic plants. Submerged and floating macrophytes are not prominent. Fallen fruit, leaves and other plant litter from the forest therefore represent the main organic input to the system and the chief source of food for many phytophagous or detritivorous members of the aquatic fauna.

The roots of riparian plants growing into the water form distinctive features in the bank and channel topography. Those of some, notably gapis, respond to submersion by branching into a luxuriant mass of rootlets. The dense mats so formed are of great importance as shelter and attachment sites for small aquatic organisms of saraca streams.

16.5.2. Micro-organisms

Under the conditions of low illumination, algae are not prominent in Malaysian forest freshwaters although there is a diverse flora. In fast-flowing forest streams, attached algae predominate, encrusting rock surfaces, submerged roots and plant stems; planktonic forms are scarcer. Of 194 algal taxa identified in the Gombak, 98 (51%) were found at the representative forest station II of Bishop (1973, Table 46), principally diatoms (Bacillariophyta) with 31 taxa, together with 16 blue-green algae (Cyanophyta), 9 green algae and desmids (Chlorophyta), 8 euglenoids (Euglenophyta) and 5 red algae (Rhodophyta). This diversity was not matched at any open country station, although there was some overlap in the species assemblages, as shown by Sørensen coefficients of association between the two habitats, falling in the range 0.32–0.44 (Bishop, 1973, Table 48, Station II *vs.* III, IV, V). Yet productivity at the forested station II, calculated by several alternative measures, was low compared with that in the lower Gombak at open country stations unaffected by excessive turbidity. For example, computations of total annual production of periphyton fell in the range 292–5876 mg m^{-2} in forest and 11 571–41 391 mg m^{-2} in open country, in the ratios 1:3–1:70 (Bishop, 1973, Table 58).

The microscopic flora of Tasik Bera is richer, with 418 taxa identified:23 blue-green algae, 24 euglenoids, 5 fire algae (Pyrrhophyta), 7 green-brown and yellow-brown algae (Chrysophyta), 61 diatoms, 293 desmids (the large number reflecting the prevailing acidity of the waters), 2 stoneworts (Charophyta) and 3 red algae . Planktonic forms are scarce in the currents of open water and stream channels, common in the still waters of pools and among the reeds and submerged water plants. Average densities are 343 cells per litre, diversity 30 species per litre. Average productivity was calculated as 1.3–1.4 mg C h^{-1}. Among planktonic algae a marked annual cycle of abundance was found, in January falling to a minimum of 11–23 cells per litre and peaking during the rainy season September–December at 30–300 times (av. 114 times) the minimum value at the five sampling stations. The increase was attributable largely to desmids and partly explained by the stirring action of currents at high water levels (Furtado & Mori, 1982).

Bacteria and fungi are also present in forest waters, and undoubtedly play important roles in the saprophytic breakdown of organic detritus. The aquatic hyphomycetes are conidial fungi growing on submerged dead leaves and twigs in fast-flowing, unpolluted streams. Fifty-five taxa, plus 5 conidial aquatic Basidiomycetes, have been identified in Malaysia, mostly from collections made at Gombak and the Cameron Highlands (Nawawi, 1985). Some are local or tropical in range; others are cosmopolitan. Many spore-types also found remain unidentified, and study of these mysterious fungi is far from complete.

Associated aquatic microconsumers may be present in considerable numbers, but the role of minute heterotrophes in the trophic dynamics of forest freshwaters is little known. The Gombak supports a fauna of mixed protozoans (chiefly Flagellata and Ciliata) and Rotifera. Water-fleas (Crustacea, Cladocera), which form an important element in the food chain of so many freshwater systems, are absent — and lacking also in the slow-flowing waters of lowland swamp forests in Johor (Johnson, 1962). A few copepods were taken in the Gombak by Bishop (1973) but only one was a true member of the stream fauna; the others were soil dwellers.In Tasik Bera, 64 microconsumer taxa have been identified : 13 protozoans, 14 water-fleas, 3 Ostracoda, 5 Copepoda and 29 rotifers. This microfauna showed no recurrent seasonal cycle of abundance.The microfauna of streams and ditches in open country and in artifical ponds may be richer. The potential diversity of rotifers, for instance, has been shown by collections from open country in Singapore, where a single pond yielded 81 species (including 3 varieties) by intermittent sampling over 20 years (Karunakaran and Johnson, 1978).

16.5.3. Decapod Crustacea

The freshwater decapod crustaceans of mainland Peninsular Malaysia form a small but important assemblage. The crabs (Brachyura) and prawns (Caridea) known at present are listed in Table 16.2, but their taxonomy is not settled and specialist study is continually adding new taxa (Ng, 1986a and b, for example). Of the five crab families, the Grapsidae and Ocypodidae are predominantly marine or estuarine, as are all alpheid prawns except *A. paludosus.*

Abbreviation or elimination of the larval life stage is a characteristic of the freshwater crabs. All potamid, parathelphusid and gecarcinucid crabs undergo direct development. Their large eggs hatch into miniature versions of the adult, which are subsequently brooded by the female under her abdomen

TABLE 16.2. The freshwater decapod crustaceans of mainland Peninsular Malaysia*.

1. Crabs (Brachyura)	2. Prawns (Caridea)
Potamidae	Palaemonidae
Johora j. johorensis	*Macrobrachium rosenbergi*
J. j. gapensis	*M. trompii*
J. j. intermedia	*M. pilimanus*
J. j. murphyi	*M. palawanense*
Stoliczia s. stoliczkanad	*M. idae*
S. s. perlensis	*M. lar*
S. leoi	*M. latidactylus*
S. rafflesi	*M. lanchesteri*
S. pahangensis	*M. sintangense*
S. chaseni	*M. malayanum*
S. cognata	*M. scabriculum*
S. tweediei	*M. javanicum*
Terrapotamon aipooae	*M. equidens*
Parathelphusidae	
Parathelphusa maculata	
Siamthelphusa improvisa	Atyidae
Somanniathelphusa sexpunctata	
Irmengardia pilosimana	*Atyopsis moluccensis*
	Caridina thambipillai
	C. weberi
Gecarcinucidae	*C. nilotica*
	C. excavatoides
Phricotelphusa sp.	*C. propinqua*
	C. tonkinensis
Grapsidae	*C.* cf. *babaulti*
Geosesarma cataracta	
G. malayanum	Alpheidae
G. nemesis	
G. perracae	*Alpheus paludosus*
G. scandens	
G. serenei	
Sesarma granosimana	
Varuna litterata	
Ocypodidae	
Potamocypoda pugil	

* Nomenclature by Mr Peter K. L. Ng

for several days before being released. The true freshwater grapsids of the genus *Geosesarma* also produce large eggs. Among *G. perracae* there are only three short larval stages, completed within the parental burrow. Other *Geosesarma* species may have similarly abbreviated development but the remaining grapsids must return to the sea to spawn, their larval phase being completely marine.

The two predominant potamid genera frequent fast-flowing hill and montane streams, sheltering under stones but not making burrows. Their ranges are largely exclusive, meeting in the Cameron Highlands whence *Johora* extends south to Singapore (including forested stretches of the Gombak) and *Stoliczia* north into Thailand. Among the grapsids, all *Geosesarma* species inhabit freshwater. One, *G. malayanum*, was recently discovered in the Endau-Rompin forest reserve living in cups of the pitcher plant, *Nepenthes ampullaria* (Ng, 1986b). These three genera are represented by a number of localized endemic species or subspecies, reflecting the loss of dispersive capacity through the brevity of the larval life. The single freshwater gecarcinucid, *Phricotelphusa* sp., is another highland species with restricted distribution, discovered in Perak by P.K.L. Ng.

The remaining grapsids are probably more accurately classified as euryhaline species. *Varuna litterata*, for example, is extremely wide ranging in the Indo-Pacific region, frequenting fresh or salt water, although ovigerous females and large males are only found in the sea. *Sesarma granosimana* has been recorded from freshwater streams in eastern Johor. The only ocypodid species that can be regarded as inhabiting freshwater, *Potamocypoda pugil*, was also recorded in a stream in swamp forest in eastern Johor.

All Parathelphusidae are lowland crabs, frequenting slow-flowing streams, usually in cultivated land, plantations, etc. *Parathelphusa maculata* occurs in Sumatra, Singapore and most of Peninsular Malaysia north to 6°N, including the lower Gombak where it does not extend into forested reaches. At higher latitudes, it is replaced by *Siamthelphusa improvisa*, extending into Thailand. The ricefield crab *Somanniathelphusa sexpunctata* infests paddies in Perak, Kedah and Perlis.

The dominant freshwater prawns of South East Asia are Palaemonidae, among which *Macrobrachium* is the principal genus, widely distributed in Asia, Africa and South America. In Peninsular Malaysia 13 species have been recognized. Best known is *M. rosenbergi* (=*M. carcinus*), the famous 'udang galah', esteemed in local gastronomy. This prawn is widely cultured and its biology is well understood. Like many *Macrobrachium* species, it lays small eggs which hatch into pelagic larvae that must complete their life cycle in the sea. Other species, e.g., *M. sintangense* and *M. lanchesteri*, have larvae that need not return to the marine environment, while the most advanced (*M. pilimanus*, *M. malayanum* and *M. trompii*) have very large eggs which hatch into well developed benthic larvae. This last group contains most of the species encountered in inland freshwaters. *M. pilimanus* predominates in fast-flowing hill streams and *M. malayanum* in neutral lowland streams; both occur in forested reaches of the Gombak. *M. trompii* favours acid peat swamp waters and in Tasik Bera (with *Caridina thambipillai*) frequents the mats of bladderwort where prawns have a mean density of 41 individuals m^{-2}. *M. lanchesteri* appears to be a recent introduction, originally from Thailand, and is now common in reservoirs, ponds, irrigation ditches and other artificial, enclosed freshwaters. *M. equidens* is normally marine, but locally common in coastal streams.

Prawns of the family Atyidae lack the large pincers present in the palaemonids, and are usually smaller in size. Although represented by fewer species, in some habitats they are more numerous than *Macrobrachium*. The largest is *Atyopsis moluccensis* (=*Atya spinipes* of authors, including Bishop), which prefers fast water and a substrate of smooth rocks and is found in forest streams with these features. Atyids are filter feeders, using their small but strongly setose (i.e., hairy) pincers as nets to trap detritus and small organisms. Members of the genus *Caridina* are widespread, best represented in lowland and coastal waters, but in some cases penetrating to the saraca streams; for instance, *Caridina weberi* has been taken in the Gombak (Norris and Charlton, 1962).

16.5.4. Insects

In general, the insects of Peninsular Malaysian forest streams and rivers are more diverse than those of temperate waters, especially dragonflies (Odonata), beetles (Coleoptera), moths (Lepidoptera) and mayflies (Ephemoptera); only stoneflies (Plecoptera) are comparatively impoverished. Although the taxonomy of many groups is incompletely known, it is clear that the insect faunas of forest freshwaters are distinctive; few species also occur in open country. Forest stream faunas are longitudinally zoned, and separable into assemblages occupying different biotopes of which sandy bottoms, leaf drifts, root-bank complexes and stones in the current are notable. Predatory forms constitute a strikingly high proportion of all assemblages. For herbivores, allochthonous foods are of major importance, either directly or after microbial transformation (Bishop, 1973). Many insects are aquatic as larvae and terrestrial (or aerial) as adults, thus transferring resources from one habitat to the other.

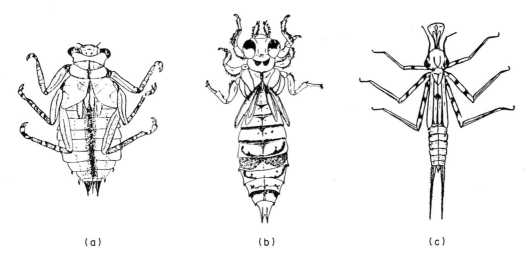

(a)   (b)   (c)

Fig. 16.3. Examples of odonate nymphs from the Gombak: (a) Libellulidae, (b) Gomphidae, (c) Coenagrionidae. After Norris & Charlton (1962, Appendix VII).

The dragonflies and damselflies (Odonata) form a major component of the fauna of all Malaysian freshwaters. The known Peninsular fauna comprises 187 species (Lieftinck, 1954), notably dragonflies of the families Libellulidae (69 species), Gomphidae (18), Aeschnidae (13) and Cordulidae (12) and damselflies, Coenagrionidae (27) and Chlorocyphidae (10 species). As larvae (Fig. 16.3) and as adults, all are predators of insects and other small animals of aquatic and terrestrial habitats, respectively. The odonates thus provide an important linkage between land and water.

Standing waters provide the larval habitat of some species of all families. Distinct communities are found in ponds or impoundments under forest cover, in the hills and lowlands. Certain aeschnid and libellulid dragonflies lay in puddles or other small, often ephemeral bodies of water. Some coenagrionid damselfly species are specialized to oviposit through small apertures, and are thus able to exploit the water that collects inside bamboo internodes (when the culm wall is cracked or pierced by boring insects, etc.) or similar sites (Lieftinck, 1954).In flowing waters, damselfly larvae typically cling to root mats, or other surface or submerged plant material, living or dead, near the banks of the water body; some (Epallagidae, Calopterygidae) also occur on stones in the riffle area of streams. The habitats of dragonfly larvae are more variable: aeschnids occur among plant roots or stable woody debris; gomphids burrow into sand or leafy or fine organic detritus; corduliids and libellulids are bottom-dwellers, usually on sandy, muddy or detritus substrates, although species of *Zygonyx* and *Onychothemis* (Libellulidae) cling to rocks in stream riffles (Furtado, 1969).Both adults and larvae maintain territories; adult males guard oviposition sites to which they conduct females during the process of courtship and mating. The adult community is fairly stable throughout the year, with slight augmentation (marking peak emergence) during drier seasons. Larval populations in flowing waters suffer irregular fluctuations, being vulnerable to disruption by floods during rainy months.

On the basis of adult communities, sets of forest stream odonates have been recognized in Selangor (Furtado, 1969). Here, *Piptospatha* streams, where torrential, are characterized by *Coelicia albicauda*, *Calicnemia chaseni* and *Indocnemis orang* (Platycnemididae), *Devadatta argyoides* (Amphipterigidae), *Euphaea ochracea* (Epullagidae) and *Rhinocypha fenestrella* (Chlorocyphidae) and, where non-torrential, by *Sundacypha petiolata* (Chlorocyphidae) and *Echo modesta* (Calopterygidae). Upper saraca streams are characterized by *Euphaea ochracea*, *Rhinocypha perforata*, *Zygonyx iris* (Libellulidae); middle saraca streams by *Neurobasis chinensis* (Calopterygidae), *Megalogomphus sumatranus* and *Gomphidia abbotti* (Gomphidae), *Onychothemis coccinea* and *O. culminicola* (Libellulidae) and lower Saraca streams by *Copoea marginipes* (Platycnemididae), *Libellago lineata* and *Rhinocypha biseriata*, *Veslalis gracilis* (Calopterygidae), *Ictinogomphus decoratus* (Gomphidae), *Onychothemis testacea*. Blackwater and slow-flowing lowland streams are characterized by *Prodasineura interrupta* and *Elattoneura analis* (Protoneuridae), *Libellago aurantica*, *Euphaea impar*.The odonates of Tasik Bera comprise 33 species, with libellulids (22 species) predominant. Odonate larvae form 8.5% of the macro-invertebrate biomass.

Larval mayflies (Ephemeroptera) are especially important in hill streams. Different groups are specialized for distinctive modes of life and occupy exclusive habitats. For example, in the Gombak and its tributary streams, where 44 taxa have been identified (Bishop, 1973), *Baetis* spp. (Baetidae) are the dominant herbivores browsing on encrusting algae on stones; *Isonychia* sp. (Siphlonuridae) is a predator frequenting the trailing aquatic roots of *Saraca* trees. In the upper saraca streams, *Isca* sp. (Leptophlebiidae) is a common detritivore, and *Epeorus* sp. and *Thalerosplines* sp. (Heptageniidae) are the common herbivores browsing on rock surfaces; in lower saraca streams, *Potamanthodes* sp. (Potamanthidae) (Fig. 16.4) is a common detritivore. In benthic habitats, *Ephemera* sp. (Ephemeridae) occurs commonly in clean unsilted situations and *Caenis* spp. among detritus and accumulations of submerged leaves. Mayflies form the dominant portion of stream drift in the Gombak, and presumably are a major food of fish.In Tasik Bera, fewer taxa were identified, but mayfly larvae were abundant, with a standing crop of 22,000 individuals m^{-2} in the bladderwort mats. The adults have vestigial mouthparts, do not feed and for their short lives subsist on stored reserves. They are none the less important prey of insectivorous vertebrates.

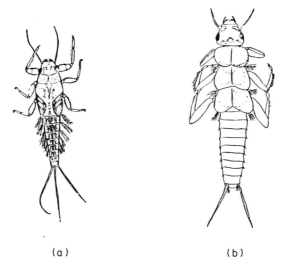

(a)   (b)

Fig. 16.4. (a) A larval mayfly (Ephemeroptera, Potamanthidae), and (b) a larval stonefly (Plecoptera, Perlidae, *Perla* sp.), from the Gombak. After Norris and Charlton (1962).

The stoneflies (Plecoptera) have generalized chewing mouthparts, and corresponding unspecialized diets. Stonefly larvae are abundant in hill streams. Out of 34 species recorded in Peninsular Malaysia, 19 have been found in the upper Gombak. Among these, *Peltoperlodes* sp. (Peltoperlidae) is a detritivore, found among submerged accumulations of leaves between boulders; *Nemoura* spp. (Nemouridae) are detritivores occurring at the base of *Saraca* root mats, and *Perla* sp. (Fig. 16.4) and *Neoperla* spp. (of the dominant family, Perlidae) are omnivores frequenting stream riffles (Bishop, 1973).

The bugs (Heteroptera) show a rich diversity. Many species are confined to standing waters but, although all are air-breathers at all developmental stages, few are resistant to pollution. The water skaters (Gerridae) and water crickets (Veliidae) (Fig. 16.5) dart rapidly over the water, supported by surface tension, each foot riding in a minute dimple. Many species are gregarious and all are predatory, often taking aerial prey that fall from forest vegetation. Of 25 species recorded in the Gombak, *Ptilomera lundbladi* (Gerridae) is common, at densities of 1–2 individuals per metre stream length (Bullock and Furtado, 1968). Somewhat similar in appearance, the water measurer *Hydrometra* sp. (Hydrometridae) spends most of its time submerged in the stiller water of pools. Also pool-dwelling are the water scorpions *Cercotmetus pilipes* and *Ranatra varipes* (Nepidae), which breathe air through their long "tails", and the back-swimmer *Enithares malayensis*. Other bugs are found among *Saraca* roots (*Helotrephes corporaali*, Helotrephidae), and in detritus along the stream bank (*Laccocoris neovicus*, Naucoridae). In Tasik Bera, 22 taxa were identified, including representatives of Gerridae, Veliidae, Hydrometridae, Naucoridae, Notonectidae and Nepidae and, numerically predominant, the water boatmen *Micronecta* spp. (Corixidae). Among the bladderwort, Heteroptera comprised 2.0% of macro-invertebrate biomass, with a mean standing crop of 0.5 g dry weight m^{-2} and 10 individuals m^{-2}.

Caddis-fly larvae (Trichoptera) are abundant in forest streams. At least 79 taxa occur in the Gombak, including Rhyacophilidae which construct no case, case-builders and those that build fixed webs and tubes on which they rely both for shelter and to ensnare their food, i.e., Philopotamidae, such as the detritivores *Chimarra* spp., and Hydropsychidae, such as the omnivores *Diplectrona* spp. (Fig. 16.6). Water beetles (Coleoptera) are also well represented in Peninsular freshwaters; standing water species have been studied by Fernando and Gatha (1963). Although adults are capable of flight, all life stages are aquatic. 30 species recorded in the Gombak include *Hydrovatus* sp. (Dytiscidae), found among sand and leaf deposits, the gregarious whirligigs *Orectochilus* and *Dineutus* spp. (Gyrinidae), and the water pennies *Eubrianix* sp. (Psephenidae) (Fig. 16.7). In Tasik Bera, beetles constitute only a minor element of the aquatic fauna.

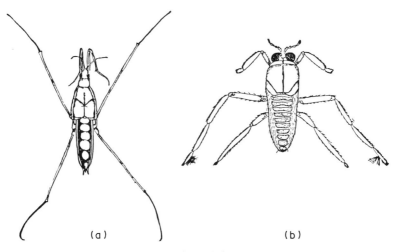

(a) (b)

Fig. 16.5. (a) A water skater (Gerridae), and (b) a water cricket (Veliidae), from the Gombak. After Norris and Charlton (1962).

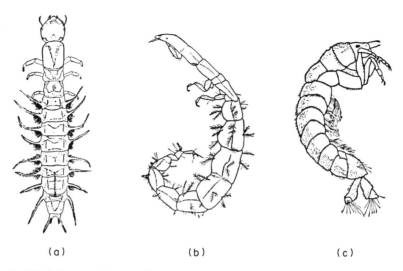

Fig. 16.6. Types of caddis fly larvae (Trichoptera) from the Gombak: (a) Rhyacophilidae, *Rhyacophilus* sp., (b) Philopotamidae, (c) Hydropsychidae. After Norris and Charlton (1962).

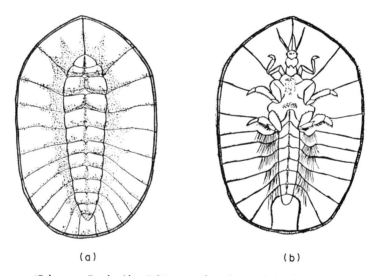

Fig. 16.7. The water penny (Coleoptera, Psephenidae, *Eubrianix* sp.) from the Gombak. After Norris & Charlton (1962). (a) Dorsal aspect, (b) ventral aspect.

Larval flies (Diptera) are abundant and important components of the aquatic fauna, but only the mosquitoes, midges (Chironomidae) (Karunakaran, 1969) and black-flies (Simuliidae) (Smart and Clifford, 1969) have been studied in detail.

Small bodies of rain-water collected in natural hollows in vegetation provide larval habitat for several Diptera, especially mosquitoes of the genus *Aedes*. The eggs of some species posses a degree of resistance to exposure and may be laid on the edge of a suitable container, to hatch later when the water level rises to submerge them. Forest mosquitoes are vertically stratified. Those that live high in the canopy exploit small water bodies in tree hollows (e.g., *Aedes 'niveus'* group). Others have been found as larvae in pockets of water held in leaf axils (*Topomyia* spp. in *Collocasia*), contained in curled fallen leaves on the forest floor (*Toxorhynchites quasiferox*), in ginger plants (e.g., *Armigeres durhami*) or in the pitchers of *Nepenthes* (*Toxorhynchites* spp., *Armigeres giveni*) (Macdonald, 1957).

The Gombak stream fauna also includes 18 black-flies and 60 chironomids, among which *Pentaneura* spp. are predators in riffle and pool substrates, *Chironomus* spp. are detritivores occurring in silted marginal waters, *Tanytarsus* spp. are abundant on all substrates and orthocladines frequent the interstices of gravel. In Tasik Bera the life cycle of *Chironomus* sp. is rapid, taking about two weeks to complete In the bladderwort mats Diptera (predominantly Chironomidae) amounted to 43% of the total standing crop of invertebrates, at a mean density of 91,145 individuals m^{-2}.

16.5.5. Acarina

The watermites (Hydrachnidia = Hydracarina) of Peninsular Malaysia have been inadequately studied. Recent collections have shown that at least 40 species in 21 genera are present; many more taxa remain unidentified (R. Wiles, unpublished). Watermites were a constant and significant component of the Gombak fauna in samples taken by Bishop (1973). Although only six species were identified, undoubtedly many more are present. In Tasik Bera, mean densities of 375 m^2 were found in bladderwort mats. Red species sometimes occurred at such high densities that the weed beds were suffused with this colour. Watermites in Tasik Bera reach peak numbers in February–April. The seasonal distribution and life histories of individual species are not known, but members of the genera *Hydrachna*, *Limnochares*, *Hydryphantes*, *Hydrodroma*, *Frontipodia*, *Limnesia*, *Atractides* and *Arrenurus* (Fig. 16.8) lay eggs in October–December. The parasitic larval stages of *Hydrachna* sp. have been found on the water bug *Ranatra*, those of *Limnochores crinata* on *Micronecta* sp. and those of *Arrenurus* sp. on odonates in October and November. Individuals of several genera collected in December survived without feeding until the following February and March (R. Wiles, unpublished data).

Fig. 16.8. The watermite *Arrenurus laticodulus*. (Photograph by R. Wiles.)

16.5.6. Molluscs

Unlike so many organisms treated in this chapter, the distribution of freshwater molluscs is only weakly affected by the presence or absence of forest cover. The strongest constraint is the amount of dissolved calcium in the water, since calcium is an essential constituent of the shell. Only in central and northern parts of Peninsular Malaysia, in the regions of limestone outcrops, do the streams and rivers contain sufficient calcium in solution to support an abundance of molluscs. Apart from calcium, other factors are important. For instance, those freshwater molluscs that respire by gills need relatively high concentrations of dissolved oxygen, whereas air-breathers can tolerate low levels. Bivalves prefer waters with a gravelly, sandy or muddy bottom, in which they can become partly buried. They are not found in places where a fast water-flow scours the stream bed, nor in still pools where thick layers of dead leaves accumulate.

The fauna is closely related to that of continental South and South-east Asia. Those species occurring in disturbed waters (including irrigation streams, ditches and reservoirs) are widespread and several may have been introduced. The forest stream fauna includes species with apparently restricted local distributions, but the taxonomy is continuously being revised and views on interspecific relationships and synonymies are unsettled.

The snails (Gastropoda) of freshwaters are represented by eight families, comprising at least 13 genera. The prosobranchs, distinguished by possession of an operculum for closing the shell and, in most cases, a true gill in the mantle cavity, comprise the Viviparidae (*Siamopaludina* and *Filopaludina*), Ampullariidae (*Pila*), Bithyniidae (*Bithynia*, *Wattebledia*), Thiaridae (*Brotia*, *Thiara* and *Melanoides*) and Buccinidae (*Anentome*). The pulmonates, lacking an operculum and, in most cases, air-breathing by means of a mantle "lung", comprise Lymnaeidae (*Lymnaea*), Planorbidae (*Indoplanorbis*, *Gyraulis*) and Ancylidae (*Ferissia*).

The viviparids feed on fine organic matter which may either be ingested directly via the mouth or filtered from the respiratory stream by means of the large gill. On the gill, particles are trapped in a mucus film which then is passed as a sticky cord along a ciliated groove on the head and directed into the right side of the mouth. The females retain their eggs within the reproductive tract until they are fully developed and emerge as young snails, complete with shell (A.J. Berry, 1963, 1974a). The ampullarids or apple snails, *Pila ampullacea* and *P. scutata* (Fig. 16.9a,b), are large and partially amphibious, capable of respiring either by gill or by air taken into the mantle cavity. Both are herbivores, feeding chiefly on green plant material rasped and torn by the radula and horny jaws at the corners of the mouth. The eggs are laid in clusters, attached to emergent vegetation. Each egg is enclosed in a gelatinous envelope, which hardens soon after laying. Fully developed young snails eventually emerge. Along with the bithyniids, the biology of which is little known, these snails inhabit slow-flowing lowland hard-water streams around limestone outcrops. They are also capable of surviving in drainage ditches and ponds, outside the forest environment, provided their basic requirement of a high concentration of dissolved calcium in the water is met. One species, *Bithynia minuta*, is an endemic cavernicole found only in streams within Batu caves, Selangor (Brandt, 1968).

Among the thiarids, *Brotia costula* and *B. spinosa* (Fig. 16.9g) live in clean, shallow, fairly fast-flowing forest streams and rivers with stony bottoms. *Thiara scabra* is tolerant of a wider range of water qualities, and has been found in brackish water and hot springs (Berry, 1963). *Melanoides tuberculata* (Fig. 16.9f) is the commonest and most wide-ranging member of the family, found in almost any kind of freshwater except fast hill streams; *M. riquetti* (Fig. 16.9d) inhabits slightly brackish water, where freshwater streams merge into mangrove swamps. All feed on organic debris and minute algae. All are apparently parthenogenetic (no males are known), and the eggs are retained in a brood pouch in the back of the head, until fully developed. The buccinids are carnivorous snails, with powerful teeth on the radula, a long proboscis which is inserted into the prey and a mobile siphon extending from the edge of the mantle cavity to sample surrounding water and so detect the presence of prey. Most are marine (the whelks) but *Antemome* sp. (Fig. 16.9e) has been collected in slow-flowing forest streams in the Taman Negara (Berry, 1963).

The pulmonates frequent still or slow-flowing calcium-rich waters. They feed directly on minute attached algae and organic particles. They have been found chiefly in ponds, drainage ditches and wet rice fields in limestone areas, and appear to be, at the most, marginally forest-adapted members of the gastropod fauna. *Lymnaea rubiginosa* (Fig. 16.10a) is of economic importance as the intermediate host of the common liver fluke of cattle, *Fasciola gigantica*. The planorbids *Indoplanorbis exustus* and *Gyraulus convexiusculus* (Fig. 16.10b,c) are common in still waters and ponds in open country and were probably introduced with fish for pond culture. The appearance of the flattened spire of the shell gives rise to the common name, 'ram's horn' snails. Finally, the ancylids have a symmetrical, shield-shaped shell with no trace of coiling. *Ferrissia javana* lives on aquatic vegetation in ponds and sluggish streams, browsing on attached algae on the aquatic plants (Fig. 16.9d).

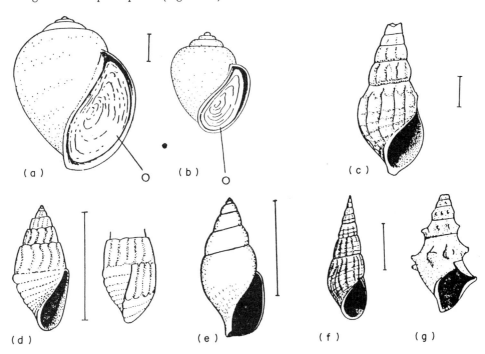

Fig. 16.9. Selected freshwater prosobranchs: (a) *Pila ampullacea*, (b) *Pila scutata*, (c) *Brotia costula*, (d) *Melanoides riquetti*, (e) *Anentome* sp., (f) *Melanoides tuberculata*, (g) *Brotia spinosa*. Scale line = 1 cm. After Berry (1963, Plate II).

The freshwater bivalves are represented by three families, the Unionidae, Corbiculidae and Sphaeridae (Berry, 1974b). The unionids, the largest, have an obligatory parasitic stage in the life cycle. Fertilization is internal and the juveniles develop in the parental gill chamber (thousands at a time) into a unique larva known as the glochidium. Released via the exhalent siphon, the glochidium can live free for a while but must attach to a cyprinoid fish and encyst to continue its development. Ultimately, it drops off and settles in the bottom sediments where it continues to grow, feeding in the normal manner. *Contradens ascia* and *Rectidens perakensis* are found in soft muds, for instance in the artificial lake at Cenderoh, Perak, or (the latter only) the lower reaches of large rivers. *Pseudodon vondembuschianus* frequents the sandy beds of forest streams. Two corbiculids, *Corbicula javanica* and *C. malaccensis*, are

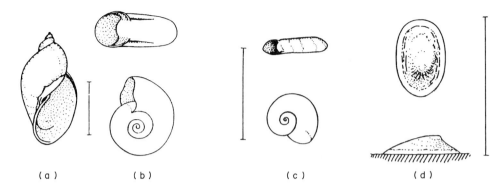

Fig. 16.10. Selected freshwater pulmonates: (a) *Lymnaea rubiginosa*, (b) *Indoplanorbis exustus*, (c) *Gyraulus convexiusculus*, (d) *Ferissia javana*. Scale line = 1 cm. After Berry (1962, Plate III).

common in clean, hard-water streams, in open country and forest. The range of the sphaerid *Pisidium javanum*, a minute pea-mussel less than 5 mm long, is not fully known, but it has been found in small numbers in the upper Gombak where the level of calcium in the water is extremely low (Berry, 1974b).

16.5.7. Fishes

Peninsular Malaysian freshwaters are rich in fish, with 382 species claimed by Mohsin and Ambak (1983, p. 247 and Table 1), representing 56 families. In comparison with the areas involved, this diversity is similar to that in other tropical river systems: for instance, 1400 species in Brazil, 408 in the Congo basin (Lowe-McConnel, 1969).Of the total fauna, 266 species are primary freshwater fish, i.e., intolerant of salt or brackish waters. Their distributions suggest a complex pattern of faunistic relations, with important centres of radiation in South China/Indochina (sharing 76 species with Peninsular Malaysia) and the Sunda region (83 species in common with Sumatra and Borneo). In particular, Peninsular Malaysia is especially rich in Cyprinidae (with 108 species, far outnumbering any other family), including the genus *Rasbora* with 22 species, five of which are endemic (Mohsin and Ambak, 1983).

Within the peninsula there is no east-west differentiation in primary freshwater fish faunas, indicating that the axial mountain ranges (Chapter 1) have not acted as barriers to distribution. Operative factors appear, rather, to reflect in part climate and in part soils and topography. Three main faunistic divisions are recognized (Fig. 16.11). The north-west division includes the islands of Penang and Langkawi, and the rivers Kedah and Muda on the mainland. The north-east/central division is the largest and includes all principal montane terrain, the rivers Kelantan, Terengganu, Perak and Pahang, and areas of freshwater swamp in the coastal lowlands. Not suprisingly, it supports the richest freshwater fish fauna. Finally, the southern division has a mainly low elevation, no limestone, and generally acidic waters. Its principal rivers, Langat, Muar, Johor, Endau and Rompin, show highly developed dendritic drainage patterns. Tasik Bera is included in this division.

Johnson's collections in the southern division emphasized that different assemblages of species are characteristic of the three lowland habitats, (i) open country including ricefields and other agricultural land, (ii) the acid blackwaters and (ii) what he termed 'lowland tree country', i.e., forest and plantations. The local tree-country fish fauna amounted to 109 species. Other studies have found 28 species in the Gombak (catchment 123 km^2), 95 in the Tasik Bera catchment (614 km^2, swamp area 61.5 km^2) and 109, including 15 endemic species, within Taman Negara (4350 km^2) (Mohd. Zakaria, 1984).

In forest waters, longitudinal zonation is marked, with progressive impoverishment of the fish fauna

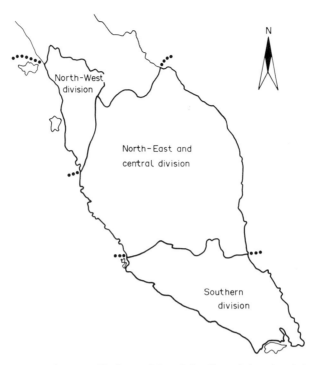

Fig. 16.11. Faunistic divisions of freshwater fish, as defined by Mohsin and Ambak (1983).

towards the upper river. In a series of sample collections reported by Mohsin and Ambak (1983), three from the middle zone in primary/secondary forest, from Pahang, Perak and Terengganu, yielded 27–31 species while five from the upper zones of large rivers in primary forest, from Selangor to Kelantan, produced only 12–17 species. In the upper zone of the Gombak, Bishop (1973) found 11 species, declining to three in the small headwater tributaries. Removal of the forest cover of lowland streams reduces the complement of fish species by an order of magnitude (Johnson, 1967a). Pollution can cause reduction on an equally dramatic scale.

The ratio 40:20:40, representing cyprinid species: catfish species: other taxa, appears to hold generally in waters of the southern division (Johnson, 1967a) including Tasik Bera (Furtado and Mori, 1982). The balance reflects the heterogenous resources of the freshwater habitat wherein detritivory, carnivory (animalivory) and omnivory are of equal importance. In the Cyprinidae, among *Rasbora* and *Puntius* (which contain more species than any other generic group) some assemblages replace each other ecologically, e.g., *Puntius lateristriga* + *Rasbora bankanensis* in torrential streams, *P. binotatus* + *R. elegans* in slower flowing waters (Johnson, 1967a). In the Gombak *Tor soro* and ikan daun *Acrossocheilus deauratus* (Cyprinidae) dominate the headwater tributaries and saraca stream; the spiny eel or tilan *Mastacembelus maculatus* (Mastacembelidae), the catfish *Silurichthys hasseltii* (Siluridae), depu *Glyptothorax major* (Sisoridae) and ikan putih *Puntius binotatus* (Cyprinidae) also occur in forested reaches but become numerically dominant in the lower stream, including forested tributaries. *Rasbora sumatrana* and ikan terbul *Osteochilus hasseltii* (Cyprinidae) and *Dermogenes pusillus* (Hemirhamphidae) occur only in the lower zone (Bishop, 1973). In all localities in the Gombak the fish fauna separates into a mid-water or bottom-feeding group comprising a detritivore and one or more predatory animalivores taking chiefly aquatic prey, and mid-water omnivores and mid-water and surface animalivores dependent on plant matter and invertebrate prey of terrestrial origin. The contribution of water plants (including algae) to the fish diet is negligible, and the population of fish dependent on aquatic prey is small by comparison with that feeding on terrestrial invertebrates falling into the river.Morphological specializations of benthic fish include ventral flattening and modifications such as the suctorial mouths of mountain loaches or susuh batu *Homaloptera* (Homalopteridae) by which they adhere to stones in torrent streams; the sisorid and akysid catfish show functionally convergent adaptations. Loaches (Cobitidae) are less specialized bottom feeders, often also frequenting the slowly turning masses of dead leaves that accumulate in pools.

Within the range of habitats provided by a large river system, there may be progressive age-related changes in the biotope selected by a given species, as well as cyclic seasonal or diurnal movements between different parts of the river channel. In the upper Kenyam river, for example, kelah *Tor tambroides* and tengas *Acrossocheilus hexagonolepis* (Cyprinidae) feed in shallow waters by night and shelter in deep pools by day, moving between the two habitats at dusk and dawn. A specific spawning site of temoleh *Probarbus jullieni* in the Tembiling river is known, a shallow, stony stretch of rapids some 70 m long where spawning occurs in late May to early June.

The temoleh attains 50 kg weight and is a predatory fish-eating carnivore when adult. Other cyprinids as juveniles may feed opportunistically on planktonic micro-organisms, attached algae, insect larvae, etc., but as adults are phytophagous. Their staple diet apparently comprises benthic algae but they also feed avidly on fruit falling into the water from riparian trees including *Ficus* spp., *Eugenia* spp., *Dipterocarpus oblongifolius, Dysoxylon angustifolium* and *Elateriospermum tapos* (Tan, 1980). In lowland streams, periodic flooding augments the foods available to fish, enlarging the area over which they can feed while simultaneously washing detritus and other foods from the forest into the water.

In Tasik Bera, the principal species are kelesa *Scleropages formosus* (Osteoglossidae), *Rasbora dorsiocellata, R. cephalotaenia, R. elegans, R. heteromorpha, R. myersi*, pelampung jaring *Puntius tetrazona*, ikan kawan *Labiobarbus festiva* and *Tor clouremis* (Cyprinidae), toman *Channa micropeltes* (Channidae), buntal *Tetraodon palembangensis* (Tetraodontidae), *Leiocassis micropogon* (Bagridae), *Xenentodon cancila* (Belonidae), temakang *Helostoma temminckii*, sepilai *Betta pugnax* and *Polyacanthus hasseltii* (Anabantidae), ikan lali *Botia hymenophysa* (Cobitidae) and *Wallagonia miostoma, Ompok leiacanthus* and anak tapah *Kryptopterus limpok* (Siluridae) (Furtado and Mori, 1982).

16.5.8. Amphibians

The forest amphibians have a particular role in the exchange of resources between the terrestrial and aquatic environments. The fauna of Peninsular Malaysia is rich, comprising four caecilians (Gymnophiona, Ichthyophiidae) and 80–90 frogs and toads (Salienta) representing the five families Pelobatidae, Bufonidae, Rhacophoridae, Ranidae and Microhylidae (P.Y. Berry, 1975; Kiew, 1984).

The caecilians are limbless, normally soil-dwelling as adults but as larvae often found in forest streams. The eggs are laid in a burrow and guarded by the female until they hatch. The young then migrate to water, although they are apparently already air-breathing. Few details of the diet are known but it is assumed that earthworms and burrowing snakes are the chief prey of adults.

Most of the frogs and toads are forest dwelling. Only 14 species are found in open country habitats such as ricefields, gardens or other deforested environments (Kiew, 1984a). Of these, the pond and ricefield frogs, *Rana erythraea, R. limnocharis* and *R. macrodactyla* (Ranidae), painted toad *Kaloula pulchra* (Microhylidae) and common toad *Bufo melanostictus* (Bufonidae) do not occur in forest. The puddle frogs, *Ooeidozyga* spp. (Ranidae), house frog *Polypedetes leucomystax* (Rhacophoridae) and burrowing froglets *Microhyla butleri* and *M. heymonsi* (Microhylidae) frequent a range of habitats, including secondary and primary forest and the forest edge. Of local occurrence are the crab-eating frog *Rana cancrivora*, found in coastal freshwater and brackish swamp and in mangrove, the swamp toad *Pseudobufo subasper*, confined to the Tasik Bera swamp forest, and the small contingent exclusive to the north-west, *Ooeidozyga lima*

(Ranidae), *Bufo macrotis* and *Microhyla ornata*.The forest fauna is altitudinally zoned, as illustrated in the collections made from pristine habitat on the northeast flank of Mt Benom (Fig. 16.12). Most frogs and toads (57 species = 77% of those listed from forest localities by Grandison, 1972, and P.Y. Berry, 1975) have been found through lowland elevations up to or just into the lower montane zone; two or three appear to frequent mid-slope elevations, and 13 are exclusively montane.

The largest genus, *Rana*, with 20 forest species, is mainly lowland in distribution; some species extend to their highest limits on Kedah peak (Gunung Jerai) where montane counterparts may be lacking (see Chapters 12, 13). Among Rhacophoridae, the genus *Philautus* is largely montane and *Rhacophorus* lowland

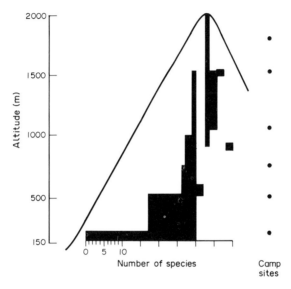

Fig. 16.12. Schematic representation of the altitudinal ranges of 39 species of frogs and toads on Mt Benom, as indicated by the collections of Grandison (1972). The stepped effect is attributable, in part at least, to the concentrated effort of collecting in the vicinity of camp sites.

(except *R. bimaculatus* and *R. leprosus*); but in other genera, including *Leptobrachium* and *Megophrys* (Pelobatidae), *Ansonia* (Bufonidae) and *Microhyla* (Microhylidae), member species replace each other successively (with overlap) over the full altitudinal range from the extreme lowlands to mountain summits.

Breeding habits reflect local exigencies of the environment at these different elevations. Thus, the one known case of direct development is a frog of montane forest, provisionally identified as *Megophrys longipes*, whose large eggs containing advanced tadpoles with legs have been found in damp moss on tree trunks on mountain summits in Perak and Selangor (P.Y.Berry, 1975). All others require free water. The Rhacophoridae are specialized to exploit small bodies of standing water, including rain puddles, the wallows of large mammals, or accumulations in sites such as the nodal cavity of a broken bamboo culm. The mating pair make a foam nest at the time of egg-laying, either on the margin of or suspended over the water. On hatching, the tadpoles emerge and enter the water; it is possible that some proceed towards metamorphosis within the foam capsule.The Microhylidae, normally ground-dwelling or burrowing in habit, also breed in small water bodies, mostly puddles; but the montane species *Metaphrynella pollicaris* makes use of accumulated rain water in tree hollows and may itself live in the canopy (Grandison, 1972). Most Bufonidae and Ranidae lay in streams. Some favour quiet pools or backwaters, others the main current. Thus, the giant frog *Rana blythi* excavates a shallow "nest" in the stony stream-bed of shallow riffles, and lays its eggs among the pebbles. Widespread in hill streams is the torrent frog *Amolops larutensis*, whose tadpoles adhere to rock or boulder surfaces in the fastest flows by means of a ventral sucker.

It is characteristic of amphibians that the diet changes progressively through the life stages. The larvae obtain their food in water, and show a range of adaptations of the mouth and feeding apparatus matching the chief component of the diet which may be plant or animal, on the water surface or within it, etc. As adults, all Malaysian amphibians are predatory animalivores. Most hunt invertebrates, but large frogs may take vertebrates: a rat, in the case of one *Rana blythi* encountered in the Gombak.

16.5.9. Other vertebrates

Excluding the marine turtles, there are 20 native Chelonia in PeninsularMalaysia, representing three families (Kiew, 1984b). Of these, the soft-shelled turtles or labi-labi (Trionychidae) are obligatorily aquatic and most water tortoises (Emydidae) are closely associated with water although often emerging to forage.The soft-shells are predominantly carnivorous, feeding mostly by predation of other aquatic animals, and are widely distributed in freshwaters from small streams to large rivers. The red-cheeked softshell *Dogania subplana* favours hill streams, hiding itself beneath rocks and large stones (Smith, 1933); but even this species may not be confined exclusively to the forest environment. The largest, *Trionyx cartilagineus*, extends into estuaries.Water tortoises are all omnivorous to some degree, ranging from mainly vegetarian spotted-shell tortoise *Notochelys platynota* to the mainly carnivorous (animalivorous) black pond tortoise *Siebenrockiella crassirostris* and Malaysian snail tortoise *Malayemys subtrijuga*. Records indicate that the Asiatic leaf tortoise *Cyclemys dentata* and spiny hill tortoise *Heosemys spinosa* frequent hill streams; others are found at lowland elevations. Most frequent forest streams and rivers, but the black pond tortoise, spotted-shelled tortoise, Malaysian snail tortoise and giant pond tortoise *Heosemys grandis* favour lowland swamps and slow-flowing streams, and the largest, the tuntong *Batagur baska* and painted river terrapin *Callagur borneensis*, are found in the lower reaches of large rivers and in estuaries.All these chelonians are rare and only the tuntong, which produces eggs of commercial interest, has been studied in natural conditions (Chapter 17). Limited data indicate low clutch sizes among small water tortoises: 2-4 eggs for the Asiatic leaf tortoise, compared with about 25 for the lowland riverine Burmese roofed turtle *Kachuga trivittata* or 10–30 for the tuntong. Such low potential reproductive rates correlate with rarity, long life expectation and hence vulnerability to environmental pressures.

Among snakes (see Grandison, 1978), two Acrochordidae are wholly aquatic. The elephant trunk snake *Acrochordus javanicus* is riverine, but not confined to forested waters. The diverse family Colubridae also includes several members more or less closely associated with freshwaters. Most specialized are the water snakes (subfamily Homalopsinae) among which *Enhydris* species have been found in forest streams, where they feed on amphibians and fish. They have modified nostrils, situated on the upper surface of the snout and closed by a valve, and are viviparous. Less modified for an aquatic existence but also drawn to streams by the availability of prey are the mock vipers *Psammodynastes pulverulentus*, a predator of amphibians and lizards, and *P. pictus*, which has been seen catching prawns from a branch hanging low over a forest stream (Tweedie, 1961), the yellow-ringed cat snake *Boiga dendrophila*, which frequents lowland forest streams and rivers and mangrove, the checkered keelback *Xenochrophis piscator* and the speckled-bellied keelback *Rhabdophis chrysargus*, seen in the water of the Gombak by Bishop (1973).

The piscivorous monitor lizard, *Varanus salvator*, ranges widely from tidal waters to small streams, including forested reaches of the Gombak.

For all forest birds, streams, pools or puddles provide water for drinking and bathing. Certain species exploit forest freshwater food resources, taking different types of aquatic prey according to dietary specializations. Notable are black bittern *Dupetor flavicollis*, Storm's stork *Ciconia stormi*, masked finfoot *Heliopais personata*, fishing eagles *Icthyophaga* spp., *Alcedo* kingfishers and forktails *Enicurus* spp.; member species of the three last-named genera show longitudinal separation in their riverine habitat. Among non-breeding visitors, several waterfowl (Anatidae) and riparian species such as Philippine banded crake *Rallina eurizonoides* and common sandpiper *Actitis hypoleucos* may follow river or stream banks into forest.Other birds are more loosely associated with the aquatic environment. Thus, black-and-red broadbills *Cymbirhynchus macrorhynchos* frequently hang their large, flask-shaped nests over water, suspended from thin stems or twigs. There is also a community of species not directly exploiting any feature of the aquatic environment itself, but nonethless closely associated with the narrow strip of riparian forest alongside watercourses. Prominent by virtue of their distinctive songs, always evocative of streamsides, are straw-headed bulbul *Pycnonotus zeylanicus* and Horsfield's babbler *Trichostoma sepiarium*.

Many mammals also require to drink free water from time to time although not necessarily descending to the ground to obtain it. Gibbons, *Hylobates* spp. drink by wetting the hand (e.g., in water in tree hollows) and sucking water from it.Some large ungulates habitually make wallows, notably pigs *Sus* spp. and Sumatran rhinoceros *Dicerorhinus sumatrensis*. On well drained slopes, the muddy pools so created can be an important resource for those other creatures able to exploit small or ephemeral water bodies, mentioned above.Mammals heavily (but not exclusively) dependent on prey from the aquatic environment are the water shrew *Chimarrogale himalayica*, the four otters, otter civet *Cynogale bennettii* and flat-headed cat *Felis planiceps*. The fur of these mammals is thick and waterproof and the feet of otters and the otter civet are webbed. There is no web ont the foot of the shrew, but a marginal fringe of stiff, short hairs performs a similar function. Moonrats *Echinosorex gymnurus* often frequent streams and feed on aquatic prey, but show no particular adaptive specializations.

ACKNOWLEDGEMENTS

We are grateful for specialist help received in the compilation of the section on molluscs from Dr A. J. Berry, that on crabs and prawns from Mr Peter K. L. Ng, and on watermites from Dr R. Wiles. The Librarian of the Freshwater Biological Association kindly provided photocopies of references not available elsewhere.

REFERENCES

Berry, A. J. (1963). An introduction to the non-marine molluscs of Malaya. *Malay. Nat. J.* 17, 1–17.
Berry, A. J. (1974a). Reproductive condition in two Malayan freshwater viviparid gastropods. *J. Zool., Lond.* 174, 357–67.
Berry, A. J. (1974b). Freshwater bivalves of Peninsular Malaysia with special reference to sex and breeding. *Malay. Nat. J.*, 27, 99–110.
Berry, P. Y. (1975). *The amphibian fauna of Peninsular Malaysia.* Kuala Lumpur: Tropical Press.
Bishop, J. E. (1973). *Limnology of a small Malayan river, Sungai Gombak.* Monographiae Biologicae 22. The Hague: Junk.
Brant, R. A. M. (1968). Description of new non-marine mollusks from Asia. *Archiv f. Molluskenkunde* 98, 213–89.
Bullock, J. A. and Furtado, J. I. (1968). Population assessment and movement in *Ptilomera dromas* (Hemiptera, Gerridae). *Archiv. Hydrobiol.* 64, 121–30.
Cheng, L. (1965). The Malayan pond-skaters. *Malay. Nat. J.* 19, 115–23.
Corner, E. J. H. (1952). *Wayside trees of Malaya.* 2nd ed. Government Printing Office, Singapore.
Corner, E. J. H. (1978). The freshwater swamp-forest of south Johore and Singapore. *Gdns Bull. Singapore* Suppl. 1.

Crowther, J. (1982). The thermal characteristics of some West Malaysian rivers. *Malay. Nat. J.* 35, 99–108.

Douglas, I. (1971). Aspects of the water balance of catchments in the Main Range near Kuala Lumpur. In J. R. Flenley (ed.), *The water relations of Malesian forests*, pp. 23–48. University of Hull, Department of Geography, Misc. Series 11.

Fernando, C. H. and Cheng, L. (1963). *A guide to Malayan water bugs (Hemiptera, Heteroptera) with keys to the genera.* Department of Zoology, University of Singapore.

Fernando, C. H. and Furtado, J. I. (1975). Reservoir fisheries resources of Southeast Asia. *Bull. Fish. Res. Studies Ceylon* 26, 83–95.

Fernando, C. H. and Gatha, S. (1963). *Guide to families of Malayan aquatic Coleoptera.* Department of Zoology, University of Singapore.

Furtado, J. I. (1969). Ecology of Malaysian odonates: biotope and association of species. *Verhandel. Intl. Verein. Limnol.* 17, 863–87.

Furtado, J. I. and Mori, S. (eds.) (1982). *Tasek Bera: the ecology of a freshwater swamp.* The Hague, Boston & London: Junk.

Goh, K. C. (1978). Human interference with the Malaysian tropical rainforest. *Malay. Naturalist* January 1978, 16–21.

Grandison, A. G. C. (1972). Reptiles and amphibians of Gunong Benom with a description of a new species of *Macrocalamus. Bull. Br. Mus. Nat. Hist. (Zool.)* 23, 45–101.

Grandison, A. G. C. (1978). Snakes of West Malaysia and Singapore. *Ann. Naturhist. Mus. Wien* 81, 283–303.

Johnson, D. S. (1961). Notes on the freshwater Crustacea of Malaya. I. The Atyidae. *Bull. Raffles Mus.* 26, 120–53.

Johnson, D. S. (1962). Water fleas. *Malay. Nat. J.* 16, 126–44.

Johnson, D. S. (1967a). Distributional patterns of Malayan freshwater fish. *Ecology* 48, 723–30.

Johnson, D. S. (1967b). The chemistry of freshwaters in southern Malaya and Singapore. *Arch. Hydrobiol.* 63, 477–96.

Karanukaran, L. and Johnson, A. (1978). A contribution to the rotifer fauna of Singapore and Malaysia. *Malay. Nat. J.* 32, 173–208.

Kenworthy, J. B. (1971). Water and nutrient cycling in a tropical rain forest. In Flenley, J. R. (ed.), *The water relations of Malesian forests*, pp. 49–59. University of Hull, Department of Geography, Misc. Series 11.

Kiew, B. H. (1984a). Conservation status of the Malaysian fauna. III Amphibians. *Malay. Naturalist* 37 (4), 6–10.

Kiew, B. H. (1984b). Conservation status of the Malaysian fauna IV. Turtles, terrapins and tortoises. *Malay. Naturalist* 38 (2), 2–3.

Macdonald, W. W. (1957). An interim review of the non-anopheline mosquitoes of Malaya. *Malaysian Parasites* 16, 1–34. Institute for Medical Research, Kuala Lumpur, Study no. 24.

Mohd. Zakaria, I. (1984). Checklist of fishes of Taman Negara. *Malay. Naturalist* 37, 21–26.

Mohsin, A. K. M. and Ambak, A. (1983). *Freshwater fishes of Peninsular Malaysia.* Penerbit Universiti Pertanian Malaysia (Malaysian University of Agriculture Press).

Nawawi, A. (1985). Aquatic hyphomycetes and other waterborne fungi from Malaysia. *Malay. Nat. J.* 39, 75-134.

Ng, P. K. L. (1986a). *Terrapotamon* gen. nov., a new genus of freshwater crabs from Malaysia and Thailand, with description of a new species, *Terrapotamon aipooae* sp. nov. (Crustacea: Decapoda: Brahyura: Potamidae). *J. Nat. Hist.* 20, 445–51.

Ng, P.K.L. (1986b). Preliminary descriptions of 17 new freshwater crabs of the genera Geosesarma, Parathelphusa, Johora and Stoliczin (Crustacea, Decapoda, Brachyura) from South East Asia. *J. Singapore Natl. Acad. Sci.* 15.

Norris, R. C., and Charlton, J. I. (1962). *A chemical and biological survey of the Sungei Gombak.* Government Printer, Federation of Malaya.

Peh, C. H. (1980). Run-off and sediment transport by overland flow under tropical rainforest conditions. *Malay. Forester* 43, 56–67.

Prowse, G. A. (1957). An introduction to the desmids of Malaya. *Malay. Nat. J.* 11, 42–58.

Sandhu, H. S., Gunaratnam, D. J., and Furtado, J. I. (1980). The impact of land use on the quantity of run-off. In J. I. Furtado (ed.), *Tropical ecology and development*, pp. 708–9. Kuala Lumpur: International Society of Tropical Ecology.

Smart, J. and Clifford, E. A. (1969). Simuliidae (Diptera) of Sabah (British North Borneo). *Zool. J. Linn. Soc., Lond.* 48, 9–47.

Smith, M. A. (1933). *The fauna of British India. Reptilia and Amphibia I. Loricata, Testudines.* Taylor & Francis, London.

Tan, E. S. P. (1980). Ecological aspects of some Malaysian riverine cyprinids. In J. I. Furtado (ed.), *Tropical ecology and development*, pp. 757–62. International Society of Tropical Ecology, Kuala Lumpur.

Completed October 1986

CHAPTER 17

Animal Conservation Strategies

Mohd Khan Bin Momin Khan

Director General, Wildlife and National Parks, Km10, Cheras Road, 56100 Kuala Lumpur, Malaysia

CONTENTS

17.1. INTRODUCTION

Man has shown a tendency to do away with wild things in his quest to create a better standard of living. The success in animal conservation in Peninsular Malaysia will depend on leadership and public understanding of the role of wildlife as a component of the environment. There is a need for adequate spiritual readiness to apply some of the power and wealth that has been acquired to keep the land both beautiful and suitable for habitation.Control of guns had little visible success in the early efforts to conserve wildlife. The modern approach to wildlife management has the same aim but is more versatile in dealing with the different species involved. Habitat manipulation, harvest or restraint of populations,

251

and the creation of refuges are some of the techniques applied. Management processes are interrelated with agriculture, forestry, mining and other land-use practices. Wildlife is a renewable resource that can be harvested indefinitely with good management. In the past much information has been collected by universities, scientific institutions and the Wildlife Department. Successful management of wildlife will depend on the accuracy of these findings.

17.2. EARLY WILDLIFE LEGISLATION

The first legislation was introduced in the Straits Settlements (Ordinance No. 111 of 1894). This was of limited scope, giving protection only to certain species of birds. It was followed by the Wild Animals and Birds Ordinance of 1904, which gave power to the Governor in Council to set a close season and to prohibit the killing or taking of any specified wild animal or bird. Under this Ordinance, towards the end of 1924 elephants *Elephas maximus*, rhinoceros and tapir *Tapirus indicus* were protected in the district of Dindings (then part of the Straits Settlements), and in 1930 the shooting of green pigeons was totally prohibited in Malacca.

In the Malay States, Selangor adopted the 1894 Ordinance of the Straits Settlements with appropriate changes. Next to follow was Pahang, with an ordinance protecting "products of state land" which *inter alia* required licensing for the capture, killing or wounding of elephants, seladang *Bos gaurus* and rhinoceros, with penalties up to $200 and/or imprisonment not exceeding one year.

In 1902 the States of Selangor and Perak published a more comprehensive enactment for the protection of wild animals and birds. This enactment was adopted by Negeri Sembilan in 1903. The outstanding feature of the 1902 enactment was the creation of the Chior Big Game Reserve. Further legislation followed in 1904, and again in 1911, when a more elaborate enactment embracing all four Federated Malay States was passed. The appointment of T. R. Hubback in 1923 as Honorary Chief Game Warden, Federated Malay States, and Honorary Adviser in the Colony of the Straits Settlements and in the Unfederated Malay States proved a landmark in the history of conservation. To this day Hubback is referred to as the "father" of the present Department of Wildlife and National Parks, and to him we owe the creation of Taman Negara (see Fig. 17.1).

It was not until 1927 that money was allocated for the salaries of State Game Wardens, the first being W. E. McNaught, appointed Game Warden, Perak, on 14 October, 1927. Negeri Sembilan followed in the following year with the appointment of H. H. Banks, and in 1929 Pahang appointed A. H. Fetherstonhaugh as Assistant Game Warden.

A second major landmark was the setting up of a Wild Life Commission of Malaya in 1930. T. R. Hubback was appointed Commissioner, with G. Hawkins as Assessor. Their report appeared in three volumes in 1932, and provides a history of wildlife conservation in Peninsular Malaysia up to that date.

The Report of the Commission made a number of important recommendations, most of which have been realized progressively. The post of Chief Game Warden was created (known today as Director General for Wildlife and National parks). The Gunong Tahan game reserve, together with adjoining areas in Kelantan and Terengganu, were made a national park (Taman Negara). The shooting, killing and taking of game is confined to open seasons for specific species of animals and birds. Commercialism of wildlife was not prohibited, as recommended, but its trade was controlled by legislation. Bird sanctuaries were established. The Protection of Wildlife Act 1972 and the National Parks Act 1980 have provided separate legislation for wildlife and for national parks, and a wildlife fund has been operating since the early seventies. The post of Honorary Deputy Game Wardens existed until recently, when such duties were taken over by officers of the department. Conservation of riverine fish, however, still comes under the purview of the Fisheries Department, and there is still no secure land tenure for national parks and wildlife.

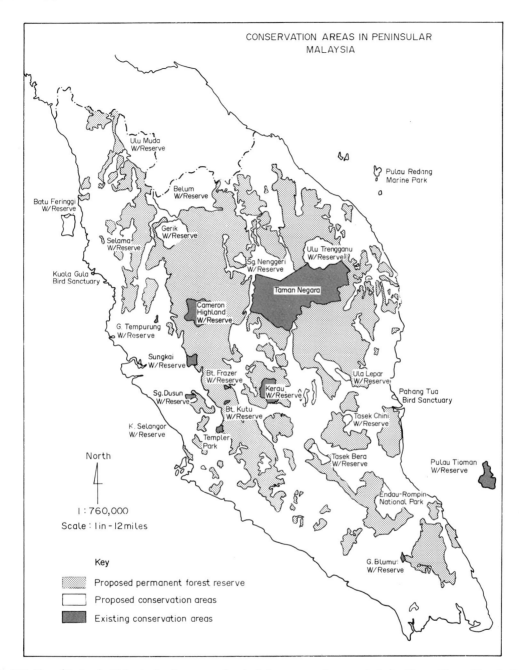

Fig. 17.1. Map of Peninsular Malaysia showing proposed and existing conservation areas, including Taman Negara, Templer Park and wildlife reserves.

In the Unfederated Malay States, the State of Johor led the field with enactment No. V of 1912, under which big game, deer and birds were protected. Licences to shoot could only be issued by the Sultan or by an officer authorized by him in writing. In Terengganu an enactment of 1923 gave protection to elephant, rhinoceros, seladang and tapir. Two years later the killing or taking of female sambar deer

Cervus unicolor and barking deer *Muntiacus muntjak* were prohibited, as were the killing and taking of the great argus *Argusianus argus* and fire-backed pheasants *Lophura* spp. An order was passed in the state of Kelantan in 1921 giving protection to elephant, rhinoceros, seladang and tapir, all of which could be shot on licence. The States of Kedah and Perlis had no conservation laws until 1956. There was, however, an Assistant Game Warden for many years in Kedah, and in 1956 his authority was extended to Perlis.

17.3. RECENT ADVANCES IN WILDLIFE PROTECTION

A big step forward was made in 1955 with the enactment of the Wild Animals and Birds Protection Ordinance by the government of the then Federation of Malaya. Adoption of the ordinance by States was, however, slow and it was not until the end of 1959 that satisfactory legislation existed over the whole of the Federation, 60 years after the passing of the first wildlife ordinance in the Straits Settlements.

The Wild Animals and Birds Protection Ordinance No.2 of 1955 provided total protection for the rhinoceroses, pangolin *Manis javanica*, gibbons *Hylobates* spp., binturong *Arctictis binturong*, slow loris *Nycticebus caucang* and tapir. Included in the schedule of Big Game were elephants, both wild cattle *Bos* spp. and the sun bear *Helarctos malayanus*. Mouse-deer *Tragulus* spp. were given protection for the first time and monitor lizards *Varanus* spp., which had been hitherto ruthlessly exploited, were declared reserved animals. Birds were listed as "Game birds" or "Birds which are neither game birds, nor totally protected birds"; all birds not included in these schedules were totally protected. Licences were introduced for taxidermists, game dealers and dealers in animals and birds, which helped to control illegal traffic in wildlife.

Ordinance No.2 of 1955 was satisfactory at the time it was passed, mainly because of the relatively large tracts of forest then still remaining. However, far too much power was given to state Game Wardens whose attitudes were not always in the best interest of wildlife conservation. In the case of totally protected species, Menteri Besar (Chief Ministers of States) had wide powers. A serious weakness was the provision to gazette Assistant Game Wardens with full powers of game wardens, which became a common practice due to lack of senior staff. The facility to compound cases involving totally protected species, and then to auction confiscated animals or to issue permits for possession, further weakened the legislation. The need for more effective powers was soon felt.

After much effort by successive chief game wardens, the Protection of Wildlife Act was passed in December, 1972, later amended in 1976 and again in 1988. The amended Act is more effective than any of its predecessors. Penalties are quite severe and will serve as a deterrent to violators of the law. The fine of $15,000, or five years imprisonment, or both, for killing the Sumatran rhinoceros *Dicerorhinus sumatrensis* (Fig. 17.2), the tiger *Panthera tigris* or the clouded leopard *Neofelis nebulosa* is adequate. The penalty for other totally protected species carries a maximum fine of $5,000, or three years imprisonment, or both. A mandatory jail sentence of not more than ten years has been introduced for possession of twenty-five or more wire snares or a fine of $5,000, or five years imprisonment, or both, for possession of less than twenty-five wire snares. Amendment of the provisions relating to the killing of animals ostensibly in defence of life, crops or property, will prevent people from taking advantage of those loopholes which are providing a defence, for example in the killing of tigers. The schedules of protected and totally protected species which were outdated have now been amended and improved. For instance, in view of the worldwide concern for otters (Lutrinae), and the perilous status of one of four otter species in Peninsular Malaysia, the whole group has been given total protection, as have certain civets (Viverridae). Birds of prey, which were in the list of "other protected species", are now totally protected.

Fig. 17.2. The carcass of a female Sumatran rhinoceros, killed by poachers at Sungai Dusun, Selangor, in 1986. Only the horn was taken. (Photograph by Hasni Abu Bakar.)

17.4. GAME RESERVES AND NATIONAL PARKS

The creation of game reserves, wildlife sanctuaries and national parks, from about 1902–1939, has proved to be most significant in setting aside protected areas for conservation (Appendix 1). Apart from Sungei Dusun which was created in 1964 (see Fig. 17.1), these are in fact the only forest areas that have remained more or less inviolate. Sustained efforts to create more wildlife reserves and national parks have not produced results. It appears that the increased scarcity of suitable development land in Peninsular Malaysia, added to the perceived rigidity of wildlife legislation, has deterred State Governments from allocating land for conservation. State Governments have to consult the responsible Federal Minister when acquiring or intending to excise any part of a legally established National Park or a wildlife reserve. This provision, while having ensured the survival of existing wildlife reserves, has inhibited the creation of new ones.

The Chior Big Game Reserve, Perak, was gazetted in 1902. For the next 20 years no new reserve was created; but then, over a short period of 18 months, six Game Reserves were established: Fraser's Hill, Kuala Selangor, Bukit Kutu, the golf course at Kuala Lumpur (all in Selangor), Serting, Negeri Sembilan, and Kerau, Pahang. In 1925, S. Lui and the Gunung Tahan game reserves (both in Pahang) were declared, three islands off Port Dickson in 1926, the Sungkai Game Reserve in 1928 and an area in Kelantan in 1929. The latter, with the Gunung Tahan Game Reserve and adjacent areas in Terengganu, became the King George V National Park (now known as Taman Negara). The Serting and S. Lui Game Reserves were revoked in 1929. Large reserves were established in Johor, through the interest of the then Ruler, H. H. Sultan Ibrahim, but their legal standing was anomalous and (as later events have shown) insecure (Hislop, 1961).

17.5. PUBLIC INTEREST IN CONSERVATION

Conservation is the main objective of a number of non-governmental organisations in Peninsular Malaysia. The Malayan Nature Society is perhaps the best known. The Society's active role in controversial issues has saved endangered species and conservation areas. An attempt to trap Sumatran rhinoceros for the national zoo might have succeeded if it had not been for the all-out effort of the Malayan Nature Society, which gained strong international and local support. Proposals to make Taman Negara serve as a dual purpose park also failed, due largely to the firm stand of the Society. Bukit Takun was saved from quarrying by the Society in the same way. The Society mounted a strong campaign in opposition to the first proposals to build a hydro-electric dam in Taman Negara, and has again been active in the face of renewed plans (1981–82). Though considerable damage was inflicted to the rhino habitat in the proposed Endau-Rompin national park, logging (carried out at an unprecedented rate) was ultimately stopped in response to protests of numerous people representing diverse interests. A subsequent scientific expedition to Endau-Rompin drew widespread public support (Kiew, 1986).

In the old days the Malayan Nature Society valued its independence and would only accept a grant after a thorough examination that it was for a worthwhile project which in no way undermined its authority and independence. It would appear that policies have changed over the years. Funds are raised through active, sustained campaigns for specific projects. The author feels that some members have used the Society to exert pressure to stop projects which are considered to be environmentally unwise. Individuals have used the Society for their own publicity and have for a long time held on to their positions. Changes every few years would bring in leaders with different ideas and accumulate expertise. Among other groups, the Friends of Templer Park Society have played an important role by averting threats to this park, with the support of other societies and the public. The Batu Caves Protection Society has been equally successful in the achievement of its objective to stop quarrying.

The World Wildlife Fund has an active Malaysian branch, which collects donations to support conservation projects. It has a mobile education unit, and has produced bulletins and posters related to conservation. It has also wholly or partly funded research projects. Money is obtained mainly through donations from Malaysian organizations and individuals.

The author feels it is tragic that World Wildlife Funds have come into conflict with locals. The situation is no different in Malaysia. Locals resent the slogan that they are not competent and are incapable to carry out wildlife projects. Based on this assumption funds were chanelled to expatriates to do wildlife projects. It is most unfortunate that locals who rightly should have received such funds and become experts in their respective field of wildlife in their own country are denied the opportunity. Given the opportunity Malaysians are as good as any expatriate. World Wildlife Fund Malaysia should pride itself in assisting Malaysians in every way possible to produce the expertise needed in wildlife conservation.

The degree of involvement by the public, Members of Parliament, and societies and institutions including international bodies, reflects the gravity of the situation. There are problems that must be solved to save species of wildlife from extermination. In the past, even when 80% of Peninsular Malaysia was under forest, problems of wildlife conservation existed. It was then that the Javan rhinoceros *Rhinoceros sondaicus* became extinct. With forest cover now so severely reduced, the situation is exceedingly critical. Extensive areas of wildlife habitat have been destroyed, and with such destruction the extermination of the native fauna. Large numbers of elephants have had to be killed in defence of crops. Crocodiles have been driven almost out of existence. The seladang is endangered. The Sumatran rhinoceros is well known to be on the verge of extinction.

Yet, Malaysia has a long history of conservation. Individuals with an understanding of natural relationships are rapidly spreading ecological thinking. Some of these individuals are active leaders of organizations working towards a better understanding of the present situation in the country. The abundance and diversity of natural resources of Peninsular Malaysia reflect its wealth, economic security and hopes for the future. The quality of life and health of the people depend to a large extent on the variety of natural resources available. A country is rich only so long as its supply of resources is greater than the needs of its people to maintain a high standard of living.

17.6. EFFECTS OF DEVELOPMENT ON WILDLIFE

The high diversity of wildlife in the forests of Peninsular Malaysia has been discussed in preceding chapters. W. E. Stevens (1968) spent two years on secondment to Peninsular Malaysia, during which he completed a report on wildlife conservation. His data on the occurrence of mammals at different elevations and in different forest types indicated that only 9% of native species can exist at any altitude, and that 53% of all species are confined to primary forest, 25% live in primary or tall secondary forest, 12% occur in primary or secondary forest but can also subsist in cultivated areas and 10% are confined to cultivated or urban areas (cf. Chapter 12).

Based on estimates of densities of primates provided by Southwick and Cadigan (1972), Khan (1978) computed losses in population due solely to forest clearance (Table 17.1). Deforestation reduces the number of mammal species to a mere handful (Harrison, 1965). Yet among mammals, as among birds (Table 17.2), the numbers of individuals per unit area may not be reduced and indeed is often increased. Clearly legal protection alone is not enough to ensure the preservation of wildlife. It is nevertheless an essential element in the overall programme. In the absence of law and its enforcement, research and management cannot bring the required results.

TABLE 17.1. Estimated losses of primates in Peninsular Malaysia between 1957 and 1975 due to the reduction of forest area from 84% (1958) to 51% (1975).

Species	Population in 1958	Population in 1975	Estimated losses	% loss
Long-tailed macaque	414,000	318,000	87,000	23.37
Pig-tailed macaque	80,000	45,000	35,000	43.75
Silvered leaf-monkey	6,000	4,000	2,000	33.33
Banded leaf-monkey	962,000	554,000	408,000	42.41
Dusky leaf-monkey	305,000	155,000	150,000	49.18
Gibbons	144,000	71,000	73,000	50.09
Siamang	111,000	48,000	63,000	56.75

TABLE 17.2. Population estimates of birds in different habitats (from McClure, 1969).

Place	Habitat	Birds per 100 acres
Kuala Lumpur	Urban and gardens	1100
Subang	Secondary forest	450
Rantau Panjang	Coconut plantation and mangrove	800
Ulu Gombak Forest Reserve	Extraction track in logged forest	400
Ulu Gombak	Virgin jungle reserve	400

Logging destroys the forest structure, and affects the survival of many species of animals, particularly those requiring aboreal niches. Mining results in the discharge of silt into rivers, causing the decline of fish and fish-dependent wildlife. Erosion also adversely affects wildlife.

Wildlife is a component of our natural resources. Good soil and water are the basic requirements for the production of crops and livestock, and wildlife is no less dependent on them. Wildlife must be included in the national plan with clear guidelines for its management in every form of land utilization. People are concerned about widespread pollution. Herbicides and insecticides in large quantities are in the hands of users with little knowledge of their function, or the possible consequences of their application. These dangerous chemicals have caused mortality among wildlife including elephants and seladang. Many rivers are now so polluted that they have become unsuitable for fish, which in turn affects fish-dependent organisms.

17.7. THE SIGNIFICANCE OF NATIONAL PARKS, WILDLIFE RESERVES AND SANCTUARIES

Originally, national parks were established for the preservation of places of scenic and cultural value or places with unique animal or plant communities. They were set aside for recreation, education and scientific study, to be passed on unimpaired to future generations. Recently it has been realized that such parks and reserves are no luxury but are in fact an absolute necessity for the future of any country. National Parks and reserves attract people from all over the world. This special type of tourist wishes to see the natural habitats of other countries. Such people have no interest in the environment of cities, from which they have just escaped.

On the other hand there are serious problems in developing countries with increasing populations. To maintain the standard of living, let alone to improve it, it is necessary to exploit natural resources such as forest timbers or minerals, and to make land available for agriculture. It is obvious that this cannot be avoided. An integrated plan for national development will include the harvesting of natural resources resulting in optimum benefit to the people with minimum waste.

Currently general trends in the establishment and management of national parks indicate a growing concern for the situation in Malaysia. A network of reserves and national parks exists in the country. Unfortunately, most only look good on paper, having in most cases no secure land tenure. The creation of national parks and wildlife reserves often was not the result of overall planning for national development.

There is now an urgent need for a serious approach to the problem. Development remains top priority. Plans for development are the first ingredients in the implementation of national projects. Establishment, development and management of national parks must be included in the overall master plan and implemented simultaneously. A system of national parks with clear objectives is indispensable, if national parks are to play the important role expected of them. A national park system that is formulated and included in the national plan will only be realised if it has the approval of the highest authority in the country and the highest decision making body of the government. In national planning, investigations are carried out to determine the suitability of land for alternative uses. Such studies must include potential national park areas if conflicts of interest are to be avoided. Permanent protection must be assured when national parks are established. National parks must be developed to serve the different interests of people. The widest subsequent use of parks, within permissible and acceptable bounds, will guarantee their perpetuation.

A capable administrative division is a prerequisite to the successful management of national parks. There is at present an acute shortage of professionally trained park managers in the region due to neglect of the field. Adequate funding is essential. Professionally qualified officers are needed for park work which includes law enforcement, research and development, and the management of visitor facilities.

It is essential for State governments to realize and recognize the value of protected areas to achieve in full their objectives. Ignorance and the lack of appreciation of the importance of protected areas have had disastrous effects. What it boils down to is the need for State governments to be aware of the consequences of the actions of agencies involved in development and to desire to create a healthy environment for the people. The existing situation is evidence of either the lack of understanding or of the low priority given to conservation against other development projects. An awareness of the value of protected areas is the first step in the formulation of a well-conceived environmental programme.

17.8. THE WAY FORWARD: PROBLEMS AND SOLUTIONS

Conservation must extend to forest reserves because these areas hold much more wildlife than national parks, wildlife reserves and sanctuaries. The area represented by national parks and wildlife reserves is small compared to forest reserves.

Research to guide management must be accelerated. The most should be made from such studies, as they are time-consuming and expensive. Long term research programmes are preferable to brief individual investigations, with little or no significance to conservation. Individual studies are often not within the priorities of an agency that manages national parks and wildlife. On the other hand, such studies undeniably help to stimulate interest and may create an opportunity for useful public criticism.

It will generally be more beneficial to rely on the relevant agency, when funds are available, to decide what studies are needed. Full reports take time to emerge, and it may take several years to publish complete results. Matching studies with the requirements of an agency will permit the prompt implementation of recommendations without the need to wait for publication.

Individual studies should be carried out by Malaysians. Conservation, after all, is needed in Malaysia. It is logical to expect Malaysians to carry out studies of Malaysian wildlife, with their easy personal contacts and knowledge of sources of advice. They are more interested in the country's wildlife, and will therefore be an asset to conservation. Local authorities on Malaysian wildlife will in time emerge, to perform much needed functions in conservation.

Variously defined wildlife management is the art of making land and water produce optimum, sustained annual crops of the best species of wildlife, consistent with the utilization of land and water for other purposes. Wildlife management in this strict sense is a relatively recent activity in Peninsular Malaysia. Game Departments have been better known as licensing agencies for hunting or trade in wildlife. Indiscriminate shooting and killing of wildlife were common, and licensees often took more game than their bag limits.

Malaysia has been extensively opening up forest since independence. At the present rate of forest clearance, it is estimated that lowland rain forest will disappear in 10–15 years. With it will be lost most of the native fauna. Inadequacies are to be found in the existing network of reserves. Many do not cover the home ranges of the species they were created to conserve. Many are far smaller than the required minimum area of 100 km^2 (10,000 ha). Buffer zones are non-existent. Surveys to demarcate boundaries are urgently needed in most national parks and wildlife reserves.

Centres for wildlife studies already exist in Malaysia, but these are too few to achieve maximum effect. There is a need to increase the number of nature study centres to educate the population and thus gain support for conservation. Programmes on conservation are shown on radio and television networks but mainly deal with foreign countries. A more relevant approach is necessary, to illustrate local problems and actions in the field of conservation. A special effort to gain the support of decision makers and politicians is undoubtedly of top priority.

Wildlife laws in parts of Malaysia are weak and enforcement is poor. The first step is to strengthen legislation. More value has to be put on the lives of animals and birds by imposing heavier fines. The support of the judiciary is indispensable, and adequate law enforcement officers equipped to perform their duties with confidence are essential. The large number of firearms in the hands of people interested in hunting presents a serious problem in the conservation of wildlife. Illegal possession of firearms carries the death penalty, but far too many persons have licensed firearms. Illegal trapping methods still cause problems. For example, each year, there are about 12 cases involving the steel-wire snare. This snare is a most efficient method of catching animals including large birds such as argus and fire-back pheasants. The sufferings to snared animals are horrific. The steel wire snare has been banned and it is an offence even to possess one.

Corruption is a serious problem that must be corrected by the payment of good salaries and the offer of attractive prospects in the wildlife service. Close supervision is necessary, with swift action against corrupt officers. Violations, no matter how small (including technical offences), should be acted upon. Minor offences may be settled out of court by offering compositions, but all serious offences should go to court and be dealt with accordingly. Smuggling occurs because of the demand and the high commercial value of many species of wildlife. To curb such smuggling, Malaysia acceded to CITES. All South East Asian countries except for Brunei have become parties to CITES. The ASEAN Agreement on nature conservation is another step in the right direction.

Apart from legislation directly relevant to the protection of wildlife and national parks, there are numerous other regulations that are not effectively enforced. In Peninsular Malaysia river terrapins *Batagur baska* are governed by the river rights laws of each State. Regulations are hardly enforced, and as a consequence the species is endangered. In pre-war days, when the legislation was strictly enforced, river terrapins were abundant, but after the war projects had to be devised to prevent their extinction.

The Land Conservation Act aims, among other things, to combat erosion. One can only assume that it is not properly enforced in view of the widespread erosion in the country. The Mining Act is ineffective in so far as silt traps and relandscaping are concerned. Forest practice still has grounds for improvement. Reafforestation is inadequate. The Environmental Quality Act is being enforced and an advance is indicated by the formation of the Department of Environment. The Department is now involved in serious efforts to impose conditions to preventing water pollution.

Yet political stability far outweighs the importance of national parks and wildlife reserves. In the face of rapid population increases there is no alternative but to exploit renewable and non-renewable natural resources. The low standard of living of rural people contributes to existing problems. Whatever the pros and cons, conservationists have to work in the light of these realities which form the challenges of the future.

Of regional significance is the inclusion of conservation in the ASEAN programme for the environment. The people involved are high-ranking officials and ministers from relevant government ministries. It is an ideal machinery for conservation, where decisions are made by people attending meetings or through them at higher levels. Under conservation, the importance of national parks, trade in wildlife, legislation, training, information exchange and wildlife management research are given prominence. Representatives from the departments of wildlife and national parks of each country participate in these meetings, workshops and field trips. Assistance and advice of international organizations like UNEP and IUCN are sought when needed.

The lack of management research has resulted in inaccurate administrative decisions. This has had serious adverse effects on wildlife. Wrongly timed hunting seasons have led to heavy mortality of gravid animals and their young. It is important to step up management research. Conservation action must be based on a thorough knowledge of the biology of animals and their role in the ecosystem. Academic research can educate public opinion and increase basic scientific knowledge. Cumulative results of a series of studies will enhance the knowledge on which management of an area can be based.

Mining presently occupies about 1% of the total land area in Peninsular Malaysia. There is no landscaping and restoration programme, because of high costs. The problem of siltation from mining is most serious on fish and fish-dependent wildlife. Existing legislation, if strictly enforced, will however prevent further silting of rivers and ensure relandscaping of mined area.

Shifting cultivation, extensive in parts of Malaysia (Chapter 18), presents a serious problem in wildlife management. While it is beneficial to some species it is detrimental to most.

17.9. WILDLIFE VALUES

The value of wildlife is often inadequately appreciated. As a general observation, outdoor recreation is a healthy, satisfying and often creative use of leisure time. The increasing number of Malaysians visiting the countryside indicates a growing spiritual need for contact with the wild.

The role of birds in the dispersal of seeds of large trees is important. Birds also feed on insects: woodpeckers on wood-boring insects, flycatchers and warblers, etc. on leaf-eating and other defoliating insects. Mammals, particularly smaller species, are important in removing large quantities of potentially injurious forest insects. Invertebrates play an important role in nutrient recycling and soil fertility. We are a long way from understanding the dynamics and balance of natural communities. Areas for study and retention as standards for comparisons are needed to measure changes and deterioration in modified ecosystems.

Maintenance of wildlife populations at sustainable, harvestable levels is desirable to meet legitimate hunting needs. The amateur hunter is one who is interested in game for the love of sport, rather than as a means of livelihood. As a group, amateur hunters are exceedingly interested in conserving their particular quarry. They are very observant during the short period they spend in the field but are inclined to accept uncritically the views of fellow sportsmen in preference to the conclusions of the scientist. This group believes strongly in legal control, but may be slow to recognize and accept new facts.

The economic contribution of wildlife lies in its utilization by commercial hunters, including people who provide goods and services to users of wildlife. Wildlife provides a multi-million dollar trade because of its high commercial value. Worldwide, the ivory trade threatens the future of elephants and the preservation of rhinoceros is difficult because of the great demand for horn and other body parts. Rural people may depend on wild animals as a food source, but town-dwellers also value wild meat. For the sake of posterity, the present generation is obliged to maintain wildlife in adequate numbers to provide for future generations. Rapid development of the country has modified much wildlife habitat, requiring great care to avoid the already upset balance of nature becoming irreparable.

The aesthete sees wildlife as something to enjoy, observe and study, but not to take for use. This group is inclined to feel that all species need full protection. The belief that in Malaysia wildlife can never be significant in terms of tourism has done harm to conservation. Yet thousands of people visit our national parks and reserves annually, to see and photograph wildlife in natural surroundings.

The wildlife officer enforces the law and controls the harvest of game. He has deep convictions as to the essential worth of his work. Yet it is clear that "value" of wildlife resources is interpreted differently by different people. To some it means use only in economic terms. To them good conservation is practised when a game population is converted into products of commercial value. To others a magnificent elephant standing in a National Park is more "useful" than if it is converted into ivory and meat. The over-generalizations of the hunter challenge the scientist's findings of fact. The hunter is impatient at the slowness of research, and the warnings of the scientist. The aesthete's point of view is that of sentiment combined with moral superiority. He regards the hunter as a menace. The hunter tolerates the aesthete as long as no attempt is made to interfere with what he considers his rights and privileges. The wildlife officer is often in the best position to appreciate and understand the conflicting attitudes of others, and thus has an opportunity to build bridges between the groups and help in the process of welding them into an effective team.

As a renewable natural resource, wildlife has many similarities with timber. Harvesting is possible, and the resource can replace itself indefinitely under management. There may be reasons for total protection of a population at a particular time, and equally good reasons for heavy cropping of population at other times. The Sumatran rhinoceros, for example, if it can be saved at all, will need the protection of individual animals and habitats. The few surviving rhinoceroses need undisturbed conditions, if they are to breed and increase. But to apply indefinite total protection to the wild boar is to invite trouble. This species has the capacity to destroy its own habitat and thus itself.

The fate of wildlife and that of other living resources are intricately tied together. Considerable care is given to the conservation of wild animals by nations practising a high level of conservation of land and its renewable resources. Nations in which wild animal populations are depleted and on the verge of extinction are usually those in which the state of soils, farmlands and forests is also precarious. Wildlife conservation, therefore, must be associated as part of the general subject of natural resource conservation. With the high human population growth in Peninsular Malaysia it seems unlikely that any area will remain unaffected by the activities of man. It would be tragic if future generations were denied the variety of animals which have accompanied and influenced our culture and history.

17.10. PRESSING ISSUES IN SPECIES MANAGEMENT

The elephant problem was tolerable before palm oil became a major industry. The loss of habitat and the liking for oil palm by elephants have resulted in the most serious problem ever to be faced. More research is required to produce economically the most effective means of elephant control. A trapping scheme solved the Jengka crop depredations, and is applicable elsewhere, where there is no available forest for elephants. Electric fences are proving effective against elephants where sufficient natural habitat remains. The Department is developing a policy of capture and translocation of problem elephants (Fig. 17.3). It is important to see the depredations of elephant in perspective, against losses from other, uncontrolled sources. With better foresight and planning, the elephant problem can be reduced or even avoided altogether.

Small residual groups of Sumatran rhinoceroses are scattered in forested areas of the peninsula (Fig. 17.4). The existence of a population in Endau-Rompin was verified by the Game Department in the mid 1970s. The core area of the proposed Endau-Rompin National Park is important as a sanctuary for them. Of top priority is patrol of the area by wildlife rangers posted at strategic places with administrative headquarters and adequate staff for active duty. Studies on a long term basis to collect additional data on population, home range, habitat requirement and to evaluate present management action are essential. More surveys are also necessary in Taman Negara to determine the distribution and status of rhinoceros in the park.

Serious studies of the seladang were started in the mid 1960s (Weigum, 1970, 1971). Further studies on distribution, numbers and movement of the species were carried out in the later part of the 1970s. Guidelines were made and now provide the basis for management which has apparently led to an increase in numbers (Khan, Sivanantham and Zolkifli, 1983). The establishment of reserves is desirable for large herds or for key small herds. Important use areas that cannot be protected may be designated as "critical seladang habitat" to be managed for wildlife production under the joint jurisdiction of the Wildlife Department and the cooperating agency. The establishment of protected areas under joint State and Federal gazettement or joint jurisdiction may necessitate concessions to allow multiple use

Fig. 17.3. An aged cow elephant, captured at Seberang Perak, where the entire herd was removed to save it after total forest clearance for development. (Photograph by Mohd Momin Khan.)

management policies. Adequate controls should be retained to minimize disturbance and habitat damage to critical areas. Surveys are necessary to locate and determine population status and home ranges. Annual concurrent surveys during the first two weeks of March will reveal population and production information. Quarterly population and age ratio counts are necessary to determine calf survival and size of breeding populations. Habitat may be improved by developing a series of small clearings in which preferred food item production is increased and maintained at a high level. The Department has also started a pilot captive breeding project (Fig. 17.5).

As a result of four years of intensive studies of the Malayan tiger more is now known about the animal (Blanchard, unpubl.). More studies are required to formulate a management plan.

17.11. A WILDLIFE PLAN

A wildlife Plan will be a major step in the management of wildlife resources in Malaysia. It will be necessary to look at the activities of man at all levels to understand the place and role of a wildlife plan. It will be the function of this plan to provide an understanding of the wildlife situations, the problems it faces and how these resources may be managed for the enjoyment of the people. Based on the needs of the people, this plan can point out the actions that must be taken to maintain or improve wildlife resources to meet such needs. The Plan will be dependent on Government policies. Present policies have to be evaluated. Impact assessment is necessary for projects affecting wildlife. Monitoring and control of pesticides are necessary to prevent damage to wildlife resources. Maintenance of existing national parks, reserves and sanctuaries in addition to creating new ones are necessary to prevent further depletion of endangered species. Services are needed for research, management extension programs and law enforcement. Harvesting of wildlife species has to be based on accurate data. Information is scarce on diseases and parasites. It is important to understand the economic values of wildlife and the effects on wildlife of land practices.

Fig. 17.4. Peninsular Malaysia's rarest large mammal, a Sumatran rhinoceros *Dicerorhinus sumatrensis*, at Sira Kemian, Ulu Perak. (Photograph by Moktar Mohamad.)

One of the major problems is the threat to riparian habitat. The density and variety of wildlife is greater in riparian habitat than most other types of habitat. Freshwater swamps are being lost as a result of drainage. Estuaries are affected by pollution and siltation. The long term effects of pesticides are not adequately known because, in Malaysia, such studies have been neglected until now. Agricultural practices using modern, mechanized methods, leave little or no habitat for wildlife. Irrigation and drainage often strip the land, and multiple or round the year cropping denies the respite of a fallow period.

17.12. THE CENTRALIZATION OF THE DEPARTMENT OF WILDLIFE AND NATIONAL PARKS

The Protection of Wildlife Act, 1972, provided the means to reorganize the Game Department. The process of centralization began in 1973. This was a giant step forward in reorganization and expansion of

Fig. 17.5. A cow seladang, originally from Ulu Lepar, Pahang, at the Department's captive breeding station. (Photograph by Mohd Momin Khan.)

the Department. Steps were taken to overcome the acute staff problem. National Parks and wildlife reserves were allocated more funds under federal jurisdiction. Funding is more readily available, and to achieve maximum benefit from national parks and wildlife reserves, priority was given to their development. Development is costly. The millions of dollars that are needed annually could not possibly be met solely from the fees of visitors to a park. State funds are spent on other important projects.

Now that centralization of the State Game Departments in Peninsular Malaysia is complete, consideration may be given to Sabah and Sarawak. The first step is an amendment to extend the Protection of Wildlife Act, 1972, to Sabah and Sarawak, followed by centralization of national parks and wildlife under one federal authority.

In the past emphasis was put on law enforcement. Law enforcement still plays a leading role in the protection of wildlife. The department handles about 5000 cases a year, about 10% of which are taken to court, the majority being settled through composition. The division of law enforcement is well developed though there is ground for improvement. Wildlife rangers have adequate knowledge of existing legislation. They have no problem in making arrests, lodging police reports or preparing case files to the extent required of them. The case files are then completed by the respective Heads of Department and processed accordingly. The most serious problem facing this division is the acute shortage of law enforcement officers. The increased provision of vehicles have helped considerably, especially in the patrol of outlying areas.

The research branch is comparatively young. Studies have commenced on the Sumatran rhinoceros, tiger, seladang, tapir (Williams, 1979), elephant (Khan, 1977) and deer (Habsah Muda, 1983). Progress so far is satisfactory with valuable data collected. Obviously a good start was made but more officers are needed to cover a number of other species. It would take 20–30 years to collect meaningful data on each species. Adequate numbers of research officers are needed to study the serow *Capricornis sumatraensis*, hornbills (Bucerotidae), pheasants and other species with a view to assist management.

An active bird ringing team has been operating for the past 15 years. This is a continuation of the many years of sustained work carried out by the University of Malaya (Chapter 13). Despite a large body of information on birds both resident and migratory, further detailed studies are essential to better our knowledge of birdlife.

A wildlife training centre has been built. Staff training is of priority. The need to produce high calibre men to perform their duties efficiently is more pressing today than ever before.

Courses conducted regularly for school teachers with a view to develop interest in nature are essential. The idea is to have teachers with some training in wildlife and its conservation. Such teachers will be needed to participate in school nature clubs that are being introduced by the Ministry of Education with the cooperation of the Department of Wildlife and National Parks. Presently the Department is making a series of wildlife films, a few of which are already completed. These films will be made in sufficient copies for utilization in schools and through television for members of the public.

17.13. CONCLUSION

Conservation in Peninsular Malaysia has evolved over periods of plenty to periods of scarcity of natural resources. During periods of plenty laws were lax resulting in wasteful utilization of wildlife. Up to the time of the first salaried game warden in 1927 wildlife laws were enforced by volunteers or officers of the land office who were involved mainly in the issue of game licences. Though it was a violation to shoot, kill or take immature game there was no standard maturity.

The incredibly low value put on the lives of animals contributed significantly to the tragic extinction of the Javan rhinoceros in 1932 and the precarious situation of the Sumatran rhinoceros, now an endangered species, the tiger and the seladang. Strong and effective legislation, long felt necessary, was slow in coming. At present, although laws appear satisfactory to curb losses from poaching and trade in wildlife, the effects of habitat loss have proved to be more disastrous than any known cause of mortality of the past. The approach to the problems has changed from an emphasis on law enforcement to a combination of research and management, extension programmes, and establishment of national parks and wildlife reserves.

National parks and wildlife reserves have so far remained more or less intact and survived pressures to exploit or excise them for various purposes. These protected areas are the most treasured for conservation. Mounting pressures to excise parts of these areas are becoming more difficult to resist. The rich timber that they hold is well known to large enterprises in this field. A number of applications in the national interest were difficult to refuse, but the loggers have proved to be dishonest and trees were cut outside approved areas.:

APPENDIX 17.14.

Existing National Parks, Wildlife Reserves and Sanctuaries in Peninsular Malaysia (see Fig. 17.1).

(i) Taman Negara

In the true sense of the word Taman Negara is the only national park in existence in Peninsular Malaysia. Created in 1938, incorporating parts of Pahang, Kelantan and Terengganu, it has an area of

1677 sq miles (4350 km^2). The foresight in setting aside this exceedingly large tract of forest as a national park is widely recognized. It is said to be the most outstanding park in Southeastern Asia.

The fauna includes most forest species, except those characteristic of coastal or estuarine habitat. Larger mammals are well represented. Several herds of elephant occur within the borders of all three component States. Seladang are well represented, with a herd each in Kelantan and Terengganu and about five herds in Pahang. Positive evidence of the Sumatran rhinoceros has been found in four localities of the park (the upper reaches of the S. Atok, Ulu Kenyam, S. Tanum and S. Sepia) together holding an estimated population of 7–12 animals. Tigers are not infrequently seen by visitors. An estimated 25% of the national tiger population is probably found in the park. Sambar are well represented; wild pig *Sus scrofa* are common. The park is noted for its tapir, often seen at salt licks. Siamang *Hylobates syndactylus*, white-handed gibbons *H. lar*, dusky and banded leaf-monkeys *Presbytis obscura* and *P. melalophos*, and both species of macaques *Macaca* occur. From time to time wild dogs *Cuon alpinus* are seen. Among birds argus pheasant is most commonly heard during its breeding season. Crested and crestless fire-backed pheasants, *Lophura ignita* and *L. erythropthalma* are frequently seen and, less often, peacock pheasant *Polyplectron malacense* and partridges. Hornbills are well represented, and most may be observed from halting bungalows during the fruit season. Since its establishment Taman Negara has survived two grave threats. First was the proposal to convert it to a "dual purpose" park and forest reserve (with the obvious motive) and more recently the Tembeling dam was close to being built. The second problem has now re-emerged. Only an enlightened public fully conscious of the absolute need for such areas can save Taman Negara. As Stevens (1968) pointed out, it is agreed that there is no need to modify the boundary of the park, but rather to have it demarcated and posted. The pleasant boat-trip between Kuala Tahan and Kuala Kenyam calls for the protection of the forest on the left side of the river.

(ii) Sungkai Game Reserve, Perak

This reserve was created in 1928 and enlarged in 1940. Its main purpose was the preservation of seladang or gaur *Bos gaurus*, which originally numbered about 30 animals. Development has reduced the reserve to about 6000 acres (2400 ha) with unfortunate consequences. The herd of seladang now ranges outside the reserve during a good part of each year. During the Japanese occupation many seladang were destroyed and after the war arsenite poisoning caused heavy mortality. Only about 8–15 animals remain. The reserve appears free from harrassment. Efforts are being made to attract the animals back into the reserve by providing pastures.

(iii) Kerau Game Reserve, Pahang

This reserve has been the site of important research, mainly primate studies (Chivers, 1980). Wildlife includes perhaps as many as 40 head of seladang, in addition to elephants, tapir, sambar and barking deer. The tiger and the clouded leopard have been recorded and there is good reason to believe other cats also occur. There are unconfirmed reports of Sumatran rhinoceros. There is now a deer farm on the border of the reserve and a base camp for the Elephant Unit just outside the reserve at Perlok. The new Kuala Lumpur-Karak highway provides improved access. Recommendations made in 1966 for a joint forest reserve function were supported by Stevens (1968), and Weber (1968) did not include the Kerau Wildlife reserve in his proposed national park system for Peninsular Malaysia.

As at present constituted, the reserve includes an undisturbed area of 30,000 acres (12,000 ha) of lowland forest. This may be the only such area under protection in Peninsular Malaysia, perhaps the most important reason for keeping it absolutely as a wildlife reserve. Logging will impair the opportunities for continued research. Knowledge of Malaysian tropical rain forest ecosystems is so scanty that it will need many decades of intensive studies to obtain the understanding on the basis of which this and other reserves can be managed.

(iv) Pahang Tua Bird Sanctuary, Pahang

This sanctuary was created to protect green pigeons *Treron* spp. which congregate in the mangroves at certain times of year, and has an area of about 3300 acres (1335 ha).

(v) Cameron Highlands Wildlife Reserve, Pahang

The whole district of Cameron Highlands was declared a reserve for deer in 1958 and in 1962 protection was extended to all animals and birds. The area is 176,000 acres (70,000 ha) of which 15,000 acres are under agriculture, 9000 acres of forest reserve and 800 acres for town sites (Stevens, 1968). It is hilly country lying above 2000 feet (610 m) elevation.

(vi) Fraser's Hill Bird Sanctuary, Pahang/Selangor

The main area (7360 acres = 2980 ha) was established within the State of Selangor in 1922; the town board area, in Pahang, was added later. The principal scientific work of bird-ringing that continued for many years has revealed the massive number of migratory birds of many species that pass through Fraser's Hill each season (Medway and Wells, 1976).

(vii) Sg. Dusun Game Sanctuary, Selangor

The primary reason for establishment in 1964 was the protection of 3–5 Sumatran rhinoceros, as recommended by Milton (1963). The rhinoceroses are known to visit the salt licks on 8–10 occasions in the reserve each year, each visit lasting a few days, but most of the time the animals are found outside the reserve. The staff situation is being improved by the addition of a number of wildlife rangers and an assistant director of wildlife. With the additional staff law enforcement can be more efficiently carried out.

(viii) Bukit Kutu Reserve, Selangor

This reserve (4800 acres = 940 ha) was established in 1922. It consists of a steep forested hill rising to 3456 feet (= 1053 m). It serves a useful purpose as a refuge and breeding area for wild animals and birds.

(ix) Templer Park, Selangor

Templer Park (2450 acres = 990 ha) was established in 1955, chiefly as a recreation area for Kuala Lumpur, and named in honour of Sir Gerald Templer. The park is very popular, especially at weekends. it is administered by the Friends of Templer Park Society, who receive grant support from the Tourist Development Corporation.:

(x) Endau-Rompin, Pahang/Johor

The massive development schemes in Pahang Tenggara and in Johor have resulted in great losses of wildlife. One of the objectives of this proposed national park would be to provide suitable habitat for wildlife made homeless. Larger mammals are well represented. The only herd of seladang still remaining in the state of Johor is found in the proposed area. There are also elephant, deer, serow, tapir, tiger, leopard *Panthera pardus*, wild pigs and barking deer, and birds including hornbills and pheasants. Without any doubt the most important species in the area is Sumatran rhinoceros, numbering 20–25 animals (Flynn and Mohd. Tajuddin Abdullah, 1983). The rivers have clear water with abundant fish.

For tourism Endau-Rompin has great potential, due to its proximity to Singapore.

The Federal Government included the Endau-Rompin proposed park in the Third Malaysia Plan and a Management plan has been prepared (Flynn, 1980). Unfortunately a FELDA scheme in the Ulu Capau area incorporates 10,000 acres (4050 ha) within the "buffer zone".

APPENDIX 17.15

Desirable additional Parks and Wildlife Reserves (see Fig. 17.1)

(i) Ulu Muda, Kedah

There is no park or wildlife reserve in the state of Kedah. A proposal has been made for the establishment of a reserve in the northeastern part of the State, particularly in order to conserve the large mammals that are concentrated in the salt lick area of the Muda river. The proposed reserve is 445 sq. miles (1153 km^2) in area, and forms the catchment of the Muda and Pedu dams. Of the salt licks, Sira Hangat is rated as one of the finest in Peninsular Malaysia, although at certain times of the year it is under water. The area holds a fairly good population of elephants and a herd of seladang numbering about 15 animals in the Sg. Teliang area. There are also populations of both species of deer, pig, tapir, tiger and other cats within unconfirmed reports of banteng *Bos javanicus* and Sumatran rhinoceros.

(ii) Grik, Perak

The efforts of the game Department to establish a reserve in Upper Perak can be traced back to the 1950s. Even then it was considered as the most important wildlife country of the State. The abundance of mammals and birds and the occurrence of a large number of salt licks in a relatively small area is most extraordinary. No other area in Peninsular Malaysia so far as is known can compare with the proposed Grik Wildlife Reserve in richness of wildlife and salt licks. Efforts were unsuccessful until 1971 when an area of approximately 210 sq miles (544 km²) was approved by the Perak Exco (State Executive Committee) as a wildlife reserve. Unfortunately, gazettement was neglected with tragic consequences. The completion of the Temenggor dam made it necessary to settle some 300 families at Air Ganda, close to the proposed reserve. Land on the east bank of the Perak river, parts of which are in the reserve, was logged in preparation for plantation. Several salt licks were destroyed. Poaching is now serious, and difficult to control because of security problems. Added to this tragedy are plans to turn the Air Cepam area, home of about 16–20 seladang, into oil palm plantation. Orang Asli peoples affected by the dam will settle at Air Dala further down river, where Sira Dala is the most important salt lick for elephants.

Maximum benefit from the reserve can only be achieved if development of Air Cepam and Gatek is stopped. Equally important is to prevent the settlement of Orang Asli at Dala. Serious crop depredation by elephant will inevitably be faced and past experience predicts that the only end will be the extermination of entire herds.

(iii) Kuala Gula, Perak

This is the sanctuary of 2,200 acres (890 ha) of mangrove habitat at the mouth of the Selinsing river. It is intended for the protection of shorebirds, the most important being the night heron *Nycticorax nycticorax* for which this is the principal nesting area in Peninsular Malaysia. Other species of birds likely to benefit include the lesser adjutant *Leptopilos javanicus*, milky stork *Ibis cinereus*, egrets *Egretta* spp., terns, white-bellied sea-eagles *Haliaeetus leucogaster*, lesser fishing eagle *Icthyophaga nana*, buzzards, brahminy kite *Haliastur indus* and a large number of migrant shorebirds. The Federal Government approved a law enforcement project by providing M\$50,000 in the Second Malaysian Plan, continued into the Third Malaysian Plan with an additional M\$50,000.

(iv) Belum, Perak

The writer is one of the few to have penetrated deep into the Belum district reaching close to the source of the Perak river. In an expedition lasting three weeks, records of wildlife were made mostly close to the banks of the river. Seladang and elephants certainly exist in good numbers. Two tigers were seen; fresh tracks and evidence of feeding indicated at least two Sumatran rhinoceroses in the area; tapir tracks were recorded daily. Both species of deer occur in good numbers; the calls of siamang and gibbon were heard and, except for the silvered leaf-monkey *Presbytis cristata* all other species of primates have been recorded; binturong have been observed.

Stevens (1968) recommended the area as a wildlife reserve and the Department recommended its establishment as a third national park for Peninsular Malaysia. It certainly qualifies in size, richness of wildlife, the beauty and purity of rivers and many other properties of an ideal national park. Because of security problems, it would be advisable at present to establish the area as a wildlife reserve with limited development for recreation where possible.

(v) Ulu Selama, Perak

An area of 86 sq miles (223 km²) is proposed as a sanctuary for the Sumatran rhinoceros, and would appear to cover the home range of these animals so far as is presently known. Visits have been made by O. Milton (in 1964), D. L. Strickland (1966), W. E. Stevens (1968), a former Chief Game Warden, J. A. Hislop and former Director of Wildlife, Bernard Thong, and the writer most recently. Fresh tracks of two animals have been confirmed recently. Elephants, tapir, barking deer and many other mammals also occur in the area. Development has reached close to the border of the proposed sanctuary. The most important salt lick, Sira Harimau, is only two miles from the nearest road. The immediate establishment of the sanctuary is of top priority.

(vi) Tasik Bera and Tasik Cini, Pahang

Tasik Bera is Peninsular Malaysia's largest freshwater lake (Chapter 16) and Tasik Cini the only other natural lake of any size. In these lakes, swamps and marshes the false gharial *Tomistoma schlegelii*, once numerous, has been seriously depleted, but more than 144 species of fishes and 215 species of birds are recorded. Of the larger mammals elephants, tigers, tapirs, deer and pigs occur in the area. Smaller animals include the mouse deer *Tragulus* spp.

(vii) Ulu Terengganu

The proposed wildlife reserve (403 sq miles = 1044 km²) lies in western Terengganu, covering the Kenyir dam with three large islands. Before impoundment the area held an estimated 20–25 seladang, 10–16 elephants and 16–25 tigers.

The State of Terengganu is famous for its tigers and the conservation of the tiger is the main objective of the proposed reserve. This proposed reserve borders Taman Negara, and the combined area will amount to 2080 sq miles (4390 km²), making it the most important conservation reserve in the country.

(viii) Nenggiri, Kelantan

The original proposal, dating from 1968, concerned an area of 143 sq miles (368 km²) which included the habitat requirements, as far as was known, of seladang, elephant, tapir, deer, pig and most of the primate species. Subsequent development has caused considerable damage, leaving only a much reduced reserve mainly for the conservation of seladang and smaller mammals.

Seladang are known to range along the Cendawan and Lalat rivers of Panggong Lalat. Observations either suggest three separate herds or one large herd that splits into smaller groups. Logging appears to have benefited the animals.

REFERENCES

Blanchard, R. F. (unpubl.) *Special Tiger Reports*. No.1. Preliminary Analysis of Tiger Mortality and Livestock Depredation in Trengganu and Kelantan, West Malaysia. No.2. Measuring Tiger Pug Marks. No.3. Preliminary Proposal for Tiger Management. Department of Wildlife and National Parks, Kuala Lumpur.

Chivers, D. J. (ed.) (1980). *Malayan Forest Primates*. London & New York: Plenum Press.

Conry, P. L. (unpubl.) *Proposals for Wildlife Reserves for Seladang in the State of Kelantan*. Report submitted to Chief Game Warden, Department of Wildlife and National Parks, West Malaysia.

Flynn, R. W. (1980). *Endau-Rompin National Park Management Plan*. Department of Wildlife and National Parks, Kuala Lumpur.

Flynn, R. W. and Mohd. Tajuddin Abdullah (1983). Distribution and number of Sumatran rhinoceros in the Endau-Rompin region of Peninsular Malaysia. *Malay. Nat. J.* 36, 219–47.

Habsah Muda (1983). Ladang ternakan rusa dan pengurusannya. *Perhilitan* 3(2), 10–17.

Hislop, J. A. Rhinoceros and Seladang —· Malaysia. Vanishing species. *Proceedings of the Conference on Conservation of Nature and Natural Resources in Tropical South East Asia - IUCN*. Publications series No.10, pp.278–83.

Hislop, J. A. (1961). Protection of wildlife in the Federation of Malaya. *Malay. Nat. J.*, Special issue: Nature Conservation in Western Malaysia, pp.136–42.

Khan, M.M.K. (1977). Studies of the Malayan elephant (*Elephas maximus*). *Malay. Nat. J.* 30, 1–38.

Khan, M.M.K. (1978). Man's Impact on the Primates of Peninsula Malaysia. *Recent Advances in Primatology* 2, 41–6.

Khan, M.M.K. and Bullock, J. A. (1972). Conservation and the Malayan Nature Journal. *Malay. Nat. J.* 25, 142–60.

Khan, M.M.K., Sivanantham, T. Elagupillay and Zolkifli bin Zainad (1983). Species conservation priorities in the tropical rain forests of Peninsular Malaysia. *Malayan Naturalist* 36(4), May 1983, 2–8.

Kiew, B. H. (1986). Progress report on Endau-Rompin expedition for the period June-November 1985. *Malay. Naturalist* 39(3&4), 3–15.

McClure, H. E. (1969). An estimation of bird population in the primary forest of Selangor, Malaysia. *Malay. Nat. J.* 22, 179–83.

Medway, Lord and Wells, D. R. (1971). Diversity and density of birds and mammals at Kuala Lompat. *Malay. Nat. J.* 24, 238–47.

Medway, Lord and Wells, D. R. (1976). *Birds of the Malay Peninsula*. Vol. V. London, H. F. & G. Witherby.

Milton, O. (1963). *Wildlife conservation in Malaya*. American Committee for International Wild Life Protection, Special Publication 16.

Stevens, W. E. (1968). *The conservation of wildlife in West Malaysia*. Federal Game Department, Seremban.

Weber, B. E. (1968). *A national park system for West Malaysia*. Federal Game Department, Seremban.

Weigum, L. E. (1970). Seladang. *Animal Kingdom* 73, 2–9.

Weigum, L. E. (1971). The last refuge. *Malay. Nat. J.* 24, 132–7.

Wells, D. R. (1971). Survival of the Malaysian bird fauna. *Malay. Nat. J.* 24, 248–56.

Whitmore, T. C. and Grimwood, I. (1976). *Conservation of Forests, Plants and Animals*. (Draft Report) for IUCN.

Wild Life Commission (1932). *Vols. I-III*. Government Printing Office, Singapore.

Williams, K. D. (1979a). Radio-tracking tapirs in the primary rain forest of West Malaysia. *Malay. Nat. J.* 32, 253–8.

Williams, K. D. (1979b). Trapping and immobilization of the Malayan tapir in West Malaysia. *Malay. Nat. J.* 33, 117–22.

Completed November 1986

CHAPTER 18

People of the Forest

A. Terry Rambo

East-West Environment and Policy Institute, Honolulu, Hawaii 96848, U.S.A.

CONTENTS

18.1. INTRODUCTION

Knowledge of prehistoric humans of Peninsular Malaysia is limited and largely conjectural (Tweedie, 1953). Hominids of the *Homo erectus* grade were present in Java, southern China and Burma; presumably these also occurred in Peninsular Malaysia, but verified fossils or cultural artifacts from the Lower or Middle Pleistocene are lacking. The Tampanian stone "tools" which Sieveking (1958) dated as Middle Pleistocene are now generally considered to be from the late Upper Pleistocene. Most of the pieces are, in any case, now thought to be the accidental product of natural processes while the few that are clearly artifacts show Hoabinhian affiliation (Hutterer, 1977).

Thus the earliest evidence of human presence is of populations belonging to the Hoabinhian cultural complex, associated elsewhere in Southeast Asia with the Pleistocene-Holocene transition period. Known from archaeological excavations of cave sites and coastal shell middens, they were hunters and foragers who exploited a broad spectrum of wild plants and animals in the coastal strand and river valleys. On the basis of excavations of similar sites in northern Thailand it has been suggested that these people were the initiators of the agricultural "revolution" in Southeast Asia (Gorman, 1971). It is inferred from artifactual and botanical evidence that early agriculture became an important activity in Peninsular Malaysia about 6000 BP.

Subsequently, in effect until the mining and plantation industries spread inland in the 19th and 20th centuries, interior Peninsular Malaysia continued to be thinly populated by man. These indigenous peoples are today known by the Malay language term "Orang Asli".

Traditional accounts have asserted that the diverse ethnic groups of Orang Asli found today are the residues of several waves of migration into the peninsula (Cole, 1945). Recently, specialists (Benjamin, 1976, 1980, 1985; Rambo, 1984; Solheim, 1980) have challenged this interpretation, suggesting that the different contemporary ethnic groups evolved *in situ* from the prehistoric population of the peninsula. In assessing the past impact of man on the forest environment, it is therefore legitimate to review the traditional activities of the Orang Asli as models.

18.2. DIVERSITY, DISTRIBUTION AND POPULATION

The Orang Asli today number some 60,000 (out of a total Peninsular Malaysian human population of some 10 million). They are dispersed over a large area and are divided into many local communities which display physical, linguistic and cultural diversity (Fig. 18.1). One of the primary problems confronting ethnologists has been to bring order to this empirical chaos by devising suitable classificatory schemes. Since the pioneering work of Skeat (1902, 1906), ethnologists have employed a tripartite system which includes Negritos or Semang, Senoi ("Sakai"), and Proto-Malays ("Jakun") as its three major ethnic categories. Although the racial category of "Negrito" is employed in the Malaysian Government Department of Orang Asli Affairs (JOA) ethnic classification (Jimin, 1968), use of biological labels for culturally defined groups is unacceptable and the older label of "Semang" is to be preferred, despite its lack of meaning to the Orang Asli themselves. "Sakai" is a pejorative Malay term meaning subject or slave. It was employed in traditional ethnic classifications both as a label for the group now referred to as "Senoi" and also as a general term for all aborigines, regardless of ethnic affiliation. In much of the older literature "Jakun" is applied to all Proto-Malays although it is actually correct only for one group in Johor, who are now more usually called "Orang Hulu".

The tripartite classification scheme is the product of 19th-century ethnological theory in which race, language and culture were seen as genetically linked characteristics and in which different ethnic groups were seen as occupying different rungs on a single evolutionary ladder. The clearest presentation was that of the Austrian ethnologist Schebesta, who classified the Orang Asli according to the now discredited *Kulturkreise* theory of W. Schmidt. The ethnic categories proposed by Schebesta (1926) are described in terms of their racial, linguistic, cultural, and evolutionary characteristics in Table 18.1. Schebesta's classification is still employed in academic descriptions of Orang Asli ethnology (Carey, 1975) as well as in administrative activities of JOA (Jimin, 1968). Revision of this scheme to recognise that race, language and culture often vary independently, and that cultural evolution is multilinear rather than unilinear, is long overdue but is a task beyond the scope of this chapter. For present purposes it will be sufficient to use the terms Semang, Senoi, and Proto-Malay. Wherever possible, more specific identification of the local group being described will be provided.

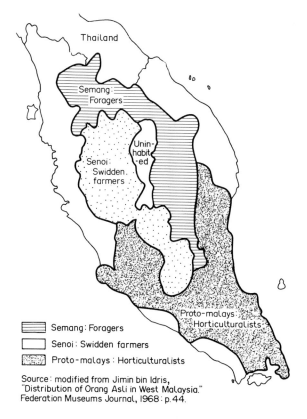

Fig. 18.1. Distribution of Orang Asli.

TABLE 18.1. Standard ethnological classification of Malayan aborigines (after Schebesta 1926).

ETHNIC GROUP	SEMANG (Negrito)	SENOI (Sakai)	PROTO-MALAY (Jakun)
RACIAL TYPE	Negrito	Australoid/Mongoloid	Proto-Malayan
PHYSICAL CHARACTERISTICS			
Stature	Very Short	Tall	Medium
Skin Color	Very Dark	Light Brown	Medium Brown
Hair Color	Black/Dark Brown	Dark Brown with Chestnut Tint	Dark Brown
Hair Form	Wooly or Frizzy	Wavy	Straight
LANGUAGE	Austroasiatic	Austroasiatic	Austronesian
CULTURAL TYPE	Nomadic Hunter-Gatherers	Shifting Cultivators	Horticulturalists
EVOLUTIONARY STAGE	Very Primitive	Intermediate	Advanced

18.2.1. Ethnicity and ecology

Concern with ethnic classification of the Orang Asli would be a matter of interest only to anthropologists, except for the fact that there is a significant relationship between ethnic identity and mode of ecological adaptation. Contrary to the once popular stereotype of all Orang Asli as primitive nomads wandering through the forest eating wild roots, fruit and game, these people actually employed distinctive adaptive strategies having profoundly different impacts on the rain forest environment (Rambo, 1979a, 1982). While there are many instances of recent changes in settlement pattern (government sponsored or spontaneous) and of experimentation with new crops and new farming methods, in terms of traditional mode of life the Semang are nomadic foragers, the Senoi are swidden (slash-and-burn) cultivators, and the Proto-Malays are mixed horticulturalists cultivating a wide range of annual and perennial plants but with a major emphasis on tree crops. A more detailed but also more unwieldy classification of ethnic groups and their ways of life was presented by Williams-Hunt (1952).

Not every group fits into the appropriate category in this classification. Many Semang engage periodically in swiddening, although only rarely does farming provide a major basis for their subsistence. Conversely, there are Senoi groups (such as some of the Semok Beri of northern Pahang) and Proto-Malay groups (e.g., the Orang Laut of the southern coastal zone of the Peninsula) who are primarily foragers. Indeed, virtually all Orang Asli engage in at least some foraging for forest products, with much of the cash income of even the horticultural Temuan Proto-Malays derived from trade in forest products such as rattan. Finally, although many Proto-Malay communities are mixed-horticulturalists, there are others (notably the Temuan of northern Selangor, and some of the Orang Hulu (Jakun) of Johor) who are shifting cultivators following a way of life much like the Senoi; and there are Senoi groups (e.g., the Mah Meri of coastal Selangor) who are settled horticulturalists and fishermen.

From the standpoint of this chapter the principal importance lies in the different impacts that Orang Asli groups following these ways of life have on the forest environment. The Semang foragers are essentially a species of large predator, part of the forest ecosystem, but hardly an ecological dominant. The Senoi swidden farmers, in contrast, exercise a much more radical influence. By periodically clearing and burning plots which are then abandoned and allowed gradually to revert to mature forest, they greatly increase the patchiness of the natural forest, thereby providing a habitat for many plants and animals that might otherwise be absent or present in smaller numbers (see Chapters 12, 13). Finally, the Proto-Malay horticulturalists destroy natural forest to replace it with synthetic agro-ecosystems (Rambo, 1982).

18.2.2. Distribution

Orang Asli are today found in all the states of Peninsular Malaysia except Penang (including Province Wellesley) and Perlis, both of which were still reportedly inhabited by Semang as recently as the mid-1800s (Fig. 18.1). It can be seen that the three main ethnic categories occupy distinct and largely non-overlapping territories. The Semang are found in a broad arc running from the foothills of the northern West coast across the peninsula along the Thai border and then down a long strip between the coastal plains and the mountains on the East coast. The Senoi occupy the Main Range (Chapter 1) with outlying communities in the lowlands of Perak, Pahang, and Selangor; and the Proto-Malays are found scattered across much of the southern third of the Peninsula.

The extent to which the three groups traditionally occupied distinctive habitats is still a subject needing more systematic empirical assessment. It appears that the Semang are generally found at elevations below 300 m, while most Senoi live at higher elevations, with an upper limit to settlement of about 1200 m — the highest elevation at which the bamboo that they use for house construction thrives (H. D. Noone, 1936). The Proto-Malays are generally found at lower elevations ranging from sea level to 300 m.

Semang bands are generally found in river valleys, building their shelters close to the river banks, a pattern also often displayed by Proto-Malay communities. Senoi settlements, and those of neighbouring Proto-Malay groups such as the Temuan of northern Selangor, are typically located on hill-tops, often at some distance from streams or rivers.

The Semang as nomadic foragers are almost invariably found inside lowland rain forest. Their temporary camps are typically constructed in small circular clearings under the shade of tall trees. The Senoi occupy clearings set in a mosaic of forest in various stages of succession. The Proto-Malays inhabit an even more variegated mosaic environment than the Senoi. Perhaps the most striking difference between the settlements of the two groups is that while the houses of the Senoi are built in the middle of large clearings exposed to the direct light of the sun, those of the Proto-Malays are usually sheltered beneath groves of domesticated fruit trees. In gross terms, the relations of the three groups to their habitats can be distinguished as follows: the Semang are people who live within the forests; the Senoi are people who destroy forest in order to use its land for their shifting cultivation and then let natural succession restore it; and the Proto-Malays are people who must first destroy the forest in order to replace it with an artificial plant community designed according to their own specifications.

18.2.3. Population

Attempts were made to enumerate the Orang Asli in the censuses of 1921, 1931 and 1947, and Williams-Hunt (1949) tried to estimate their numbers using aerial photographs. However, reliable statistics were not obtained until the JOA Census of 1969, which showed a total of 53,000 (Carey, 1975). The as yet incompletely analysed Malaysian national census of 1980 enumerated approximately 60,000 Orang Asli (Benjamin, 1985). Table 18.2 presents population figures for each aboriginal group in 1969 and 1980. For 1980, the Semang are the least numerous with just over 2,000 members, the Senoi the most numerous with almost 40,000, while the Proto-Malays number some 19,000 individuals. Reliable historical data are lacking, but it appears that, with the exception of the Semang, whose numbers have evidently either declined slightly or remained stable for at least the past 50 years, Orang Asli have generally been increasing since the 1920s, although many suffered a temporary drop during World War II and the subsequent emergency (see Gomes, 1979, for a review of the demographic literature). There is no reason to fear that the Orang Asli are in danger of biological extinction — long-term preservation of their unique cultures is more problematic.

Measurement of population densities reflects problems in delineating the territories actually exploited by Orang Asli communities. Rough estimates of different types of land use (Table 18.3) based on the 1969 census imply that the Orang Asli exert at least some impact on 7 million hectares out of a total forested area in 1966 of 9 million hectares. Only some 100,000 hectares, however, are affected by agricultural activities, primarily swidden farming.

18.3. ORANG ASLI IMPACT ON THEIR ENVIRONMENT

Given the many centuries during which man has been in Malaysia, it is reasonable to assume that the forebears of the Orang Asli, even the nomadic (Semang) foragers, have influenced the character of the

TABLE 18.2. Population of Orang Asli groups

Group/tribal division	1969 Census[a]	1980 Census[b]
Semang		
Kintak	120	100[c]
Kensiu	100	100[c]
Jahai	700	950[c]
Mendrik	120	125
Batek	500	800[c]
Lanoh	260	224
Total	1,800	2,300[c]
Senoi		
Temiar	9,929	11,593
Semai	15,506	18,327
Jah Hut	2,013	2,442
Che Wong	272	200[c]
Semok Beri	1,406	2,078
Mah Meri	1,198	1,356
Semelai	2,300	2,582
Temok	100	350[c]
Total	32,724	38,930[c]
Proto-Malay		
Jakun (Orang Hulu)	9,100	9,799
Temuan	7,650	9,312
Kanak	50	34
Laut	1,500	n.d.
Selitar	300	n.d.
Total	18,600	19,150[c]
TOTAL	53,124	60,380[c]

Notes:
[a] Source: Carey 1975, reporting results of special JOA Census of Orang Asli in 1969.
[b] Source: Benjamin 1985, reporting preliminary results of the Malaysian Census of 1980.
[c] Figures ending in zeros are estimates.

rain forest ecosystem (Rambo, 1979b). It is a very real problem, however, to identify and measure any effects. As already noted (Section 18.2.1) the different groups have affected the forest ecosystem in different ways: the Semang function essentially as predators within the natural forest ecosystem, the Senoi alter its successional state, and the Proto-Malays permanently alter its structure. It is now worth examining in greater detail Orang Asli impact on specific ecosystem components, looking first at plants, then animals, and finally soil, water, and the atmosphere.

18.3.1. Forest plants as food sources

Forest plants are important to the Orang Asli as sources of food, materials for personal and ritual use and, perhaps most significantly, trade goods. A wide range of plant parts may be exploited for food,

TABLE 18.3. Orang Asli land use

Population[a]	Foraging/ collecting/ hunting	Swidden farming[b]	ha/person Tree orchards[c]	Irrigated paddyland	Total agricultural land	million ha/whole ethnic group Foraging collecting	Agriculture	Total area used
Semang (2,000)	900	—	—	—	—	1.8	—	1.8
Senoi (33,000)	100	2.5	—	—	2.5	3.3	0.08	3.4
Proto-Malay (19,000)	100	1	0.5	0.1	1.6	1.9	0.03	1.9
All groups (54,000)	—	—	—	—	—	7.0	0.1	7.1

Notes:
[a] Population figures are from the 1969 census (Carey 1975).
[b] Includes area under fallow, assessing an average rotation period of 10 years.
[c] Includes fruit and rubber orchards.
NB: All figures are approximations at best, especially with regard to foraging/collecting/hunting areas. Size of Semang territories is from Endicott 1974. Size of Senoi swidden plots is derived from Dentan (1965) and Cole (1959). Size of Proto-Malay agricultural plots is from Gomes (1982).

including leaves, tubers, fruit and seeds. Burkill (1966) made numerous references to Orang Asli consumption of forest plants but no comprehensive list of food species used by any specific community has been compiled. Foo (1972) recorded 57 species eaten by a sample of Orang Asli communities belonging to all three major ethnic categories. This is undoubtedly an underestimate, as Dentan (1971) listed more than 60 species of cultivated plants used for food by the Semai Senoi alone, and Dunn (1975) found 65 species of fruit trees exploited by the Temuan of Ulu Selangor.

Wild yams (*Dioscorea* spp.) of at least ten different species are the main wild source of carbohydrates for the Batek and other Semang and are an emergency food of all Orang Asli. Exploitation by the Batek is heavy with a woman collecting an average of 14 pounds (6.4 kg) of roots in a six-hour day (Endicott, 1977). The supply of mature plants within convenient walking distance of the camp is quickly exhausted and a period of from one to two years is required for sufficient new roots to reach harvestable size before it is worthwhile for the Semang to return again to search the area (Endicott, 1974).

Fruits, particularly durian *Durio zibethinus*, are probably the second most important source of carbohydrates for the Semang, and are also heavily exploited by all other groups, both for personal consumption and for sale to outsiders. The durian is not native to Peninsular Malaysia but has been widely spread throughout the forest edge by man. Orang Asli effectively cultivate the trees, assiduously clearing undergrowth around them in the fruiting season to make it possible to find the dropped fruit and in effect removing competitors. Trees are privately owned by the individuals who first plant or claim them, and rights are inherited. Fruit of the tampoi *Baccaurea griffithii* is avidly sought, both to eat and to make into wine. It is commonly reported that the Semelai are the only Orang Asli group to make alcoholic beverages but this is a misconception. Both Temuan Proto-Malays and Temiar Senoi have been observed by the author to place ripe tampoi fruit into bamboo tubes and allow it to ferment for several days into a mild alcoholic drink.

In addition to deliberate planting (e.g., durian), Orang Asli may play a passive role in dispersing the seeds of fruit trees in the forest. Ridley (1893) reported that rambutan *Nephelium* sp. seeds swallowed whole by Proto-Malays are later deposited intact in their faeces. Dentan (1965) added that such "automatically fertilized" seeds grow into groves around deserted Semai campsites. I have observed Temuan eating wild rattan fruit and casually discarding the seeds whole walking along forest tracks.

Leguminous seeds, particularly petai *Parkia speciosa*, keredas *Pithecellobium* spp. and perah *Elateriospermum tapos* are eaten by many Orang Asli groups and are also of great significance in the trade of forest products. At least some groups today systematically cultivate petai trees: Temiar observed by the author at Pos Poi in Perak were collecting volunteer seedlings in the forest and transplanting these into their swiddens after the rice harvest.

Lianas of unidentified species are used as sources of drinking water by Orang Asli travelling through forest where streams are lacking. A length of stem is first cut as high as a man can reach and then close to ground level.

As the preceding discussion has shown, the line between exploitation of wild plants and cultivation, especially of fruit trees, is often unclear among the Orang Asli. There is, however, a large number of species that are clearly domesticated cultigens, invariably planted in swiddens, home gardens or orchards (see Appendix).

18.3.2. Plants in material culture

It has been observed, with more than a little truth, that to speak of Orang Asli material culture is to speak of bamboo and the thousand-and-one uses to which it is put. Their houses, their blowpipes, their cooking pots, even the knives used to sever their umbilical cords at birth, are all made from bamboo (see Chapter 5). So great is the role of this group of plants in their life that, rather than being called the forest people, they might equally well be labelled the people of the bamboo.

Culturally, bamboos are often of uncertain status, sometimes wild, sometimes deliberately planted, and sometimes in the shadowy middle ground, cared for, even cultivated by the Orang Asli, but not actually domesticated. The Jahai Semang, for example, say that they burn clumps of "wild" bamboo in order to facilitate harvesting by eliminating dead branches and leaves and also to promote the growth of new shoots.

A wild species, *Bambusa wrayi* or buloh seworr as it is commonly known, is the source of the lining tubes for Semang and Senoi blowpipes. Having internodes of up to 2 m long, *Bambusa wrayi* is strictly montane in distribution (see Chapter 5). Each known clump in the Main Range is considered to be within the territory of a particular Senoi or Semang community, and access is rigorously controlled by that community. As a consequence, a considerable trade in bamboo lengths and finished blowpipes traditionally developed between the communities fortunate enough to have access to a clump of *Bambusa wrayi* and those living at lower elevations and in the more southerly parts of the mountains where it is absent (R. O. D. Noone, 1954).

Orang Asli living beyond the reach of the trade in *Bambusa wrayi* stems must make do with inferior substitutes, usually making their blowpipes by carefully piecing together two shorter internodes of *Schizostachyum jaculans*. The Semok Beri, however, construct their blowpipes by binding together two half-round strips of hard wood, in which they first cut two grooves and then laboriously polish the bore (Endicott, 1975). These weapons are shorter and heavier than the bamboo version, resembling in outward appearance the Dayak blowpipes of Borneo, except that spears are not fitted to the muzzles. Recently, the author observed Semok Beri using aluminium tubes purchased at the market in Pekan as the inner liners for bamboo covered blowpipes.

The dart quiver for the blowpipe is invariably made of bamboo. The Semang use quite small sections, often only 4 or 5 cm in diameter, whereas the Proto-Malays often use very large diameter sections, exceeding 10 cm across.

Bamboo stems are used as cooking vessels. Sections of large diameter green stems are filled with rice and chopped up pieces of game and then set directly on the fire. Usually, cooking is completed before the green stem chars through. Large stems are also used as water containers. Every Semang shelter will have two or three sections of bamboo standing on end to provide convenient water for drinking and cooking.

The full list of plants used by Orang Asli for some purpose or another is virtually endless, being, as Dunn (1975) pointed out, essentially coterminous with the total inventory of higher plant species in the Malaysian rain forest. The following discussion is therefore intended to be illustrative rather than comprehensive.

Leaves of mengkuan *Pandanus atrocarpus*, and other pandanus are woven into mats, bags, pouches, etc. The leaf of the bertam palm *Eugeissona tristis* is the principal source of thatch for houses. Traditionally, men's loincloths and women's skirts were made of cloth produced by pounding together strips of bark from the ipoh tree *Antiaris toxicaria*, using a grooved wooden beater. Semang women wore unique skirts made from fungus rhizomorphs.

Resin or damar collected from dipterocarps, e.g., kerning gondol *Dipterocarpus kerri*, was traditionally used for torches and as a fire starter by the Semelai and Proto-Malay groups (Gianno, 1983) as universally in Malaysia in past times.

A number of forest species are well known as sources of poison. The latex of the ipoh tree is the main ingredient used by the Semang in the poison for blowpipe darts while the Proto-Malays employ the sap of the creeper *Strychnos* sp. The roots of lianas (*Derris* spp.) are the source of tuba, used to poison fish in jungle streams. Gianno (1983) reported that *Derris* is cultivated by the Semelai.

As forest people, it is not surprising that the Orang Asli employ plants in many of their ritual activities. Their ritual medical practitioners make use of a wide variety of species in their curing rituals. Foo (1972) listed 11 plants used by the Semelai of Tasek Bera alone. The National Museum has initiated a systematic collection of ethnobotanical data on traditional Malaysian medicinal plants but the results have not yet been published.

18.3.3. Trade in plants and plant products

Most Orang Asli are, and have been for centuries, if not millenia, deeply involved in trade relations with their "civilized" neighbours (Dunn, 1975). In the case of the Semang, it can plausibly be argued that their way of life, far from being a survival from ancient pre-agricultural times, represents a modern adaptation to a niche created by the trade in forest products (Rambo, 1984).

The Orang Asli have at one time or another collected virtually every forest product for which a market existed among the coastal Malay and Chinese communities. A comprehensive list was given by Dunn (1975). Wild fruits and leguminous nuts are seasonally important. Resins and gums, particularly wild rubber of jelutong *Dyera costulata* and damar, were major trade items before World War II but have since been largely displaced by cheaper synthetics.

Bertam palm leaves are sold for thatching. Temiar in Kelantan derived considerable income in the early post-World War II period from timber, felling riverside trees and floating the logs down to sell on the coast. The Orang Laut of coastal Johor cut mangroves both for sale as fire wood and to make charcoal (Williams-Hunt, 1952).

Medicinal plants are also collected and sold to non-Orang Asli. I have observed Jahai Semang selling unidentified roots reputed to be aphrodisiacs to Chinese traders for high prices. Incense wood, particularly gaharu *Aquilaria* sp., is sporadically collected by Proto-Malay groups.

The most important item in the forest product trade today is rattan, particularly rotan manau *Calamus manan*, the large cane used in furniture making (see Chapter 4). A single stem, approximately 2 m in length and 5 to 8 cm in diameter will bring the collector as much as US$1. So lucrative has the trade become in recent years that some Temuan in Ulu Langat have abandoned cultivation of their irrigated rice fields in order to devote more time to rattan collecting. In economic terms this was highly rational since a single hour spent in rattan collection yielded enough cash to purchase rice that would have required 12 hours work to grow themselves (Ali-Rachman, 1980). Unfortunately, over-exploitation is rapidly depleting stocks of wild rattan even though the Temuan reportedly replant young stems disturbed in the course of their collecting activity (Dunn, 1975). In the face of growing scarcity of wild rattan some Temuan in Ulu Langat have begun to experiment with deliberately planting rattan seeds in their gardens.

18.3.4. Animals as food

Wild animals are of significance to the Orang Asli primarily as a source of food. They are also kept as pets and are a commodity in trade with outsiders. Virtually any species of mammal, bird, reptile, amphibian or fish, some insects, and other invertebrates such as molluscs, may be consumed by Orang Asli. Leeches, scorpions and spiders appear to be universally taboo, while each ethnic group has its own special prohibitions, especially at times of illness. Bolton (1972) presented detailed food taboo lists for several groups.

Animals are taken by many methods, of which blowpipe hunting is the best known. This unique weapon has come to symbolize the Orang Asli to outsiders. The blowpipe, however, is effective only against relatively small game and is used primarily to take tree dwelling mammals and birds, such as squirrels, monkeys or hornbills. Larger, ground dwelling mammals such as pigs, deer, serow *Capricornis sumatraensis*, etc., have traditionally been caught with spear traps, dead falls or nooses. Some Semang and Senoi groups also hunted large game with bows and arrows before these were displaced by the shotgun (Rambo, 1978). Small mammals are also trapped by snares and nooses, and small birds captured with bird lime sticks or shot, usually by small boys, with catapults.

It is difficult to assess the impact of hunting on wild animal populations. Dentan (1965) recorded that a Semai Senoi group killed 102 animals belonging to at least 25 species in 146 days. One third were rats, and large mammals such as pigs and deer accounted for only seven individuals. Zainuddin (1977) reported that ten Temuan Proto-Malay families at Ulu Langat killed 92 animals in a 10-day period.

Fish are caught by a variety of methods: traps, hook and line, using hand-thrown nets purchased from the Malays or introducing tuba poison into streams. Women and children frequently hand-catch small fish from under rocks. Systematic data on fish catches are unavailable, although ten Temuan families in Ulu Langat reportedly captured 191 fish in a 10-day period (Zainuddin, 1977).

Traditionally, Orang Asli hunting may only have depressed local populations of some species. Firearms, which were first introduced in the late 1800s and proliferated during the Emergency, have increased hunting efficiency, allowing Orang Asli to exterminate local populations of some of the rarer, slower breeding terrestrial mammals such as the rhinoceros, tapir and seladang. Other species, such as pigs and deer, appear to be more resilient although they quickly become scarce in the immediate neighbourhood of Orang Asli settlements possessing shotguns. The Orang Asli are not conservationist in orientation, and if the opportunity presents itself they will kill any creature they consider to be edible with no evident regard for preservation of future breeding stocks. Nesting female hornbills and their eggs and young, for example, are captured by Batek Semang by the simple expedient of cutting down the hollow tree in which they are sealed (Endicott, 1979). The Temuan of Ulu Langat continue to kill serow, even though they realise that these are becoming increasingly rare.

18.3.5. Domesticated and wild animals as pets

Dogs, which are used for hunting and guard duty, are kept by virtually all Orang Asli. Little attempt is made to raise domesticated animals for food. Chickens and more rarely ducks are kept by Senoi and Proto-Malay households, and the eggs eaten on rare occasion, and goats have been introduced to Senoi villages by government development schemes. Raising livestock is inhibited, however, by a cultural taboo against killing and eating animals that one has cared for. It is permissible to sell such an animal to someone else whose plans to consume it are known, but many Orang Asli feel uncomfortable about doing so.

Orang Asli frequently capture juveniles of wild species and expend much attention on these pets — on occasions lactating women have been known to suckle wild piglets and other baby mammals. Species reported to be among pets are squirrels, monkeys, pigs, otters, civets, bamboo rats and typical rats, flying fox, tortoise, a variety of birds, and even tigers (Williams-Hunt, 1952). Most animals are allowed to escape back into the forest when they become full grown but monkeys may be kept on chains for years. Wild jungle fowl *Gallus gallus* chicks are raised together with the village chickens and become integrated into the flocks (Dentan, 1965).

18.3.6. Trade in animals and animal products

Wild animals figure in the forest product trade, with monkeys, particularly the coconut-picking pig-tailed macaque *Macaca nemestrina*, bringing good prices in the coastal centres. Perhaps the most important animal product for trade purposes is honey, which the Semang collect in large quantities. The Batek employ smoke to stun the bees and then cut the whole nest off the tree branch (Endicott, 1977).

18.3.7. The Orang Asli as prey

The Orang Asli are exploited in turn by some wild animals. Most feared are tigers. Elephants are also considered highly dangerous, although it is their destruction of crops in the swiddens that causes the greatest problem. Greater damage is probably actually done by the Malaysian wood rat *Rattus tiomanicus* and several species of munias and weaver finches. Guarding the ripening padi from these predators is a major labour cost in Temuan wet rice cultivation in Ulu Langat. Other rats, including the house rat *Rattus diardii* and the little Burmese rat *R. exulans*, also prey on stored grain (Dentan, 1965).

Leeches and biting insects are a source of discomfort. Cockroaches may frequent Orang Asli houses in great numbers. It is microorganisms such as amoebae, intestinal worms, and malarial plasmodia that exact the heaviest toll from the Orang Asli, a toll made less bearable by the lack of any really effective traditional remedies. It is particularly ironic that the modifications of the forest ecosystem that the Orang Asli make to ensure their livelihood at the same time greatly expand the habitat for many disease organisms. Clearing of swiddens creates ideal breeding conditions for the main malaria vector, *Anopheles maculatus*, as well as providing favourable conditions for the spread of mite-borne scrub typhus (Polunin, 1962).

18.3.8. Orang Asli impact on other ecosystem components

Impacts of Orang Asli activities on soil, water, and atmospheric components of the forest ecosystem are limited. Soils are probably the most strongly affected component, largely as a result of swidden farming. Principal impacts on the soil are erosion, nutrient concentration and export, and pollution by disease organisms. Erosion is promoted by clearance of swidden plots, where, after burning of the slash, the soil surface is exposed to the full force of the rains. The rate of soil loss is particularly high among Senoi groups who prefer to clear fields on steep slopes due to the better fire caused by updrafts on these sites. Even critics of shifting cultivation admit, however, that the small size of plots and their location in the midst of large tracts of uncleared forest, serves to localize run-off with sediments being redeposited in surrounding vegetated areas. The contribution of erosion from Orang Asli swiddens to the very high sediment loads carried by many Malaysian rivers at flood time must consequently be small (Hill, 1982).

Orang Asli settlements serve as points of concentration for large quantities of products extracted from many square kilometres of surrounding forest and, as a result, it appears probable that soil nutrient levels there become higher than normal in the course of years of occupation. This association is evidently recognized by the people themselves since it is reported that Senoi often choose abandoned house sites to grow tobacco, a particularly nutrient-demanding crop.

Along with concentration of nutrients, the soil in long inhabited settlements also tends to become heavily polluted with disease organisms and parasites. Studies by Dunn (1972) show that the nomadic Semang have a much lower intensity of helminth infection than do the settled Senoi and Proto-Malays. The trade in forest products may also have long-term implications for the nutrient balance of the forest ecosystem. The trade is definitely unbalanced with large quantities of organic materials moving out of the forest while much smaller quantities are returned. No quantified data are available on the loss of nutrients due to trade but, given the long history and considerable magnitude of the commerce in forest products in Malaysia, a detailed assessment of its implications might be usefully undertaken.

Aside from the question of agricultural erosion adding to stream sediment loads to which reference has already been made, Senoi and Proto-Malay groups on the west coast of the Peninsula have in the past carried out small scale tin washing in streams and rivers, presumably causing some change in sediment loads in the process. Such activity is rare nowadays.

The Orang Asli affect water quality primarily through their introduction into streams of faecal matter, with consequent disease contamination. The Semang, and children of all groups, excrete into streams relying on the movement of water for their sewage system. Given the small populations involved, organic pollution is minimal and, given the generally low nutrient availability in Malaysian streams, may be beneficial by providing increased food supplies to fish. On the negative side, however, is the introduction of disease organisms, particularly internal parasites, which may infect downstream consumers of the water (Dunn, 1972). Use of commercial insecticides to poison fish in place of the traditional tuba, have serious consequences on the limnology.

The Orang Asli exert relatively little influence on the atmospheric component of the forest ecosystem. Clearance of swiddens profoundly alters the micro-climate, both increasing ambient temperatures and lowering relative humidity, but these effects are restricted to a small area (Rambo, 1985). Burning of the swiddens, of course, contributes both gases and particulate matter to the atmosphere. Given the relatively small area they burn each year, the Orang Asli hardly qualify as major polluters, although on a global basis swidden burning has been implicated as a source rivaling industrial activity as a generator of particulate pollution of the atmosphere.

18.4. CONCLUSIONS

Looking at the total range of Orang Asli impacts on the rain forest ecosystem, it is possible to summarise these into four major categories:

(a) Direct selection

The question of whether or not the Orang Asli have been a significant selective agent revolves around two issues: how long they have been present in the ecosystem, and how intensive a pressure they have exerted. As was discussed above (Section 18.1), it appears that human populations, presumably ancestral to the Orang Asli, have been present in Peninsular Malaysia since at least the Late Pleistocene. It seems probable, however, that numbers have always been small and, given the low level of traditional technology, human impact on natural resources correspondingly limited. Orang Asli hunting and collecting may have kept local populations of wild animals and plants smaller than would have been the case in a pristine environment but it is doubtful if selection pressure has been exercised. Only with regard to large, rare, slow-breeding terrestrial mammals such as the rhinoceros, tapir and seladang, can near extinction be attributable to increased hunting pressure; but this only occurred after the diffusion of firearms made killing these beasts easier and much less dangerous to the hunters. Highly valued plants in the forest product trade may have been so over-exploited as to be locally eliminated, as is occurring with rattans today.

The clearest evidence of selection, however, is with regard to adaptation of the human population itself. Senoi populations dwelling in malarial habitats show very high levels of abnormal Haemoglobin E, a mutant red-blood cell type believed to give resistance to malarial infection similar to that provided to African populations by sickle cell. Proto-Malays and members of recent immigrant populations do not display such high frequence of HbE (Lie-Injo and Ganesan, 1977).

(b) Dispersal

The Orang Asli certainly play a role in dispersing plant seeds, particularly those of edible fruits. It is possible that, in the process, a certain amount of selection for size or flavour has occurred, so that wild-growing durian and other fruits are different than they might have been without the human presence.

(c) Domestication

The Orang Asli do more than simply disperse wild plant propagules; they also provide a certain amount of care for individual fruit trees, e.g., clearing away brush and eliminating competitors, so that these trees have a better survival chance. Wild seedlings are also transplanted into swiddens, creating groves where cross pollination presumably occurs at an accelerated rate compared to the scattered situation of completely wild trees.

(d) Habitat modification

It is by doing things that intentionally or accidentally modify their environment that the Orang Asli exert the greatest impact on the forest ecosystem. Clearance of swidden fields is particularly significant in this regard, by helping to create a mosaic environment with scattered patches in varying successional stages. It has been argued that Orang Asli swidden farming is the necessary condition for the survival of grazing mammals such as seladang (Wharton, 1968). H.D. Noone (1936) made the similar claim that Semai swidden farming created a favourable habitat for birds in the Cameron Highlands. Certainly, the population of birds observed in a Jahai Semang swidden in Kelantan appeared much denser than in the surrounding mature forest. Diversity was much reduced, however, and the species resident in the swidden were opportunistic types of the kind also common in the suburbs of Kuala Lumpur (Rambo, 1985). Swidden clearance is also, as was noted above, the necessary condition for the successful breeding of *Anopheles maculatus*, the principal mosquito vector for malaria in upland Malaysia.

By changing the forest ecosystem to achieve their own goals, the Orang Asli unknowingly also affect the survival of other species. Some of these in turn influence the survival of the Orang Asli. The process is an endless dialectic between society and nature, in which each is continually transformed and causing transformation in the other. Thus, attempts to understand the structure and dynamics of the rain forest ecosystem of Peninsular Malaysia are necessarily incomplete without including consideration of the role played by the Orang Asli, the people of the forest.

18.5. APPENDIX

Millet *Setaria italica* and Job's tears *Coix lachryma-jobi* are thought to have been the traditional grain crops of the Senoi, only being displaced by upland rice in the last hundred years (Hill, 1977). Temiar at Pos Poi, Perak, who had abandoned growing millet only a few years before the author's visit, said that they preferred millet's taste but rice gave a much higher yield. They now grow only rice because it is impossible to grow both crops together in the same area.

Rice *Oryza sativa* of several different varieties is now the principal grain crop of most Senoi and all Proto-Malays. Most varieties are dry land types but the Proto-Malays plant a limited amount of wet rice. In recent years traditional varieties of the latter have been wholly displaced by the improved strains distributed by the Ministry of Agriculture.

More important as a source of calories to the Senoi and the Proto-Malays than the culturally preferred grains is the tuber of manioc *Manihot esculenta*, commonly referred to in Malaysia as "tapioca". Although nutritionally inferior to rice, manioc is a much more reliable producer, giving assured yields even in years when all other crops fail. It can also be left in the ground for prolonged periods after reaching maturity and thus requires no special harvesting or storage practices.

Literally dozens of other annual crop species are planted by Orang Asli farmers. Particularly important are maize *Zea mays*, cucumbers, squash and gourds, eggplant, beans, and amaranths. Numerous perennials are also systematically planted and cultivated, including bananas, chili peppers, sugar cane, pineapples, papaya, ginger, and lemon grass. The betel nut *Areca catechu* which is chewed together with the betel leaf *Piper betle*, lime and, sometimes, tobacco, is a central element in Orang Asli culture. Senoi and Proto-Malays grow both the areca palm and the betel vine in their gardens while the Semang purchase their supplies from neighbouring Malay villages.

The Proto-Malays also engage in extensive arboriculture, planting and tending many different trees, including durians, rambutans, jack fruit, mangosteens and various citrus species. Large, relatively pure stands of rubber trees are also maintained by some Proto-Malay communities. Although initially established through deliberate planting, these are maintained by self-seeding with the owners simply thinning out undesired seedlings.

Noteworthy when considering the current assemblage of cultivated crops is the American origin of many key species. It is difficult to conceive of what the diet of the Senoi and Proto-Malays was like before chilis, maize, papaya and pineapple were introduced by Europeans in the 1600s. It is even more difficult to envisage how the Senoi, given their heavy current reliance on manioc as a staple food, were able to survive at anywhere near their present population density prior to its introduction. This has led Cole (1959) to suggest that the Temiar were essentially nomadic foragers living much like the Semang do today until they adopted manioc cultivation. Perhaps hardest of all to imagine, for anyone familiar with the present day Orang Asli love of smoking, is how they lived before the introduction of tobacco!

ACKNOWLEDGEMENTS

This chapter is of necessity in large part based upon research by other investigators. I have relied particularly heavily on the detailed monographs of Geoffrey Benjamin, Robert K. Dentan, F. L. Dunn, and Kirk Endicott. The continuing value of the earlier studies by Walter W. Skeat and Father Paul Schebesta also deserves recognition. I. H. Burkill's monumental *Dictionary of the Economic Products of the Malay Peninsula* remains the primary source for identification of indigenous terms for plants and animals.

My own field research was carried out during my tenure as a lecturer in the Department of Anthropology and Sociology at the University of Malaya. I am grateful to the University of Malaya for the generous financial support it provided under its staff research grants (Vote F). Several colleagues and students, especially Mazidah Zakaria and Alberto G. Gomes, produced valuable help in the course of this work. Staff of the Department of Orang Asli Affairs under the direction of Dr. Baharon Azhar bin Raffie'i were unfailingly helpful in arranging my field work. An earlier draft of this chapter was critically reviewed by George W. Lovelace. Karl L. Hutterer provided information of what is currently known about the prehistory of Malaysia.

REFERENCES

Ali-Rachman, M. A. (1980). *Energy Utilization and Social Structure: An analysis of the Temuan Orang Asli of Peninsular Malaysia*. Unpublished M.A. thesis. Kuala Lumpur, University of Malaya Department of Anthropology and Sociology.

Benjamin, G. (1976). Austroasiatic subgroupings and prehistory in the Malay Peninsula. In *Austroasiatic Studies*. Eds. P. N. Jenner, L. C. Thompson & S. Starosta. Oceanic Linguistics Special Publication No. 13. Honolulu, The University Press of Hawaii, Vol. I., Pages 37–128.

Benjamin, G. (1980). *Semang, Senoi, Malay: Culture-history, kinship, and consciousness in the Malay Peninsula*. Canberra, Department of Prehistory and Anthropology, Australian National University (typescript).

Benjamin, G. (1985). Between isthmus and islands: Notes on Malayan palaeo-sociology. *Paper presented at the 12th Congress of the Indo-Pacific Prehistory Association*. Penablanca, Philippines, 26 January–2 February, 1985.

Bolton, J. M. (1972). Food taboos among the Orang Asli in West Malaysia: a potential nutritional hazard. *Am. J. Clinic Nutr*. 25, 789–99.

Burkill, I. H. (1966). *A Dictionary of the Economic Products of the Malay Peninsula*. 2 Volumes, reprinted 2nd edition. Kuala Lumpur, Ministry of Agriculture and Cooperatives.

Carey, I. (1975). *The Orang Asli: The Aboriginal Tribes of Peninsular Malaysia*. Kuala Lumpur, Oxford University Press.

Cole, F. C. (1945). *The Peoples of Malaysia*. New York, Van Nostrand.

Cole, R. (1959). Temiar Senoi agriculture: A note on aboriginal shifting cultivation in Ulu Kelantan, Malaya. *Malayan Forester* 22, 191–207, 260–71.

Dentan, R. K. (1965). *Some Senoi Semai Dietary Restrictions: A Study of Food Behavior in a Malayan Hill Tribe*. Unpublished Ph.D. dissertation, Yale University.

Dentan, R. K. (1971). Some Senoi Semai planting techniques. *Economic Botany* 25(2), 136–59.

Dunn, F. L. (1972). Intestinal parasitism in Malayan aborigines (Orang Asli). *WHO Bulletin* 46(1), 99–114.

Dunn, F. L. (1975). *Rain-forest Collectors and Traders: A Study of Resource Utilization in Modern and Ancient Malaya*. Kuala Lumpur, Monographs of the Malaysian Branch, Royal Asiatic Society, No. 5.

Endicott, K. (1974). *Batek Negrito Economy and Social Organization*. Unpublished Ph.D. dissertation. Cambridge, MA, Harvard University.

Endicott, K. (1975). A brief report on the Semaq Beri of Pahang. *Federation Museums J.* 20, 1-23.

Endicott, K. (1977). Some features of the economy of the Batek Negritos of Peninsular Malaysia. *Paper presented at the 48th ANZAAS Congress*, Melbourne, 1 September, 1977.

Endicott, K. (1979). *Batek Negrito Religion: The World-view and Rituals of a Hunting and Gathering People of Peninsular Malaysia*. New York, Oxford University Press.

Foo, E. L. (1972). *The Ethnobotany of the Orang Asli, Malaysia, with Special Reference to Their Foodcrops*. Ed. by T. Koonlin. Kuala Lumpur, Botany Unit, School of Biological Sciences, University of Malaya.

Gianno, R. (1983). Resin classification and technology in the Western Malesian forest. *Paper presented at the XI International Congress of Anthropological and ethnological Sciences Meeting*, Vancouver, 20–25 August, 1983.

Gomes, A. G. (1979). The demography of the Orang Asli. *Federation Museums J.* 24, N.S., 75–90.

Gomes, A. G. (1982). *Ecological Adaptation and Population Change. Semang Foragers and Temuan Horticulturalists in West Malaysia*. Honolulu, Hawaii, East-West Environment and Policy Institute Research Report No. 12.

Gorman, C. (1971). The Hoabinhian and after: subsistence patterns in Southeast Asia during the late Pleistocene and early recent periods. *World Archaeology* 2(3), 300–18.

Hill, R. D. (1977). *Rice in Malaya: A Study in Historical Geography*. Kuala Lumpur, Oxford University Press.

Hill, R. D. (1982). *Agriculture in the Malaysian Region*. Budapest, Akademiai Kiado.

Hutterer, K. L. (1977). Reinterpreting the Southeast Asian Paleolithic. In *Sunda and Sahul*. Ed. J. Allen, J. Golson & R. Jones, 31–71. London, Academic Press.

Jimin bin Idris (1968). Distribution of Orang Asli in West Malaysia. *Federation Museums J.* 13, 44–8.

Lie-Injo, L. E. & Ganesan, J. (1977). Biochemical genetic characteristics of Malaysians. *Malaysian Nature J.* 31(2), 75–80.

Noone, H. D. (1936). Report on the settlements and welfare of the Ple-Temiar Senoi of the Perak-Kelantan watershed. *J. Federated Malay States Museum*, 19 Part 1, 1–84.

Noone, R. O. D. (1954). Notes on the trade in blowpipes and blowpipe bamboo in North Malaya. *Federation Museums J.* 1–2, 1–18.

Polunin, I. (1962). The effects of shifting agriculture on human health and disease. In *Symposium on the Impact of Man on Humid Tropics Vegetation*. Goroka, Papua New Guinea, 1960. Canberra: UNESCO Science Cooperation Office for Southeast Asia.

Rambo, A. T. (1978). Bows, blowpipes and blunderbusses: ecological implications of weapons change among the Malaysian Negritos. *Malayan Nature J.* 32(2), December, 209–16.

Rambo, A. T. (1979a). Human ecology of the Orang Asli: a review of research on the environmental relations of the aborigines of Peninsular Malaysia. *Federation Museums J.* N.S. 24, 41–71.

Rambo, A. T. (1979b). Primitive man's impact on genetic resources of the Malaysian tropical rain forest. *Malaysian Applied Biology* 8(1), 59–65.

Rambo, A. T. (1982). Orang Asli adaptive strategies: implications for Malaysian natural resource development planning. In *Too Rapid Rural Development*. Eds. C. MacAndrews & L.S. Chia, pp. 251–99. Athens, Ohio University Press.

Rambo, A. T. (1984). Why are the Semang? Ecology and ethnogenesis in Peninsular Malaysia. *Paper presented at the East-West Environment and Policy Institute - University of Michigan Center for South and Southeast Asian Studies Conference on Ethnic Diversity and the Control of Natural Resources in Southeast Asia*. Ann Arbor, MI, 22–24 August, 1984.

Rambo, A. T. (1985). *Primitive Polluters: Semang Impact on the Malaysian Tropical Rain Forest Ecosystem*. Ann Arbor, MI, University of Michigan Museum of Anthropology Anthropological Papers No.76.

Ridley, H. N. (1893). On the dispersal of seeds by mammals. *J. Straits Branch Royal Asiatic Society* 25, 11-32.

Schebesta, P. (1926-28). The jungle tribes of the Malay Peninsula. *Bull. School of Oriental Studies (London)* 4, 269–78.

Sieveking, A. (1958). The palaeolithic industry of Kota Tampan, Perak, Northwestern Malaysia. *Asian Perspectives* 2, 91–102.

Skeat, W. W. (1902). The wild tribes of the Malay Peninsula. *J. Royal Anthropological Institute* 32, 124–41.

Skeat, W. W. and Blagden, C. O. (1906). *Pagan Races of the Malay Peninsula*. 2 Volumes. London, Macmillan & Co.

Solheim, W. G. II. (1980). Searching for the origins of the Orang Asli. *Federation Museums J.* 25, N.S., 61–76.

Tweedie, M. W. F. (1953). The stone age in Malaya. *J. Malayan Branch Royal Asiatic Society* 26(162), Part 2, October, 1–90.

Wharton, C. H. (1968). Man, fire and wild cattle in Southeast Asia. *Annual Proceedings, Tall Timbers Fire Ecology Conference* No. 8, 107–67.

Williams-Hunt, P. D. R. (1949). A technique for anthropology from the air in Malaya. *Bull. Raffles Museum* Series B. No. 4, 44–69.

Williams-Hunt, P. D. R. (1952). *An Introduction to the Malayan Aborigines*. Kuala Lumpur, Government Printer.

Zainuddin, A. R. (1977). *The Temuan ecosystem*. Unpublished Graduation Exercise. Department of Anthropology and Sociology, University of Malaya.

Completed October 1986

Species Index

Subject Index

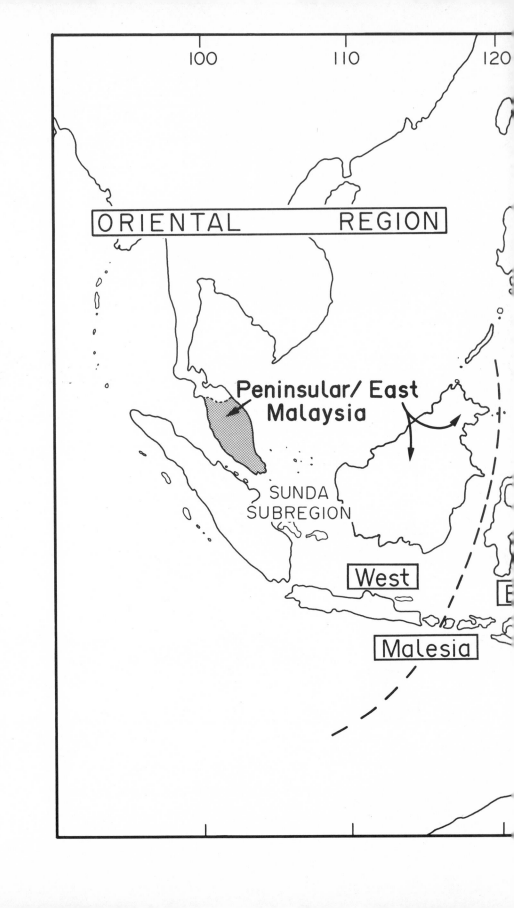